CHEMISTRY RESEARCH AND APPLICATIONS

ARGON

PRODUCTION, CHARACTERISTICS AND APPLICATIONS

CHEMISTRY RESEARCH AND APPLICATIONS

Additional books in this series can be found on Nova's website under the Series tab.

Additional e-books in this series can be found on Nova's website under the e-book tab.

CHEMICAL ENGINEERING METHODS AND TECHNOLOGY

Additional books in this series can be found on Nova's website under the Series tab.

Additional e-books in this series can be found on Nova's website under the e-book tab.

CHEMISTRY RESEARCH AND APPLICATIONS

ARGON

PRODUCTION, CHARACTERISTICS AND APPLICATIONS

BOGOS NUBAR SISMANOGLU
HOMERO SANTIAGO MACIEL
MARIJA RADMILOVIC-RADJENOVIC
AND
RODRIGO SAVIO PESSOA
EDITORS

nova publishers
New York

For permission to use material from this book please contact us:
Telephone 631-231-7269; Fax 631-231-8175
Web Site: http://www.novapublishers.com

Additional color graphics may be available in the e-book version of this book.

Library of Congress Cataloging-in-Publication Data

Library of Congress Cataloging-in-Publication Data

Argon : production, characteristics and applications / editors, Bogos Nubar Sismanoglu, Homero Santiago Maciel, Marija Radmilovic-Radjenovic, Rodrigo Savio Pessoa.
pages cm
Includes bibliographical references and index.
ISBN 978-1-62618-204-2 (hardcover)
1. Argon. 2. Argon--Industrial applications. I. Sismanoglu, Bogos Nubar.
TP245.A7A74 2013
665.8'22--dc23

2013006215

Published by Nova Science Publishers, Inc. † New York

CONTENTS

PREFACE

The aim of this book is to present a compilation of a number of reports specially written by notable scientists from around the world to present their recent research and novelties involving the noble gas argon. The inert argon gas has low thermal conductivity which combined with its low cost favors its use in the generation of laboratory plasmas at low and high pressure, for a range of applications, like industrial, environmental preservation, medical and general researches, to name a few.

The well-known emission spectrum from ultraviolet to infrared of the argon atom is by far the most commonly used in the optical and electrical characterization of gas discharges. This characterization sheds light to design the best plasma system configuration in order to enhance its applications.

Argon is a noble gas that is present in the Earth's atmosphere at approximately 1% and is produced industrially by the partial distillation of liquid air. It is a by-product of the oxygen and nitrogen production. Argon is the most commonly used gas in research, industry, medicine (dermatology, ophthalmology), analytical chemistry, nanotechnology and plasma technology due to its excellent thermal performance and cost-efficiency in comparison to other noble gases.

Chapter one presents some characteristics and production methods of argon and, finally, provides various applications in diverse areas of human knowledge. Argon isotopes ^{40}Ar-^{39}Ar are applied to geochronology and thermochronology to date the age of ancient materials, such as rocks and fossils. Low temperature atmospheric-pressure argon plasmas are relatively new contributions for diagnostics and medical applications, whereas electrical and optical studies of these discharges types are in course nowadays.

In the geology of ore deposits, petrography, mineralogy, geochemistry and chemical technology, argon is used for measurement of rock permeability, the parameter that governs many processes of heat and mass transfer in the Earth crust as, for example, development of ore-forming systems and contaminant transport by ground water from geological repositories of high level radioactive waste.

The used method is described in chapter two. The simplest technique for laboratory measurement of rock permeability is the steady state method where pressure gradient is constant in time. In spite of low viscosity, filtration of measurable volume of argon through a tight (low permeable) rock sample calls for a long time. Advantages of argon are relatively low cost and availability.

The technique allows accomplishing high-accuracy determinations of permeability within the range of values of 10^{-22}–10^{-15} m^2 both under normal conditions and at elevated PT-parameters. The method was approved in measurement permeability of main types of rocks of the uranium ore field and the areas of significant radioactive contaminations of the underground medium, related to radioactive waste disposal.

In order to improve the oxide scale adherence on alloys in various corrosive environments at high temperatures, chapter three focuses on the effect of argon annealing on sol gel coated specimens.

On a chromia forming 330Cb alloy (Fe-35Ni-18Cr-1Nb-2Si), the influence of a lanthanum sol-gel coating on the oxide scale adherence has been studied at 900°C, in air. Argon annealing of lanthanum sol-gel coatings have been performed at various temperatures in order to find the best conditions to insure the scale adherence. The research shows that the coating argon annealing at 1000 °C leads to the preferential formation of $LaAlO_3$ instead of La_2O_3. This coating procedure leads to an alumina scale formation showing the best adherence under thermal cycling conditions at 1100 °C. It is demonstrated that argon annealing of coated alloys can largely improve the oxide scale adherence.

Coherent extreme ultraviolet radiation can be produced by high-order harmonic generation of energetic femtosecond laser pulses in argon gas. In chapter four, it is shown that using 800 nm driving pulses and argon gas, the energy conversion efficiency is about 10^{-5} and the absorption limit reaches at the photon energy of ~ 45 eV. The high-harmonic generation in semi-infinitive gas cell have been considered to obtain high efficiency, highly spatial coherence and a good beam profile. The paper demonstrates that the developed source can be used for a multiple-harmonic-order coherent diffractive imaging with spatial resolution below 100 nm.

In chapter five, some important applications of argon in microwave plasmas sustained at atmospheric pressure are shown. In natural state, argon is an inert gas unable to induce chemical reactions with other reactants. However, in plasma state, argon becomes active due to the different and highly energetic species produced from its ionization such as electrons, ions, excited neutral particles and photons. All these species play an important role in the kinetics of the plasmas generated with this gas and determine the capabilities of the plasma when interacting with other substances.

The spectral lines emitted by this argon plasma can be utilized to develop methods that allow for measuring the density of the particles in the plasma (electrons, ions and molecular species) and their energy (temperatures). In the technological field, argon and argon-based plasmas are used in applications covering the sterilization of medical instruments, hydrogen production, synthesis of carbon nanostructure, analytical chemistry, etc.

The production and application of argon cluster ion to nanotechnologies (nanofabrication) and surface analysis is reviewed in chapter six. Gas cluster ion consists of gaseous atoms such as argon aggregated by Van der Waals forces and the cluster typically contains hundreds to thousands of atoms or molecules. The average level of kinetic energy per atom forming the cluster ion is equal to the value of the energy obtained by accelerating the cluster ion divided by the number of constituent atoms of the cluster ion (called cluster size).

Furthermore, when a cluster ion strikes the surface, chemical reactions such as sputtering or etching intensively occurred within a very thin layer by the multiple collisions among the numerous atoms constituting the cluster ion. Such nature of the bombardment by the gas cluster ions has a potential to change drastically an ion-beam technology. In the next chapter some aspects of the multi charged argon beam transport are considered. The basics of the ion beam physics and paraxial ion beam optic is presented. Short survey of magnetic ion-optical elements and their transfer matrices is given. The basic properties of the analyzing section consisting of a magnetic quadrupole lens and a bidirectional analyzing magnet are described. Results of beam transport calculations along the beam transport line from the ion source up to the experimental chamber are considered. The results of the charge state analysis by a magnetic analyzer of the argon ions, produced by ECR type ion source, are presented in terms of beam current distribution over the different ion species.

Chapter eight contains the results of the detailed theoretical and simulation studies (modeling) of the breakdown voltage in direct current, radio and microwave frequency argon discharges, without and in the presence of magnetic field. The emphasis was made on the role of the secondary emission process on the breakdown voltage. Calculations were performed using a Particle-in-cell/Monte Carlo collisions (PIC/MCC) code with the secondary emission model adjusted to include the energy and angular dependence of the electron yield at large separations as well as the secondary electron emission enhanced by field emission in microgaps.

The results presented here indicate that PIC/MCC code with the improved secondary emission models provides a good physical description of plasma for various gap sizes, not only in the presence of the electric field, but also under the simultaneous action of both electric and magnetic field leading to a satisfactorily agreement with the experimental results. The obtained results could be useful for determining minimum ignition voltages in microplasma sources as well as the maximum safe operating voltage and critical dimensions in other microdevices.

Chapter nine presents the chemistry studies of low pressure electrical discharges generated with argon gas. In the semiconductor industry, this gas is present in most of the processing steps of a chip or a micro device. Among the most common equipments that use argon-based plasma environment it can be cited the magnetron sputtering reactor, designed for the deposition of thin films, and the reactive ion etching reactor, designed for etching process of materials at micro/nano scale level. Moreover, nowadays there are microplasmas reactors that allow generating plasmas at pressures of the order of tens of Torr. Experimental diagnostic tools such as Langmuir probe and optical emission spectroscopy are used to determine plasma/gas parameters namely electron density, electron temperature, argon ion density, metastable species, gas temperature, etc. Moreover, some main results of plasmas simulations with collisional-radiative model are presented. A revised set of collisional-radiative reactions are presented and correlated with the experimental results. Finally, for the case of magnetron sputtering plasma, it is presented a brief discussion about the influence of argon gas incorporation in the properties of the grown thin films by this technique.

The purpose of chapter 10 is to investigate about characteristics of the cathode sheath generated in an argon low pressure hollow cathode discharge (HCD). The theoretical model for the cathode sheath is derived considering it to be non-collisional. The plasma parameters and the floating potential profile along the discharge axis were inferred from the current-voltage characteristics of a single Langmuir probe positioned at the inter-cathode space of the HCD. By using the probe data the floating potential profile was determined to verify the position of the plasma-cathode sheath interface and to promote a qualitative discussion between theoretical and experimental results.

The next two chapters are brief reports where in the first one is presented a discussion about argon ion beam milling that is used for sample preparation, and the second contains a discussion of the use of argon in medical research.

Bogos Nubar Sismanoglu
Homero Santiago Maciel
Rodrigo Sávio Pessoa
Marija Radmilović-Radjenović

August 2012
São José dos Campos

In: Argon: Production, Characteristics and Applications ISBN: 978-1-62618-204-2
Editors: B.N. Sismanoglu, H.S. Maciel et al.

Chapter 1

ARGON PROPERTIES, PRODUCTION AND RECENT APPLICATIONS

B. N. Sismanoglu[1*], R. S. Pessoa[1,2], R. Caetano[1], Y. D. Hoyer[1], H. S. Maciel[1,2] and K. G. Grigorov[1,3]

[1]Technological Institute of Aeronautics (ITA-DCTA), São José dos Campos/SP, Brazil
[2]IP&D, University of Paraiba Valley (UniVap), Brazil
[3]Space Research and Technology Institute, Sofia, Bulgaria

ABSTRACT

Argon is a noble gas that is present in the Earth's atmosphere at approximately 1% and is produced industrially by the partial distillation of liquid air.It is a by-product of the oxygen and nitrogen production. Argon is the most commonly used gas in research, industry, medicine (dermatology, ophthalmology), analytical chemistry, nanotechnology and plasma technology due to its excellent thermal performance and cost-efficiency in comparison to other noble gases. This chapter presents some characteristics and production methods of argon and, finally, provides various applications in diverse areas of human knowledge.Argon isotopes $^{40}Ar/^{39}Ar$are applied to geochronology and thermo-chronology to date the age of ancient materials, such as rocks and fossils. Low temperature atmospheric-pressure argon plasmas are relatively new contributions for diagnostics and medical applications, whereas electrical and optical studies of these discharges are in course nowadays.

1. INTRODUCTION

Argon (Ar) chemical element, the third major component of air, is a colourless, odourless, tasteless, non-flammable, non-reactive, nontoxic, inert gas and is one of the most abundant of the rare gas (noble gases – weak ability to chemically interact with other elements) of the periodic table and the most frequently used by human being. The unreactive

* E-mail address: bogos@ita.br.

and monatomic Ar gas was effectively discovered by Lord Rayleigh and Sir William Ramsay in 1894 [1, 2]. Argon is isolated on a large scale by the fractional cryogenic distillation of liquid air, to produce oxygen for the industry, the research and for medicine purposes. Argon gas condenses at -185.8°C and forms a crystalline cubic solid at -189.4° C. Itslow thermal conductivity and known electronic properties enables its use in the plasma technology.

Table 1. Basic data of the chemical element Ar [12]

Property	Value
Atomic number (Z)	18
Atomic mass (a.m.u.)	39.95
Periodic table	series - noble gases; group - 8; period - 3
Electron configuration	[Ne]$3s^2 3p^6$
Atomic radius (Å)	0.88
Atomic volume (cm^3/mol)	28.5
Number of electrons/protons	18
Number of neutrons	22
Covalent radius (Å)	0.98
Crystal Structure	cubic face centered
Electron configuration	$1s^2\ 2s^2p^6\ 3s^2p^6$
Heat of fusion	1.188kJ/mol
First ionization potential (eV)	15.76 (provide the measure of the strength between the outer electrons of the atom shell and the nuclear charge)
Melting point (°C)	- 189.35
Boiling point (°C)	-185.35
Property	Value
Thermal conductivity (W/mK) at 273K	0.0165
Density: kg/m^3 at 273K	1.784
Heat capacity (J/mol.K)	20.8
Water solubility (mg/L) at 20°C and 1 bar	62
Triple point temperature (K)	83.81
Isotopes at atmospheric pressure	^{40}Ar (99.5%), ^{36}Ar (0.35%), ^{38}Ar (0.06%)
Color	colorless

In addition of beingvery cost-efficient,at low pressure discharges appear to be red colored and at high pressure,such as atmospheric pressure microplasmas and plasma jets, emits blue. Among the others applications of the Ar gas, the following can be counted: (i) for lighting, to measure the radiation levels in "Geiger" counters. Ar fills are used to reduce heat loss in sealed units. Insulating glass technology uses Ar (low-conductivity and density greater than the air) between glass panes as a good source of insulation, which improves window performance by reducing conductive and convective heat transfers. This gas is the most commonly used as fill, due to its excellent thermal performance and cost-efficiency in comparison to other gas fills [3]; (ii) to manufacturing stainless steels, in order to avoid the formation of nitrides, especially when processing metals such as titanium [4]; (iii) as an inert filler gas for lamps (± 90% is Ar) and plasma chambers (arc plasmas, plasma jet, torch plasmas, radio-frequency and direct-current high pressure plasmas) [5]; (iv) in metallurgy, for arc-welding metals and plasma-arc cutting [5, 6]; (v) in the plasma surface technology, for growing crystals of silicon and other semi-conductors [7]; (vi) in medicine, argon blue lasers have many purposes, including, but not limited to, forensic medicine, general surgery, for cryosurgery and ophthalmic surgery [8]; (vii) it is used for preventing the wine to turns into vinegar during the production process [9]; (viii) in the food industry [10]; (ix) in microfabrication processes to produce etching and sputtering of semiconductors [7]; (x) by museums to protect documents, material and books from air oxidation [11]. Basic data on the chemical element Ar can be found out in the web site of NIST (National Institute of Standards and Technology)[12].

Argon atom has a net positive charge, making this element non-reactive. However, Khriachtchev et al. [13], at the University of Helsinki, found a stable novel argon compound named argon fluorohydride (HArF), synthesized within the matrix of cold argon, at 27K. According to these authors [13]: *"The noble gases have a particularly stable electronic configuration, comprising fully filled s and p valence orbitals. This makes these elements relatively non-reactive, and they exist at room temperature as monatomic gases. Pauling predicted in 1933 that the heavier noble gases, whose valence electrons are screened by core electrons and thus less strongly bound, could form stable molecules."* In the paper, the authors report that *"the photolysis of hydrogen fluoride in a solid argon matrix leads to the formation of argon fluorohydride (HArF), which we have identified by probing the shift in the position of vibrational bands on isotopic substitution using infrared spectroscopy. Extensive ab initio calculations indicate that HArF is intrinsically stable, owing to significant ionic and covalent contributions to its bonding, thus confirming computational predictions that argon should form a stable hydride species with properties similar to those of the analogous xenon and krypton compounds reported before."*[13]. The HArF, meanwhile, is weakly stable at this very low temperature. Moreover, Argon-containing ions and excited state complexes, such as ArH^+ and ArF*, respectively, are also known to exist [14, 15]. Theoretical calculations have predicted several Ar compounds that should be stable [16], but for which no synthesis routes are currently known. The purpose of this chapter is to present a brief overview about the main production processes of argon gas and some of its main applications.

2. Argon Production

The production of argon is an important economic factor in the industrial gas industry. Generally argon is a by-product of cryogenic air separation [17]. However, a number of additional processing steps are necessary to produce a required purity of argon. One of the critical purity requirements is the concentration of contained nitrogen. Many applications of argon demand that it be essentially free of nitrogen. In the past, the production of high purity argon involved a number of processing steps to produce a crude argon stream which was then upgraded in a refinery. Argon processing starts with the low pressure column of a cryogenic air separation plant. A low grade argon stream is withdrawn from an intermediate point in the low pressure column. The low grade argon stream is then fed into an argon column where it is separated into an overhead crude argon stream containing about 97.5 percent argon and a bottom stream which is returned to the low pressure column. The overhead stream also typically contains about 1.5 percent oxygen and about 1.0 percent nitrogen.The crude argon stream from the top of the argon column is then warmed to about ambient temperature, at which time hydrogen is added and the mixture is compressed and sent to a Deoxocatalytic furnace where the oxygen is removed. Then the combusted argon is cooled, dried and further cooled to essentially liquefaction temperature. The cold argon stream is sent to the refinery column where the excess hydrogen and remaining nitrogen are removed. Normal production provides an argon product stream containing less than 5 ppm nitrogen or oxygen [18]. Nowadays, the cryogenic process is comprised of unit operations that compress, purify, and separate the air feed into the required gaseous and liquid oxygen, nitrogen, and argon product flows. The air feed is compressed prior to the removal of the primary impurities of H_2O and CO_2 via adsorption. The air is then cooled by heat exchange with exiting product and waste streams and then sent to a set of mass and energy integrated distillation columns where the air is separated into oxygen, nitrogen, and argon. Liquid products are sent to storage and the gaseous products are re-warmed by the feed streams and compressed to the required final product pressure(s). A typical flow sheet for high purity oxygen, nitrogen, and argon production is shown in Figure 1. The exact compressor, expander, and column configurations are determined by the product purities, flows, and pressures required [19]. As stated earlier, one of the main challenges in the high purity argon production is reducing the nitrogen content of the feed to an argon column. German Patent № 1 048 936, describes a means for reducing the nitrogen content of the feed to an argon column. The suggested process increases the number of trays used in the section of the low pressure column, between a feed from, the argon condenser and the point where the argon column feed is withdrawn. The use of additional trays in the low pressure column, for the purpose of reducing the nitrogen content of the feed to the argon column, imposes a pressure drop penalty which increases the air compressor discharge pressure and therefore the energy requirements. Further, the increase in pressure level reduces relative volatility within the columns, resulting in a lowering of argon recovery [18].

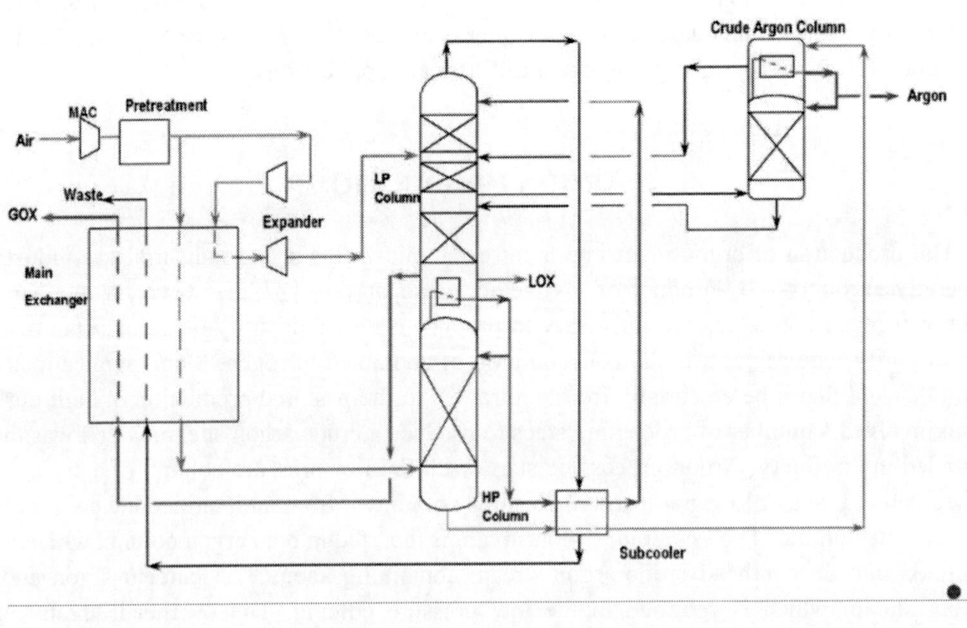

Figure 1.Schema of simplified process for cryogenic production of oxygen and argon [19].

The use of structural packing in order to increase the number of the equilibrium stages in low pressure column between the argon feed condenser and the point where the column is withdrawn was suggested by Bianchi et al. and protected by U.S. patent N°. 5,133,790. The additional rectification in the lower pressure column is provided by the incorporation of structured packing rather than by trays. This reduces the nitrogen concentration substantially while maintaining the argon concentration at or near its maximum, enabling the production of nitrogen-free argon directly. The use of structured packing, rather than trays, avoids the energy penalty and reduced argon recovery [18, 20]. Full scale testing of the system proposed by Bianchi et al., (which utilizes structured packing throughout the low pressure column), revealed that it is difficult to achieve low nitrogen levels in the argon column feed. Attempts were made to achieve low nitrogen levels using a single bed of packing between the feed from the argon condenser and the point where the argon column feed is withdrawn. The performance was not satisfactory [18, 20]. In the next topics, we separate someinformation of US patents and papers about the argon production. These are listed chronologically, showing the evolution of the cryogenic technology in the rare gas production and the main processes steps itself.

2.1. Process for Recovering Argon

This is a process for recovering argon in which a crude argon column is supplied with an argon feed gas containing a larger proportion of oxygen from a double rectifier and is cooled at its head portion with or without liquid air from a sump of the lower column of the double rectifier aiming the producing of argon reflux in the crude argon column. In the process, argon gas is compressed to a pressure sufficient for evaporating liquid oxygen fed from a condenser of the double rectifier, is then pre-cooled in a heat exchanger, and thereafter liquefied by heat exchange with the liquid oxygen. Then the head portion of the crude argon column is cooled with the liquefied argon, which is then returned in a gaseous state for further compression to thereby circulate the argon gas. For more details see reference [21].

2.2. Cryogenic Rectification Method for Producing Refined Argon

This patent presents a cryogenic rectification method employing rectification over a defined height within the lower pressure column of a double column system which reduces the nitrogen concentration of argon, nitrogen, oxygen mixture while maintaining the argon concentration at or near its maximum enabling production of nitrogen-free argon directly from the argon column system [20].

2.3. Cryogenic Distillation of Air Using Multiple Distillation Columns to Produce Argon, Nitrogen, and/or Oxygen

The present invention relates to a process and apparatus for the cryogenic distillation of air using multiple distillation columns to produce argon, nitrogen, and/or oxygen.

A process and system for producing an argon-enriched stream comprising a multiple column air separation system, a first crude argon column, and a second crude argon column is provided. The improved process involves removing a vapor stream from the upper portion of the first argon column; reducing the pressure of the vapor in a pressure reducing device; and then flowing the reduced pressure vapor to a second argon column which operates at a lower operating pressure relative to the first argon column [22].

2.4. Column Configuration and Method for Argon Production

This invention relates to the production of argon and more particularly to a low pressure column configuration of a cryogenic air separation system which provides an argon-rich supply and substantially nitrogen free argon distillation column. It consists of a system for low nitrogen level column enabled by a lower pressure in the feed point at the cryogenic air separation system by using of two structured packing beds of about equal height in the lower pressure column driving to a mixing and redistribution of the liquid air between them. Another objective of this invention is to provide improved argon production system wherein the low pressure feed point is largely nitrogen free.

The following citation summarizes the main points of this invention:

> "To produce an argon product with a low level of included nitrogen (typically 10 ppm), a low level of nitrogen must be achieved in a section of the low pressure column for the feed to the argon column of a cryogenic air separation system. This is accomplished by use of two beds of structured packing of about equal height in the low pressure column, with mixing and redistribution of liquid between them. The packed beds are located in the column section between the feed from the argon column condenser and the point where the argon column feed is withdrawn." [23]

2.5. Cryogenic Argon Production System with Thermally Integrated Stripping Column

A cryogenic rectification system for producing argon employing an argon stripping column which receives a feed in its upper portion from an associated cryogenic air separation plant, and which is re-boiled by another fluid taken from the cryogenic air separation plant [24].

Summary of the invention [24]:

> "The above and other objects, which will become apparent to those skilled in the art upon a reading of this disclosure, are attained by the present invention, one aspect of which is:
>
> A method for producing argon comprising:
>
> (A) introducing feed air into a cryogenic air separation plant and producing by cryogenic rectification within the cryogenic air separation plant a fluid comprising nitrogen and argon;
>
> (B) passing said fluid comprising nitrogen and argon from the cryogenic air separation plant into the upper portion of an argon stripping column as argon stripping column feed;

(C) passing argon stripping column feed down the argon stripping column against up flowing vapor and producing nitrogen-richer fluid in the upper portion of the argon stripping column and argon-richer fluid in the lower portion of the argon stripping column;

(D) vaporizing a portion of the argon-richer fluid by indirect heat exchange with re-boiling fluid taken from the cryogenic air separation plant to produce said up flowing vapor; and

(E) recovering another portion of the argon-richer fluid from the lower portion of the argon-stripping column as product argon.

Apparatus for producing argon comprising:

(A) a double column cryogenic air separation plant comprising a higher pressure column and a lower pressure column; and means for passing feed air into the cryogenic air separation plant;

(B) an argon stripping column having a bottom re-boiler;

(C) means for passing fluid from the lower pressure column of the cryogenic air separation plant into the upper portion of the argon stripping column;

(D) means for passing fluid from the cryogenic air separation plant to the bottom reboiler and from the bottom re-boiler to the cryogenic air separation plant; and

(E) means for recovering product argon from the lower portion of the argon stripping column.

As used herein, the term "column" means a distillation or fractionation column or zone, i.e. a contacting column or zone, wherein liquid and vapor phases are countercurrently contacted to effect separation of a fluid mixture, as, for example, by contacting of the vapor and liquid phases on a series of vertically spaced trays or plates mounted within the column and/or on packing elements such as structured or random packing.

The term "double column", is used to mean a higher pressure column having its upper end in heat exchange relation with the lower end of a lower pressure column.

Vapor and liquid contacting separation processes depend on the difference in vapor pressures for the components. The high vapor pressure (or more volatile or low boiling) component will tend to concentrate in the vapor phase whereas the low vapor pressure (or less volatile or high boiling) component will tend to concentrate in the liquid phase. Partial condensation is the separation process whereby cooling of a vapor mixture can be used to concentrate the volatile component(s) in the vapor phase and thereby the less volatile component(s) in the liquid phase. Rectification, or continuous distillation, is the separation process that combines successive partial vaporizations and condensations as obtained by a countercurrent treatment of the vapor and liquid phases. The countercurrent contacting of the vapor and liquid phases is generally adiabatic and can include integral (stagewise) or differential (continuous) contact between the phases. Separation process arrangements that utilize the principles of rectification to separate mixtures are often interchangeably termed rectification columns, distillation columns, or fractionation columns. Cryogenic rectification is a rectification process carried out at least in part at temperatures at or below 150 degrees Kelvin (K).

As used herein, the term "indirect heat exchange" means the bringing of two fluids into heat exchange relation without any physical contact or intermixing of the fluids with each other.

As used herein, the terms "upper portion" and "lower portion" mean those sections of a column respectively above and below the mid-point of the column.

As used herein, the term "stripping column" means column operated with sufficient vapor up flow relative to liquid down flow to achieve separation of a volatile component from the liquid into the vapor in which the volatile component becomes progressively richer upwardly.

As used herein, the term "cryogenic air separation plant" means a plant, comprising at least one column, which processes feed air and produces at least one of product nitrogen and product oxygen."

2.6. Method and Installation for Air Distillation with Production of Argon

An air distillation methodfor argon production comprising an air distilling apparatus in particular with double column and at least one column for producing impure argon is resumed bellow and protected by patent claims [25]:

"A plant for the production of argon by air distillation, which comprises:

(A) a feed line for supplying air to be distilled;

(B) a heat exchanger through which said feed line passes;

(C) an air distillation apparatus having an inlet fluidly connected to said feed line, and an outlet;

(D) at least one impure-argon production column in fluid communication with said air distillation apparatus; and

(E) at least one branch line for sending at least a portion of argon extracted in excess into said heat exchanger or into said distillation apparatus."

2.7. Process and Plant for Producing Argon by Cryogenic Distillation

In this process for separating the argon-enriched gas, the gas is purged of oxygen in the first column and then a stream is removed from the top of the first column is sent into a stripping column [26].

Summary of the invention [26]:

"It is an object of the invention to provide a process for separating an argon-enriched gas by cryogenic distillation, comprising the following steps: the step of sending an argon-enriched gas stream to a first column; the step of separating the gas by distillation in the first column into an oxygen-enriched liquid and an argon-enriched gas; the step of at least partially condensing the argon-rich gas in an overhead condenser of the first column in order to form reflux; the step of removing another argon-rich stream containing at most 1000 ppm oxygen and 1000 ppm nitrogen from the first column and of sending it to the top of a second column; the step of warming the bottom of the second column by means of a warming gas; and the step of withdrawing, as final product, an argon-rich fluid from the lower portion of the second column.

The stream sent to the top of the second column may be a gas or a liquid.

Preferably, the warming gas is a fraction of the argon-enriched gas which feeds the first column.

In certain cases, the first column is in two sections, one section being fed with the argon-enriched gas and the other having the overhead condenser.

Normally, the first column is fed from the low-pressure column of a double column. Preferably, an overhead gas of the second column is sent either to the top of the first column or to the overhead condenser of the first column.

It is another object of the invention to provide a plant for separating an argon-enriched gas by cryogenic distillation, comprising: a first column having an overhead condenser; a second column having a bottom re-boiler; means for sending an argon-enriched gas to a first level of the first column; means for withdrawing an argon-rich fluid at a second level of the first column; means for sending the argon-rich fluid to the top of the second column; means for sending a warming gas to the bottom re-boiler; means for withdrawing an argon-rich final product from the bottom of the second column, characterized in that, between the first and second levels, there are at least 100 theoretical trays.

The argon-rich fluid sent to the top of the second column may be a gas or a liquid.

Preferably, thc plant comprises means for sending a portion of the argon-enriched gas to the bottom re-boiler as warming gas.

Optionally, the first column is constructed in two sections. Preferably, the plant comprises means for sending an overhead gas from the second column either to the top of the first column or to the overhead condenser of the first column."

2.8. Additional Techniques for Production of High Quality Argon

Some important papers and works are presented to show additional techniques concerning the production of high quality argon for industrial and research applications. The paper entitled "First Large Scale Production of Low Radioactivity Argon from Underground Sources"reports on the first large-scale production of low radioactivity argon from underground gas wells [27]. According to the research:

"Low radioactivity argon is of general interest, in particular for the construction of large-scale WIMP (Weakly Interacting Massive Particles) dark matter searches and detectors of reactor neutrinos for non-proliferation efforts. Atmospheric argon has an activity of about 1 Bq/kg from the decays of ^{39}Ar; the concentration of ^{39}Ar in the underground argon we are collecting is at least a factor of 100 lower than this value. The argon is collected from a stream of gas from a CO^2 well in southwestern Colorado with a Vacuum Pressure Swing Adsorption (VPSA) plant. The gas from the well contains argon at a concentration of 400-600 ppm, and the VPSA plant produces an output stream with an argon concentration at the level of 30,000-50,000 ppm (3-5%) in a single pass. This gas is sent for further processing to Fermilab where it is purified by cryogenic distillation. The argon production rate is presently 0.5 kg/day. We had established with previous sampling campaigns that some underground gas wells contain argon with low levels of ^{39}Ar with respect to the atmospheric concentration. We have demonstrated the practicality of extracting underground argon from gas streams containing minute concentrations of argon (a few hundreds of parts per million) on the scale required for the development of large-scale dark matter detectors. We have developed a technique suitable for processing large low rates of gas, capable of concentrating the argon phase by more than a factor of 10 in a single pass. This technique has been demonstrated on a primarily CO^2 stream, but it is applicable to a variety of naturally occurring gas streams. We have shown that it is possible to concentrate traces of argon from an underground stream into a crude argon product stream with argon concentrations up to 5%. We have demonstrated that this plant is already capable of producing 0.5 kg per day of low radioactivity argon, with a duty factor better than 30%. Operation of the plant is cost-effective, and the plant will be used to produce a few hundred kilograms of low radioactivity argon for WIMP dark matter searches."

Where demand for argon cannot be met as a by-product of oxygen production, ammonia-plant purge gas can provide an alternative source of feed gas [28]:

"By using Petrocarbon Technology, Costain Oil, Gas and Process Ltd have recently commissioned their first such plant on the AMC plant at Verdigris in Oklahoma. The cryogenic process has an overall argon recovery exceeding 97%, meeting CGA grade E specification (less than 7.5 ppm total impurities). The liquid argon production of 33 t d^{-1} is stored in a 210 m^3 horizontal vacuum-insulated tank and can be pumped to either road trailer or railcar for distribution. The cryogenic process also produces hydrogen which is recycled at 92% purity and 94% recovery to the ammonia-plant synthesis loop thereby boosting ammonia production [28]."

Anovel membrane/cryogenic hybrid scheme is presented wherein crude argon from a cryogenic air separation unit is fed to an oxygen selective membrane unit to remove a substantial portion of the oxygen [29]. From the authors' work:

"The oxygen enriched permeates flow from the membrane unit is returned to the crude argon distillation column of the cryogenic air separation process. The non-permeate stream is enriched in argon and can be further purified in a catalytic unit to produce an oxygen-free argon stream. The proposed process makes use of the synergy between the two separation units whereby, the cryogenic unit offers high recovery and the membrane provides purification leading to improved argon recoveries at higher argon concentrations. Calculations show that this process, in conjunction with an oxygen removal catalytic process, provides an economical alternative for the production of pure argon as compared to the conventional process using just a cryogenic unit and a catalytic unit to remove oxygen."

In other work, heats pump in a ternary distillation process with a side rectifier to produce argon from air distillation is discussed [30]:

"A ternary feed mixture ABC can be separated into individual components through the use of a main distillation column with a thermally linked side rectifier. To enhance such a separation, a heat pump can be implemented to transfer heat from the condenser at the top of the side rectifier to the reboiler at the bottom of the main column. In this paper, one such heat pump is described and applied to an air distillation system separating the ternary mixture containing nitrogen, oxygen and argon. The separation is performed by a conventional double column with a crude argon side column. When this system is operated at an elevated pressure to obtain higher product pressures, the separation of oxygen and argon becomes very difficult and leads to reduced argon recovery. The proposed heat pump enhances the separation by providing a supplementary crude argon condensing duty through the vaporization of a liquid oxygen stream from the bottom of the low pressure (LP) column. This scheme improves the liquid/vapor ratio (L/V) in the bottom section of the LP column and, more importantly, increases the vapor feed to the crude argon column. This increased feed rate leads to a substantial increase in argon recovery for the elevated pressure air distillation process."

The paper entitled "Production of Argon from an Oxygen–Argon Mixture by Pressure Swing Adsorption" [31] reports on pressure swing adsorption technique. According to the researchers:

"Kinetic selectivity of four potential adsorbents for the separation of oxygen–argon mixture by pressure swing adsorption (PSA) was compared based on adsorption equilibrium and kinetics measured in the linear range. In the chosen adsorbent, Takeda II carbon molecular sieve (CMS), the measurements were extended up to 6 atm pressure. Appropriate models were developed that explained the measured data and gave necessary parameters for subsequent use in PSA simulation. PSA experiments were conducted to concentrate argon from a 95:5 (mole ratio) oxygen–argon mixture using a single-bed, three-step cycle operated in 0–3 atm pressure range. PSA simulations were conducted using a bi-dispersed adsorbent model and a dual transport resistance with strongly concentration-dependent thermodynamically corrected transport parameters in the CMS micro-pores. It was a kinetically controlled separation with a high proportion of faster component in the feed mixture. The performance of the single-bed, three-step vacuum cycle at different operating conditions was systematically studied using the PSA simulation model validated with experimental results."

3. Argon Applications

3.1. Geochronology

Argon is a mixture of stable isotopes and the argon method is applied to geochronology to date the age of ancient materials, because argon has been found in rocks and fossils. Geochronology is the science of determining the age of rocks, fossils, and sediments, within a certain degree of uncertainty inherent to the applied method. A variety of dating methods are used by the geologists leading to the elaboration of different schemes, classifications and terminologies.The approach using several methods simultaneously is preferred for achieving better results (by consilience)[32]. The geochronological scale is a periodic scale using the year as a basic unit. Apparent ages obtained in geochronometry are referred to as radiometric or isotope dates. For older rocks, multiple annual units are normally written in thousands of years (ka) or million years (ma); Holocene and Pleistocene dates are normally quoted in years before 1950 (years BP (before present)) or more recently have been quoted as b2k (i.e. before the year 2000) [33]. The $^{40}Ar/^{39}Ar$ method is based on the decay of potassium to the inert gas argon which becomes physically trapped in the crystal lattice on formation. A reliable age is dependent upon the argon being held in place in substantial parts of the crystal. The commonly used step heating method, which involves progressive degassing of the samples up to melting point and analysis of the argon from each step, provides a way of looking at argon loss from different parts of the lattice and enables well-preserved parts of the crystal yielding crystallization ages to be distinguished from those which have suffered argon loss [34]. A wide variety of geochronological tools or methods can be employed to estimate quantitative and qualitative dating of rocks and sediments. Techniques exist to date all geological materials, from billions of years to historical records. $^{40}Ar/^{39}Ar$ provides isotopic ages ranging from 10,000 years to billions of age with 0.1-0.5% precision. $^{40}Ar/^{39}Ar$ provides isotopic ages that are used to derive thermal histories of geologic materials. Samples of any age throughout Earth's history may be dated accurately and precisely by the $^{40}Ar/^{39}Ar$ method, but a practical minimum age is about 10,000 years. This technique, particularly when applied to various minerals within a single sample and used in conjunction with other geochronological techniques, can provide not only the age of formation, but also its post-formation thermal and alteration history (based on argon "closure" temperatures in various minerals, the technique can delineate thermal history over a range of approximately 100-550°C)[35]. The $^{40}Ar/^{39}Ar$ dating method has become a widely used method in the field of isotope geochronology. It is one of the most widely applicable and precise methods in geochronology and has been successfully used for dating the oldest and youngest rocks on earth, for dating lunar samples, bulk rocks and single crystals, and for dating rocks in situ using a variety of lasers systems. However, the increasing number of ages generated by the $^{40}Ar/^{39}Ar$ dating technique, more and more requires a consistency in the calculation of these ages and particularly their errors [36, 37]. The reference age of monitor minerals used and the uncertainties in the ^{40}K decay constants are main sources of systematic errors in $^{40}Ar/^{39}Ar$ dating, and the atmospheric correction on ^{40}Ar and the J-value determination are important sources of analytical errors [38, 39].

There are five isotopes of the noble gas argon which have to be measured for Ar/Ar dating.Naturally occurring argon is comprised of 0.337 ± 0.0003 at.%^{36}Ar, 0.036 ± 0.001 at.% ^{38}Ar and of the major isotope ^{40}Ar (99.6 ± 0.0003 at.%). ^{36}Ar, ^{38}Ar and ^{40}Ar are stable isotopes. Unstable ^{39}Ar and ^{37}Ar are produced from Ca and K during neutron irradiation. ^{39}Ar decays to ^{39}K by beta-emission with a half-life of ~ 269 years. Because of this slow decay rate ^{39}Ar can be treated as stable during the short time it is involved in the analysis. ^{37}Ar decays with a half-life of ~ 35 days which has to be considered during Ar/Ar analyses [40]. Thermochronology, the study of the thermal history of rocks, provides an important record of the vertical motions of bodies of rock over geological timescales, enabling us to quantify the nature and timing of tectonic processes. Isotopic age data constrain the age of rocks and minerals, but in many cases they are interpreted without a proper understanding of the relationship between the age measured and the physical processes within the Earth. Potassium-Argon (K-Ar) dating, and the associated $^{40}Ar/^{39}Ar$ method of analysis, are among the most widely used thermochronological methods. Potassium is a major element that makes up about 1.5% of the Earth's crust and is abundant in common rock-forming minerals such as K-feldspars, amphiboles and micas. Potassium has three naturally occurring isotopes: ^{39}K and ^{41}K, which are stable and together make up about 99.9% of the natural abundance of K, and the radioactive isotope 40K. Naturally occurring calcium, another abundant and major rock-forming element in the Earth's crust, is dominated by the isotope ^{40}Ca (the natural isotopic abundance of which is 96.9%). Therefore, there are commonly large quantities of ^{40}Ca bound up in mineral structures, which are indistinguishable from the radiogenically produced ^{40}Ca. Moreover, the diffusion behavior of Ca is complex due to its reactivity. For these reasons, the decay reaction of ^{40}K to ^{40}Ca is not generally used as a dating method, although some studies have utilized this reaction in particularly favorable conditions. Argon, although a secondary constituent of the Earth's atmosphere (~ 1% by weight), is sufficiently rare in rocks that the radiogenic component can be precisely measured. In most circumstances it can be assumed that Ar was not present in the crystal structure at its formation. Because Ar is a noble gas, it is unreactive and its diffusion behavior is relatively simple, making it a particularly effective thermochronometer [41].

3.2. Argon Plasmas

High purity argon gas (≥ 99.999%) is commonly used in the low and high pressure plasma technology. Its applications ranging from pure research to specific fields such as medicine (dermatology, ophthalmology, for example), semiconductor industry, analytical chemistry, nanotechnology, among others [6, 7]. In the sequel, we present some of the latest applications of plasma technology, where special focus is given to the promising field of plasma medicine [8]. In the paper entitled "Low-temperature sterilization using gas plasmas: areview of the experiments and an analysis of the inactivation mechanisms", Moisan et al. reviewed the literature on plasma sterilization [42]. Using an ionized gas (plasma) to achieve sterilization is an alternative to the conventional sterilization means because the sterilization of heat-sensitive materials is then possible and the sterilizing agents are thoroughly innocuous. The Moisan review is major issue on thc plasma sterilization methods andthe respective roles of UV photons and reactive species such as atoms and radicals [42]. A new, low power capacitively coupled plasma (CCP) micro torch (30 W, 13.56 MHz, 0.5 $L.min^{-1}$ Ar flow) was investigated in conjunction with commercially available micro-spectrometers for the simultaneous multielemental analysis by atomic emission spectrometry of liquid samples without desolvation [43]. Compared to the traditional Inductively Coupled Plasma (ICP) mass spectrometers, the microtorch has a simple construction, runs at low argon flow and can be integrated in a portable system suitable for in-situ simultaneous determination of the consisting elements. In laboratory the analytical performance of a medium power (275 W) argon CCP torch with a Mo tubular electrode and single or double ring electrodes used in the multielemental analysis by atomic emission spectrometry (CCP-AES) of pneumatically nebulized liquid samples was investigated. The results were the basis for building a simultaneous spectrometer with photodiode array as detector for the determination of alkali and earth-alkali elements as a substitute to the old methane-air flame photometer [43]. Argon plasma applications in medicine with a special focus on dermatology and ophthalmology has been growing to meet the demands. The recent tremendous progress in understanding the physical plasma phenomenon, together with the development of new plasma sources have put growing focus on the application of plasmas in health care [44]. Active plasma components, such as molecules, atoms, ions, electrons and photons, reactive species, ultraviolet radiation, optical and infrared emission and heat have the ability of activating, controlling and catalyzing reactions and complex biochemical procedures [44]. Thermal and non-thermal (i.e. cold) plasmas - both already widely established in the medicine - have been used for various therapeutic applications. Particularly in dermatology, plasma applications hold big potential, for example, in wound healing, such as efficient disinfection or sterilization, therapy of various skin infections or tissue regeneration. The works of Heinlin et al. gives an overview on potential plasma applications in medicine - including the recent research on skin diseases - and summarizes possible interactions between plasmas and living tissue [44].

Recently,Isbary and coworkers showed that a 2-min treatment with either of two cold atmospheric argon plasma devices is a safe, painless and effective technique to decrease the bacterial load in chronic wounds [45]. The 5-min argon cold atmospheric plasma treatment has already demonstrated efficacy in reducing bacterial numbers in chronic infected wounds *in vivo*. They investigated a plasma treatment to assess safety and the reduction in bacterial load, regardless of the kind of bacteria and their resistance level in chronic wounds, achieving a significant reduction in bacterial load [45]. In other work, Lam et al. used argon plasma coagulation for the treatment of hemorrhagic radiation colitis, at the regional cancer center (Cross Cancer Institute) in Edmonton, Alberta, Canada [46]. The argon plasma coagulation used in this work is non-equilibrium plasma because the argon gas temperature is almost at room temperature, remaining the heavy active species cold, whereas the electron temperature (electrons mean translational energy) reaches a few thousands of Kelvin degrees. Low temperature atmospheric argon plasmas (radio-frequency, direct current or microwave-driven plasmas) are relatively new method and contribute vastly to the medical diagnostics and applications.Studies of the electrical and the optical parameters of these discharges are in course nowadays. The group of Ermolaeva and coworkers [47] developed a microwave generator for medical applications. Using optical emission spectroscopy, this group studied the optical parameters of the argon discharge and observed the non-equilibrium of the plasma, making so the plasma parameters effective for the specific medical use.

CONCLUSION

In this chapter we summarized the main argon properties, the production methods and some recent applications in the fields of geochronology, thermochronology, medicine and plasma technologies. Basic data of this chemical elementwas also presented. We reviewed the argon production from the past, where the production of high purity argon involved number of processing steps to produce a crude argon stream which was then upgraded in a refinery, to nowadays, where the cryogenic process is comprised of unit operations that compress, purify, and separate the air feed into the required gaseous and liquid oxygen, nitrogen, and argon product flows.

Additionally, a list of some important US patents and papers about the argon production complement this section. The mentioned patents and procedures were listed chronologically, showing both the evolution of the cryogenic technology in the rare gas production and the main processes steps itself. The argon production methods are: process for recovering argon, cryogenic rectification method for producing refined argon, cryogenic distillation of air using multiple distillation columns, column configuration and method for argon production, cryogenic production system with thermally integrated stripping column, method and installation for air distillation and process and plant for argon producing by cryogenic distillation. Additional techniques for production of high purity argon were examined and presented in this chapter.

About the applications of argon gas, we showed that the isotope argon method can be applied to a geochronology to date the age of ancient materials because argon is found in rocks and fossils.The $^{40}Ar/^{39}Ar$ dating method has become a widely used method in the field of isotope geochronology. By the other hand, the thermochronology study the thermal history of the rocks and provides important information on the vertical motions of the rocks over geological timescales, enabling us to quantify the nature and timing of the tectonic processes. In this section, some recent papers were listed in the geochronological and thermochronological areas. Finally, in the last topic of this chapter we listed a number of works that deals with argon gas to produce low and high pressure plasmas for a range of applications. The high purity argon gas is the most commonly used in the plasma technology and besides the pure research, applications in the fields of medicine, semiconductor industry, analytical chemistry and nanotechnology are actively being developed at the moment. It should be noted that low temperature atmospheric-pressure argon plasma opens new contributions for these applications, whereas electrical and optical studies of these type discharges are in course nowadays.

ACKNOWLEDGMENT

This work was supported by the following Brazilian agencies projects: CAPES/PNPD (process nº 02765/09-8), CAPES/PVE (process nºBEX9796/12-6), FAPESP/CNPq/PRONEX (process nº 2011/50773-0) and FAPESP (process nº 2012/13064-4).

REFERENCES

[1] *Lord Rayleigh - Biography*. Nobelprize.org. 28 Jul 2012 http://www.nobelprize.org /nobel_prizes/physics/laureates/1904/strutt-bio.html.

[2] Argon: An Overview, WikiProject Wikipedia-Books, site: http://en.wikipedia.org /wiki/Book:argon, access: 07/20/2012.

[3] A.A. Pek, *On the dynamique of juvenile solutions,* Nauka, Moscow (1968).

[4] V.I. Malkovsky, A.A. Pek, P.L. Arseniev, and D.N. Topor, *Izvestya RAN, FizikaZemli,* 12, 57 (1988).

[5] W. A. Krivsky, *Metallurgical and Materials Transactions B*4 (6), 1439-1447 (1973).

[6] Fridman, *Plasma Chemistry*, Cambridge University Press, Cambridge (2008).

[7] M. A. Lieberman, A. J. Lichtenberg, *Principles of Plasma Discharges and Materials Processing*, 2th edition, New York: Wiley (2005).

[8] M. Laroussi, M. G. Kong, G. Morfill, *Plasma Medicine: Applications of Low-Temperature Gas Plasmas in Medicine and Biology*, Cambridge University Press, Cambridge (2012).

[9] http://www.eoearth.org/article/Argon?topic=49557, access 07/20/2012.

[10] K.C. Spencer, D.J. Humphreys, ACS Symposium Series 836, 270-291 (2003).

[11] N. Valentin, *International Biodeteriorationand Biodegradation* 32 (4), 263-278 (1993).

[12] http://webbook.nist.gov/cgi/cbook.cgi?ID=C7440371andMask=200, access 07/20/2012.

[13] L. Khriachtchev, M. Pettersson, N. Runeberg, J. Lundell, M. Räsänen, *Nature*406, 874–876 (2000).

[14] D. Petsalakis, G. Theodorakopoulos, *J. Phys. B: At. Mol. Opt. Phys.*27, 4483 (1994).

[15] H. Akashi, Y. Sakai, T. Tagashira, *J. Phys. D: Appl. Phys.* 27, 1097 (1994).

[16] Cohen, J. Lundell, R. B. Gerber, *Journal of Chemical Physics* 119(13), 6415-6417 (2003).

[17] J. T. Bernstein, *Cryogenic argon Production*, presented in: Modern Air Separation Plant Technology Conference (1999).

[18] D. P. Bonaquist and M. J. Lockett, United Stated Patent No. 5, 857, 357, Jan. 12, (1999).

[19] D. R. Vinson, *Computers and Chemical Engineering* 30, 1436–1446 (2006).

[20] J. R. Bianchi, D. P. Bonaquist and R. A. Victor, United Stated Patent No. 5, 133, 790, Jul. 28, (1992).

[21] H. Ogada, United Stated Patent No. 4, 575, 388, Mar. 11, (1986).

[22] F. De Bussy and Jean- Pierre Tranier, United Stated Patent No. 5, 682, 767, Nov. 4, (1997).

[23] D. P. Bonaquist and M. J. Lockett, United Stated Patent No. 5, 857, 357, Jan. 12, (1999).

[24] D. P. Bonaquist, N. J. Linch and M. J. Roberts, United Stated Patent No. 5, 916, 261, Jun. 29, (1999).

[25] F. De Bussy, F. S. and B. Saulnier, United Stated Patent No. 6, 269, 659, Aug. 7, (2001).

[26] B. Saulnier and J. P. Gourbier, United Stated Patent No. 6, 574, 988, Jun. 10, (2003).

[27] H. O. Back, F. Calaprice, C. Condon, E. de Haas, R. Ford, C. Galbiati, A. Goretti, T. Hohmann, An. Ianni, B. Loer, D. Montanari, A. Nelson and A. Pocar, Cornell University Library, http:// www. arxiv.org/.

[28] M. J. Healy, *Gas Separation and Purification*6, 191 (1992).

[29] R. Agrawal, S. R. Auvil, J. S. Choeand D. W. Woodward, *Gas Separation and Purification*4, 75 (1990).

[30] R. Agrawal, D. W. Woodward, T. F. Yee, *Gas Separation and Purification*8, 37 (1994).

[31] Xu Jin, A. Malek and S. Farooq, *Ind. Eng. Chem. Res.* 45 (16), 5775 (2006).

[32] P.R. Renne, K.R. Ludwig, and D.B. Karner, *Progress and Challenges in Geochronology. Science Progress* 83,107 (1998).

[33] http://www.inqua-saccom.org/, accessed in Aug. 1, 2012.

[34] http://www.inqua-saccom.org/, accessed in Aug. 2, 2012.

[35] http://geology.cr.usgs.gov, accessed in Aug. 2, 2012.

[36] A. P. Koppers, *Computer and Geosciences*28, 605 (2000).

[37] S. Scaillet, *Chemical Geology*162, 269 (2000).

[38] P. R. Renne, C. C. Swisher, A. L. Deino, D. B. Karner, T. C. Owens and D. J. Depaolo,*Chemical Geology (Isotope Geoscience Section)*145, 117 (1998).

[39] K. Min, R. Mundil, P. R. Renne and K. R. Ludwig, *Geochemica et CosmochimicaActa*64, 73 (2000).

[40] Scherer, $^{40}Ar/^{39}Ar$ dating and errors, Weingasse 13, 09599 Freiberg, http://www.geo.tu-freiberg.de/.

[41] J. Braun, P. van der Bek and G. Batt, *Numerical Methods for the Interpretation of Thermochronological Data*, Cambridge University Press, http:// www.thermoeurope.eu/.

[42] M. Moisan, J. Barbeau, S. Moreau, J. Pelletier, M. Tabrizian, L'H. Yahia, International Journal of Pharmaceutics 226 (1–2), 1–21 (2001).

[43] T. Frentiu, D. Petreus, M. Senila, A. I. Mihaltan, E. Darvasi, M. Ponta, E. Plaian and E. A. Cordos, *Microchemical Journal*97, 188 (2011).

[44] J. Heinlin, G. Isbary, W. Stolz, G. Morfill, M. Landthaler, T. Shimizu, B. Steffes, T. Nosenko, J. Zimmermann, S. Karrer, *J. Eur. Acad. Dermatol. Venereol.* 25(1), 1-11 (2011).

[45] G. Isbary, J. Heinlin, T. Shimizu, J. L. Zimmermann, G. Morfill, H. –U. Schmidt, R. Monetti, B. Steffes, W. Bunk, Y. Li, T. Klaempf, S. Karrer, M. Landthaler, W. Stolz, *British Journal of Dermatology*, 167, 404 (2012).

[46] M. C. W. Lam, M. Parliament, C. K. W. Wong, *Case Rep. Gastroenterol.* 6, 446-451 (2012).

[47] S. Ermolaeva, O. Petrov, N. Zigangirova, M. Vasiliev, E. Sysolyatina, S. Antipov, M. Alyapyshev, N. Kolkova, A. Mukhachev, B. Naroditsky, *Science for Peace and Security Series A: Chemistry and Biology*, Part 2, 163 (2012).

In: Argon: Production, Characteristics and Applications
Editors: B.N. Sismanoglu, H.S. Maciel et al.
ISBN: 978-1-62618-204-2

Chapter 2

USE OF ARGON FOR MEASUREMENT OF ROCK PERMEABILITY

***V. I. Malkovsky*[1,2,*]*, A. V. Zharikov*[1,3] *and V. M. Shmonov*[3]

[1]Institute of Geology of Ore Deposits, Petrography, Mineralogy, and Geochemistry, Russian Academy of Sciences, Moscow, Russia

[2]D. Mendeleyev University of Chemical Technology of Russia, Moscow, Russia

[3]Institute of Experimental Mineralogy, Russian Academy of Sciences, Moscow, Russia

ABSTRACT

Permeability is one of the main parameters governing many processes of heat- and mass transfer in the Earth crust as, for example, development of ore-forming systems and contaminant transport by ground water from geological repositories of high level radioactive waste. These processes depend to a considerable degree on transport of energy and substances by fluids – gases or liquids which circulate in the Earth crust. For that reason velocity of the fluid flow is of prime importance for these processes. Permeability of weakly fractured rocks can be determined in laboratory experiments on relatively small rock samples. The sample of cylindrical shape is manufactured from the examined rock. The flank surface of the sample is sealed and two bases of the cylinder are in contact with a fluid under different pressure. Permeability is determined from volume of a fluid which flows through the sample under the pressure gradient during the experiment. The fluid volume should be large enough for due precision of the measurement. As a result, measurement of tight rock permeability can be time consuming process. The measurement duration decreases with a decrease of dynamic viscosity of the fluid. For that reason, it is reasonable to select a weakly viscous fluid for these measurements. In addition to that any interaction between the fluid and the rock should be absent. The noble gases meet the both requirements. Advantages of argon are relatively low cost and availability. The simplest technique for laboratory measurement of rock permeability is steady state method where pressure gradient is constant in time. In spite of low viscosity, filtration of measurable volume of argon through a tight (low permeable) rock sample calls for a long time. For that reason, so called "pulse decay" method was developed by W. Brace and his colleagues in 1968. Amount of the filtered

* E-mail address: malk@igem.ru.

gas in this method is determined from pressure drop in the closed reservoir connected with the inlet cross-section of the rock sample. As a result of Knudsen effect, permeability depends on gas density. Therefore, determination of permeability in the whole range of fluid density (from rarefied gas to liquid) calls for measurements at different gas pressures. Therefore even this method is time consuming. We modified the "pulse decay" method what reduced radically its duration and increase the accuracy of measurements.

The technique allows to accomplish high-accuracy determinations of permeability within the range of values of 10^{-22}–10^{-15} m^2 both under normal conditions and at elevated *PT*-parameters. The modified method is based on an original technique which does not call for measurements at different pressures of argon. The method was used for measurement of permeability of many types of rocks. A next modification of this technique leads to development of the "method of alternating flows" for measurement of nonisotropic permeability of layered rocks. Argon is used as a gas in the experiments. The method was approved in measurement permeability of main types of rocks of the Streltsovskoye uranium ore field and the areas of significant radioactive contaminations of the underground medium, related to radioactive waste disposal (South Ural and Eastern Siberia).

1. INTRODUCTION

Permeability is one of the main parameters governing the processes of fluid heat-mass transfer in the Earth crust [1]. Such processes substantially control to a marked extent, for an example, development of ore-forming systems [2; 3], and contaminant transport by ground water from geological repositories of high level radioactive waste [4]; operation rate of geothermal heat sources [5]. Since characteristics of these processes depend significantly on rock permeability, determination of this parameter is of great importance.

In the laboratory measurements of rock permeability, as a rule, cylindrical samples are used. In the most simple technique (so called steady state method) the lateral walls of sample are sealed off, and constant pressure of flowing media (gas or water) p_{in} and p_{out} ($p_{in} > p_{out}$) are held upon the end faces (Figure 1 a). Under the assumption that the velocity of filtration throughout a sample is constant and normal to the end faces, the permeability value is determined from the Darcy law:

$$k = \frac{\mu l G}{S(p_{in} - p_{out})}, \quad (1)$$

where l and S are the length and cross sectional area of the sample, μ is the dynamic viscosity of flowing medium, G is the flow rate. Its value in these experiments is given usually as $G=v/t_e$, where v is the volume of flow issuing from the outlet section of the sample, t_e is the experiment duration. Such approach supplemented by special equipment and technique for accurate measurements of low v has been successfully realized by Morrow and Byerlee [6] and Morrow and Lockner [7]. The fluid flows in the sample through connected small pores and microfractures. This flow is very slow and, as a result, filtration of a volume which can be measured with sufficient accuracy is time consuming process. As a consequence, it is reasonable to select such fluid for these measurements dynamic viscosity of which is low. In addition to that any interaction between the fluid and the rock should be absent. The noble gases meet the both requirements. Advantages of argon are relatively low cost and availability.

Brace with the colleagues [8] pioneered another technique permitted to avoid the necessity to measure small volumes of the filtered fluids in the experiments. The end faces of a sample with sealed lateral surface were connected with the closed reservoirs filled with certain medium (liquid or gas) under various pressures p_{in} and p_{out} (Figure1 b). Due to the medium filtration through the sample the pressures in the both reservoirs gradually level off. Permeability values were determined as the best fit from matching the data of time-dependent pressure in the both reservoirs obtained by measurements and calculated from a solution of the mass balance equation. For this purpose graphic [8], analytical [9, 10] and numerical [11] methods were developed. The technique was called pulse decay method. It is very widely used now for measurements of low permeable rocks (e.g., in [12–16]). The advantage of the method is that it is based on measurements of pressure variation in the reservoirs what is easier than measurement of ultra-small volumes of the filtered fluid.

The method of oscillating pore pressure is also in use [17, 18]. With this method, the permeability is determined on the base of data on attenuation and of phase's retardation of a sinusoidal wave of pore fluid pressure when it propagates through a sample.

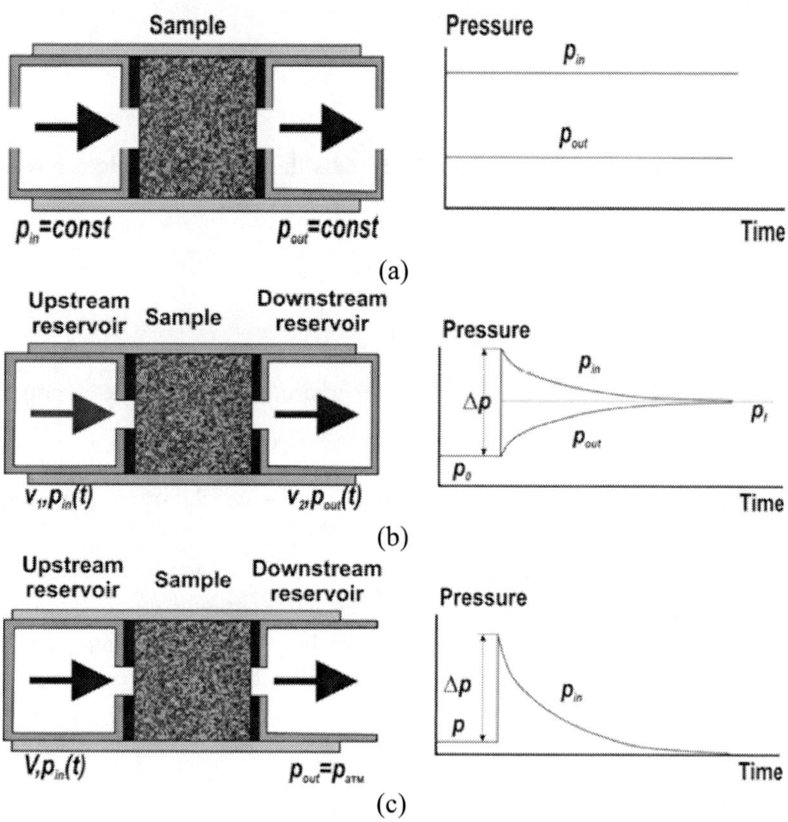

Figure 1. Concept of permeability determination methods and dependences of the pressure p_{in} change in the upstream reservoir during measurements: a – steady state, (b) – conventional pulse decay, c – modified pulse decay method.

As it is mentioned, gas is often used in the measurements as a flowing medium (argon [8; 19] or nitrogen [16]). However, it was found that permeability values obtained in the experiments with liquids and with gases can differ considerably. This effect was explained [20] as caused by increasing frequency of molecular wall collisions relative to intermolecular collisions when molecular mean free path becomes comparable to sample pore sizes what leads to deviation from Poiseuille flow in sample pore channels. To correct for this effect the dependence of the gas permeability (k) on the gas pressure (p) was proposed:

$$k = k_w\left(1 + \frac{b}{p}\right), \tag{2}$$

where k_w is the water permeability of the sample and b is the constant characterizing pore size.

Since b characterizes the ratio of pore channel dimensions to the molecular mean free path, b permits to estimate the average radius of the pore channels in the rock sample [21, 22] which is one more argument in favor of permeability measurements with use of gas as a flowing medium. It should be taken into account that b is not a rock property because this parameter depends also on temperature and dynamic viscosity of the gas. The estimation from [21] for the equivalent average radius of the pore channels r

$$r = \frac{4}{3}\frac{\pi\mu\bar{v}}{b}$$

was extended in [23] to the case when flow channels in the rock sample are represented by microcracks

$$\delta = 2\frac{\pi\mu\bar{v}}{b}$$

where δ is the equivalent average radius of the microcracks, $\bar{v}$ is the average molecular velocity of the gas.

When gas is used as a flowing medium, k_w and b have been usually derived from the results of the several measurements of the gas permeability carried out under various average gas pressure $p_{av} = (p_{in} + p_{out})/2$.

A relative drawback of this technique is a necessity to carry out several experiments under various p_{av} to obtain a single k_w value. When the technique proposed in [8] is in use, analytical description of the process (i.e. time variations of $p_{in}(t)$ and $p_{out}(t)$) was derived assuming that the distribution of gas pressure along the length of the sample satisfies the equation

$$\frac{\partial^2 p}{\partial x^2} = 0. \tag{3}$$

Here x is the longitudinal coordinate, $x=0$ and $x=l$ represent the inlet and outlet faces of the sample.

The equation (3) for gas flow through a cylindrical rock sample with sealed off sidewalls can be derived as follows. The components of the flow velocity vector satisfy the Darcy's law [24]

$$\vec{v} = -\frac{k}{\mu}(\nabla p - \rho\vec{g}) \tag{4}$$

where ρ is gas density, $\vec{g}$ is acceleration due to gravity. Since the gas density is low, the term of gravity force can be neglected. Pressure distributions at the end faces of the sample are uniform. Therefore the flow in the sample is one-dimensional.

The expression for only one non-zero velocity component from (4) takes the form

$$v_x = -\frac{k}{\mu}\frac{\partial p}{\partial x} \tag{5}$$

If porosity of the rock φ is uniform in the whole sample, the components of the velocity vector satisfy the continuity equation [24]

$$\frac{\partial}{\partial t}(\rho\varphi) + \mathrm{div}(\rho\vec{v}) = 0$$

This equation in the case of 1-D flow takes the form

$$\frac{\partial}{\partial t}(\rho\varphi) + \frac{\partial}{\partial x}(\rho v_x) = 0 \,. \tag{6}$$

Substitution of (5) for the velocity in (6) gives

$$\frac{\partial}{\partial t}(\rho\varphi) - \frac{\partial}{\partial x}\left(\frac{\rho k}{\mu}\frac{\partial p}{\partial x}\right) = 0 \,. \tag{7}$$

If the porosity is small and relative variations of ρ, k and μ are weak the equation (7) is reduced to (3). The experiments were carried in [8] at such temperatures and pressures that argon can be considered as an ideal gas. Therefore, the density of the gas is proportional to the pressure. As it follows from (2), k is also pressure-dependent. Hence, variations of ρ, k and μ can be neglected only at low relative variations of the pressure. As a result, the equation (3) is valid for the gas filtration only in the case when the relative difference of pressures at the end faces of the sample is little, that is

$$\frac{p_{in}(t) - p_{out}(t)}{p_{av}} << 1$$

where $p_{av}(t) = (p_{in}(t) + p_{out}(t))/2$ ($p_{av}(t)$ remains practically constant during the experiment because $p_{in}(t)$ decreases and $p_{out}(t)$ increases with time). Since the variations of p do not exceed $p_{in}(0) - p_{out}(0)$, ρ, k and μ are approximately constant in this case. Average value of k which can be determined in the experiment by means of the best fit method corresponds to the average value of the pressure in this experiment. The parameters k_w and b characterize the pressure dependence of k. With the pulse decay method in its conventional variant from [8], k_w and b can be obtained if k is measured, at least, at two different average pressures. But if the relative changes of p during the experiment are significant, i.e.

$$\frac{p_{in}(t) - p_{out}(t)}{p_{av}} \cong 10^0,$$

it is possible that k_w and b can be determined from the data of a single experiment. This can be achieved by a simple modification of the pulse decay technique [8], when only the inlet section of the sample is connected with a closed reservoir filled with gas at some large (in comparison with atmospheric) pressure, and the outlet section of the sample remains to be open (Figure 1 c).

In this case, the gas pressure distribution along the length of the sample does not satisfy the equation (2), and all the time-dependencies of the gas pressure in a reservoir differ from ones determined in [8]. Technique for theoretical determination of $p_{in}(t)$ for the modification of the pulse decay method will be presented below.

2. Theoretical Model

Let us consider the system, schematically shown in Figure 1 c. A side surface of a sample is sealed off. One of the end faces of the sample is connected with a closed reservoir of volume V filled with argon. The second end face of the sample remains to be open, that is the pressure p_{out} at the outlet section of the sample during the experiment is equal to atmospheric pressure p_f. A step increase of the argon pressure in the upstream reservoir up to the value of $p_f + \Delta p$ is introduced at a start time. We assume that the porosity, dimensions and the matrix structure of the sample depend weakly on the pressure within the considered range of the gas pressure, and the process of argon filtration through the sample is governed by the Darcy law (5)

The experiments are carried out at pT-parameters from the range ($T \cong$ 300–500 K and $p \cong$ 0.1–10 MPa) where the state equation for ideal gas is valid for argon. Since the sample is isothermal during the whole experiment, $\rho = Ap$ where A is temperature-dependent coefficient. We substitute the expressions for the density and the expression (2) for k into (7). As a result, we obtain

$$\varphi \frac{\partial p}{\partial t} = k_w \frac{\partial}{\partial x}\left[\frac{p}{\mu}\left(1 + \frac{b}{p}\right)\frac{\partial p}{\partial x}\right]; \quad t > 0, \quad 0 < x < l. \tag{8}$$

Boundary conditions for p can be written as

$$x = 0, \quad p = p_{in}(t); \quad x = l, \quad p = p_f \tag{9}$$

where $p_{in}(t)$ satisfies the equation of the mass balance in the reservoir

$$\frac{dp_{in}}{dt} = \frac{S p_{in} k_w}{V\mu}\left(1 + \frac{b}{p_{in}}\right)\frac{\partial p}{\partial x}\bigg|_{x=0} \tag{10}$$

The initial conditions for equation (8) can be written as

$$t = 0, \quad p(x) = \begin{cases} p_f, x > 0; \\ p_f + \Delta p, x = 0. \end{cases} \tag{11}$$

The boundary problem (8–11) describes completely the gas flow through the sample.

Let us consider a simplified model of the process assuming that storage due to compression of the gas in the pore volume of the sample can be neglected (at $\varphi << 1$), and μ =*const.* Then, we will obtain from equation (8)

$$\frac{\partial}{\partial x}\left[p\left(1 + \frac{b}{p}\right)\frac{\partial p}{\partial x}\right] = 0 \quad \text{or} \quad (p + b)^2 = c_1 x + c_2,$$

where c_1 and c_2 are the constants of integration, which are determined from the boundary conditions (9). Substitution of the explicit expression for $p(x)$ into the equation (10) gives the general Riccati equation:

$$\frac{dp_{in}}{dt} = \frac{1}{2}\frac{Sk_w}{V\mu l}\left(p_f - p_{in}\right)\left(p_{in} + p_f + 2b\right) \tag{12}$$

Solution of the Cauchy problem (12), (11) can be written as

$$p_{in}(t) = \frac{p_f(1 + \tau) + 2b\tau}{1 - \tau}, \quad \tau = \frac{\Delta p}{\Delta p + 2p_f + 2b} \cdot \exp\left\{-\frac{\left(p_f + b\right)Sk_w}{V\mu L}t\right\}. \tag{13}$$

Therefore, the function $p_{in}(t)$ where the k_w and b are parameters is determined. The value of p_{in} is measured in N time points $\{t_i, i = 1, ..., N\}$ during the experiment. The measured values of p_{in} in these time points are $\{p_{in}^{(\exp)}(t_i), i = 1, ..., N\}$. The parameters k_w and b can be determined as values which minimize the function

$$\Phi(k_w,b)=\sum_{i=2}^{N}\left[p_{in}^{(\exp)}(t_i)-p_{in}^{(calc)}(t_i;k_w,b)\right]^2 \tag{14}$$

where $p_{in}^{(calc)}(t_i;k_w,b)$ are the calculated values of p_{in} in time points $\{t_i, i=1,..., N\}$ at specified values of the parameters k_w and b.

Values of $p_{in}{}^{(calc)}$ can be calculated with use of the expression (13), but it should be taken into account that this expression was obtained under assumption that storage due to compression of the gas in the pore volume can be neglected (i.e. the left part of the equation (8) is taken to be zero). The systematic error associated with this assumption increases with the increasing porosity. Since the same assumption is used in the conventional pulse decay method (CPDM) presented in [8], the problem of evaluation of this systematic error is considered in several studies on CPDM [9–11]. It is shown in [11] that if the ratio of the effective sample pore volume is less than 0.25, then the systematic error in permeability determination caused by this assumption does not exceed 10%. This result followed from comparison between numerical solution of the problem in general formulation with the approximate analytical solution obtained under assumption that the left part of the equation (8) is zero. However this result was obtained for CPDM where the both end faces of the sample are connected with the closed reservoirs. This is not obvious for estimation of corresponding systematic error in the case of the modified pulse decay method (MPDM) when the outlet section of the sample remains to be open to the atmosphere. One should compare in this case the values of k_w and b obtained with use of approximate analytical solution (13) with the values obtained on the basis of numerical solution of the problem (8–11) in general formulation. That is why two methods for calculation of $p_{in}^{(calc)}$ in (14) (and respectively two variants of k_w and b determination from the condition of Φ minimum) are considered. Using the first method, $p_{in}^{(calc)}$ is calculated from expression (13), i.e. from the analytical solution of the simplified problem. With the second method, $p_{in}^{(calc)}$ is determined from the results of numerical solution of the boundary problem (8–11) obtained with regard to the real nonzero value of the sample porosity and the pressure dependence of argon viscosity μ.

The numerical solution of the problem (8–11) was obtained through integration of the equation (8) by the Crank-Nickolson method [25].

Let us partition the interval $0<x<l$ into N equivalent segments $\{x_j \le x \le x_{j+1}, j=1,..., N-1\}$ where

$$x_j=\Delta_x(j-1),\quad \Delta_x=l/(N-1)$$

We denote

$$t_n=\Delta_t(n-1),\quad p_j^n=p(t_n,x_j)$$

where Δ_t is step of integration with respect to time.

Values $\{p_j^n, \quad j=1,\ldots,N\}$ at n = 1 can be determined according to the initial condition (10) as

$$p_1^1 = p_f + \Delta p, \quad p_j^1 = p_f, \; j = 2,\ldots,N \tag{15}$$

With thc Crank-Nickolson method, numerical integration of the equation (8) is governed by the expression

$$p_j^{n+1} = p_j^n + \frac{k_w \Delta_t}{2\Delta_x^2}\left\{\frac{\left(p_j^n + p_{j+1}^n + 2b\right)}{\mu_j^n + \mu_{j+1}^n}\left(p_{j+1}^{n+1} - p_j^{n+1} + p_{j+1}^n - p_j^n\right) - \frac{\left(p_{j-1}^n + p_j^n + 2b\right)}{\mu_{j-1}^n + \mu_j^n}\left(p_j^{n+1} - p_{j-1}^{n+1} + p_j^n - p_{j-1}^n\right)\right\}, \quad j = 2,\ldots,N-1 \tag{16}$$

where $\mu_j^n = \mu\left(p_j^n, T\right)$, T is the absolute temperature of the sample which remains constant during the experiment.

Values of p_1^n, p_N^n at n = 2,… can be determined from the boundary conditions (9) as

$$p_1^n = p_{in}(t_n), \quad p_N^n = p_f; \quad n = 1,\ldots \tag{17}$$

where $p_{in}(t_n)$ is obtained from numerical integration of the equation (10) by finite difference method according to the expression

$$p_{in}(t_{n+1}) = p_{in}(t_n) + \frac{Sk_w\Delta_t\left(p_2^n + p_1^n + 2b\right)}{2V\left(\mu_1^n + \mu_2^n\right)\Delta_x}\left(p_2^n - p_1^n + p_2^{n+1} - p_1^{n+1}\right), \quad n = 1,\ldots \tag{18}$$

where $p_{in}(t_1 = 0) = p_f + \Delta p$ in accordance with the initial condition (15) and boundary condition (17).

Initial distribution of the pressure in the sample is determined from the condition (15). If distribution of the pressure at some t_n is known, it can be calculated at the next time point t_{n+1} from the expressions (16)–(18). As a result, we obtain pressure distribution in the sample at any time point and, in particular, pressure at the inlet section of the sample $p_{in}(t)$.

If we restrict ourselves to the first two terms of virial decomposition, the expression for the dynamic viscosity for argon in the considered region of p,T-parameters takes the form [26]

$$\mu = \mu_0\left[1 + C_1(T)b_0\rho + C_2(T)(b_0\rho)^2\right] \tag{19}$$

where

$$\mu_0 = \mu_0^0 \, b_0^{-2/3} \sqrt{T} C_0(T),$$

$$C_0 = \sum_{i=-2}^{5} d_{0,i} \left(\frac{\varepsilon}{k_B T}\right)^{i/2}, \quad C_1 = \sum_{i=-3}^{8} d_{1,i} \left(\frac{\varepsilon}{k_B T}\right)^{i/2}, \quad C_2 = \sum_{i=-3}^{6} d_{2,i} \left(\frac{\varepsilon}{k_B T}\right)^{i/2},$$

b_0 is the second virial coefficient for the gas of hard spheres, ε is the depth of the potential well, k_B is Boltzmann's constant.

$$\mu_0^0 = 1.489\times10^{-6},\ b_0 = 1.2037\times10^{-6}\ \mathrm{m^3/kg},\ \varepsilon / k_B = 119.2246\ \mathrm{K}.$$

As it follows from [26], the gas law for argon in the considered region of p,T-paremters ($T \cong 300$–500 K and $p \cong 0.1$–10 MPa) can be written with sufficient accuracy as

$$p = \rho RT$$

where $R = 208.13$ J/(kg×K).

Coefficients in the expansions for C_1, C_2, and C_3 from [26] are given in the table 1.

Calculation of μ_j^n in numerical integration of (8) immediately with use of the expression (19) is not necessary because this decreases speed of the computing. It is more efficient from this point of view to generate the table $\{P_i,\ \mu(P_i,T),\ \ i = 1,\ldots, I\}$ at the beginning of the calculations where T is the temperature of the sample and

$$P_i = p_f + \frac{\Delta p}{I-1}(i-1).$$

Then the values of μ_j^n can be calculated in the integration procedure through piece-linear interpolation of the tabular data. The more is I the higher accuracy of the interpolation.

Minimum of the function Φ (i.e. desired quantities of k_w and b corresponding to the minimum of the function) is sought by the gradient method with check of convergence.

Table 1. Coefficients $d_{i,j}$.

i 1	−3	−2	−1	0	1	2
0	–	-6.1787×10^{-4}	4.92685×10^{-2}	1.44483	−1.98282	4.19488
1	-3.1675×10^{-4}	1.95736×10^{-2}	-4.7945×10^{-1}	5.8226	4.094×10^{1}	1.6204×10^{2}
2	1.02584×10^{-4}	-6.5734×10^{-3}	1.67771×10^{-1}	−1.99835	1.46332×10^{1}	-4.8085×10^{1}
i 1	3	4	5	6	7	8
0	−6.5354	4.6801	−1.22045	–	–	–
1	-3.4186×10^{2}	3.3898×10^{2}	–	-3.1319×10^{2}	2.54454×10^{2}	-6.519×10^{2}
2	9.22737×10^{1}	-9.8389×10^{1}	5.38523×10^{1}	-1.1669×10^{1}	–	–

Let us examine accuracy of k_w determination. In the absence of systematic errors (e.g. caused by argon leakage in the experimental apparatus), accuracy of permeability estimation depends primarily on accuracy of $p_{in}^{(exp)}(t)$ measurements. For estimation of k_w determination error, let us consider the necessary conditions of Φ extremum

$$\frac{\partial \Phi}{\partial k_w} = 2\sum_{n=2}^{M}\left[p_{in}^{(calc)}(t_n;k_w,b) - p_{in}^{(exp)}(t_n)\right]\frac{\partial p_{in}^{(calc)}}{\partial k_w}(t_n;k_w,b) = 0,$$

$$\frac{\partial \Phi}{\partial b} = 2\sum_{n=2}^{M}\left[p_{in}^{(calc)}(t_n;k_w,b) - p_{in}^{(exp)}(t_n)\right]\frac{\partial p_{in}^{(calc)}}{\partial b}(t_n;k_w,b) = 0,$$

which may be written in general form as

$$F_1(k_w,b;\{p_n, n=2,\ldots,N\}) = 0, \quad F_2(k_w,b;\{p_n, n=2,\ldots,N\}) = 0$$

where $p_n = p_{in}^{(exp)}(t_n)$.

Then k_w variations at small deviations of p_n can be expressed approximately (up to the terms of the second infinitesimal order) as:

$$\frac{|\delta k_w|}{k_w} \le \frac{\sum_{i=2}^{N}\left|\frac{\partial F_2}{\partial p_i}\frac{\partial F_1}{\partial b} - \frac{\partial F_1}{\partial p_i}\frac{\partial F_2}{\partial b}\right|}{k_w\left|\frac{\partial F_1}{\partial k_w}\frac{\partial F_2}{\partial b} - \frac{\partial F_2}{\partial k_w}\frac{\partial F_1}{\partial b}\right|}|\delta p|. \tag{20}$$

where δp is the accuracy of $p_{in}^{(exp)}(t)$ measurements.

3. Experimental Arrangement and Procedure

For implementation of the method presented below an experimental setup was specially developed and built. It allows to measure rock sample permeability in the range of values from 10^{-15} to 10^{-22} m^2 under temperature up to 600 °C and confining pressure up to 50 MPa. The main unit of the setup is an experimental cell (Figure 2 a), where a sample under study is placed. Cylindrical sample (2) jacketed by shrinking or gold tube (3) is placed between two obturators (4, 5). Water incoming through the inlet (9) produces confining pressure. Gas enters through the obturator (5) flows through the sample and through the obturator (4) passes to the atmosphere. Rubber rings (7, 8), nut (10) and flange (6) are used for obturators sealing. The cell case (1) and obturators (4, 5) are made of stainless still.

Figure 2b shows the scheme of the setup. The experimental cell is placed into the high-pressure vessel *(11)*. The water transferring the confining pressure to the sample is compressed by the screw press *(12)* and supplied to the high-pressure vessel along the main line through the valve *(13)*. Due to the water pressure, the sealing jacked is pressed to the side surface of the sample and prevents from argon leakage along the sidewall of the sample. The confining pressure is measured by pressure gauge *(14)* (0.6 class) with an error of ±0.36 MPa. The sample is heated by electric furnace *(15)*. For filtration through the sample, 99.92-purity argon is used. The gas flows from reservoir *(16)* along capillaries through MO standard pressure gauge *(17)* (0.16 class), valves *(20)* and (21), and upstream reservoir with piezometric pressure transducer (*19),* which is intended to perform measurements with an accuracy of ±0.01 MPa, at the inlet cross section of the sample. Argon goes directly to the atmosphere through the opposite cross section of the sample. There is also upstream reservoir *(18)* in the gas line of the apparatus. It is possible to change the volume of the upstream reservoir by connecting or closing this reservoir and, thus, set the optimal filtration rate of the gas through the sample. The confining pressure, gas pressure, ambient temperature, and temperature of the sample and upstream reservoir are recorded by the electronic system and written by computer *(22)*. Then, these data are processed by the special original software developed by the authors, which implements the above permeability-determining method.

The thorough preparation of the samples strongly effects on accuracy of experimental results. Samples with diameters of up to 25 mm and heights of up to 21 mm made with a diamond tool were used in the experiments. All samples were cleaned by alcohol in an ultrasonic bath to eliminate contaminations arising during the manufacturing process. Then, the samples were thoroughly dried in a muffle cabinet. The rate of the samples heating varied from 2 to 3 °C/min to exclude rock cracking. The estimates of the time of drying as a function of the rock permeability and drying temperature with consideration for capillary effects are given in [27]. The temperature selected for drying (105 °C) was maximal until the rock structure decompaction and clay mineral degradation do not take place. Depending on the permeability of the samples, the drying time varied from 2 days to 1 month.

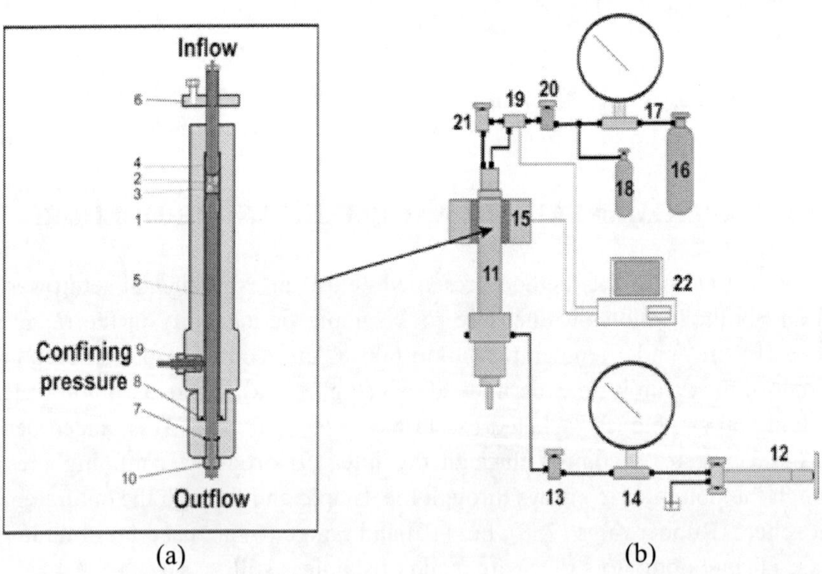

(a) (b)

Figure 2. Scheme of (a) the experimental cell and (b) setup for measuring permeability of the rock samples: a – (1) – case; (2) - sample; *(3) sealing* tube (3); *(4, 5) obturators; (6)* - flange; *(7, 8)* rubber rings; (9) inlet; *(10) – nut; b - (11)* high-pressure vessel; (12) screw press; *(20, 21)* valves; *(14, 17)* pressure gages; *(15)* electric furnace; *(16, 18)* reservoirs for gas; *(19)* upstream reservoir with the pressure transducer; and (22) computer.

The diameter and height of the samples were determined with accuracy of ±0.005 mm. The weight of the dried samples was measured by analytical balance with accuracy of ±0.0002 g.

To reveal formations of microcracks at the periphery before the tests the manufactured samples were examined with a scanning electron microscope. The observations have shown that there were no sample failures. In particular, the assurance of preserving the sample integrity is "the full" absence of the argon filtration through low-permeable samples within a day with the sensitivity of the measuring system of 0.002 cm^3.

Upon completion of the setup arrangement, a stepwise increase (up to 3.0 MPa) of gas pressure in the upstream reservoir is performed. Then the pressure starts to decrease due to filtration through the sample.

At the beginning, the process runs considerably fast then becomes more and more slow. Its rate depends on sample permeability and dimensions, test and setup parameters comprising upstream reservoir volume, as well. As it is shown in [28], for the accurate determination of the sample permeability, it is necessary to perform pressure variation in the reservoir in a sufficiently large range. So, start (P_{start}) and finish (P_{fin}) pressures in the experiments were chosen following the expression:

$$\frac{P_{start} - P_{fin}}{P_{start} - 0.1} \geq 0.5$$

The use of the upstream reservoirs of various capacities (5, 10, 100, and 1000 cm^3) allowed to measure permeability in a wide range of values, to obtain the representative data (the typical number of experimental points on the curves of the dependence $p_{in}(t)$ was as large as 700–900) and to optimize test duration (varied from several hours to several days). As a rule, the experiments were performed on samples with a 25-mm diameter and 8-mm height. In this case, e.g. when the sample with a permeability of 3.5 x 10^{-15} m^2 was studied, the pressure in the reservoir with a 1000-cm^3 volume decreased from the initial value of 0.5 MPa to 0.1 MPa for 300 s. When the sample with permeability of 5.7 x 10^{-22} m^2 was studied and the upstream reservoir with the 5-cm^3 volume was used, the initial pressure of 12 MPa decreased to 2.4 MPa approximately for 3.5 days.

4. Results and Discussion

In the course of the method testing with use of the arrangement and procedures described above a set of experiments was carried out. The rock samples with different values of porosity and a priori different permeability were selected for the tests. Experiments were performed at room temperature and a confining pressure of 2.0 – 5.0 MPa. Characteristics of the samples and the results of measurements are given in the Table 2.

Firstly, the software was tested: the numerical solution of the filtration equation and the experimental data were compared. Comparison showed a good agreement: Figure 3 shows an example of the data obtained for the granite sample, which demonstrates good coincidence of the experimental curve and numerical solution. Analogous results were also obtained for the other samples both with high and low values of porosity and permeability.

Table 2. The results of permeability measurements

Sample	Rock	φ	$\lvert\delta k_w\rvert / k_w$	k_w, m^2		b, MPa		$k_w^{(Kl)}$	$1-\frac{k_w^{(Kl)}}{k_w^{(num)}}$
				$k_w^{(an)}$	$k_w^{(num)}$	$b^{(an)}$	$b^{(num)}$		
Porf-1	porphyrite	0.0021	0.154	$2.48 \cdot 10^{-20}$	$2.427 \cdot 10^{-20}$	2.499	2.568	$3.8 \cdot 10^{-20}$	-0.541
Porf-2	porphyrite	0.0026	0.376	$6.974 \cdot 10^{-20}$	$6.764 \cdot 10^{-20}$	1.757	1.835	$8.5 \cdot 10^{-20}$	-0.257
Porf-3	porphyrite	0.0027	0.286	$3.011 \cdot 10^{-20}$	$2.897 \cdot 10^{-20}$	1.266	1.361	$2.6 \cdot 10^{-20}$	0.103
Porf-4	porphyrite	0.0033	0.117	$1.111 \cdot 10^{-19}$	$1.069 \cdot 10^{-19}$	1.182	1.259	$1.1 \cdot 10^{-19}$	-0.029
Sand-1	sandstone	0.1370	0.155	$2.707 \cdot 10^{-15}$	$2.543 \cdot 10^{-15}$	0.4501	0.5088	$2.0 \cdot 10^{-15}$	0.214
Sand-2	sandstone	0.0998	0.118	$1.687 \cdot 10^{-15}$	$1.616 \cdot 10^{-15}$	0.6529	0.7014	$1.6 \cdot 10^{-15}$	0.010
B-1	basalt	0.1990	0.191	$1.009 \cdot 10^{-16}$	$9.417 \cdot 10^{-17}$	1.007	1.117	$1.0 \cdot 10^{-16}$	-0.062

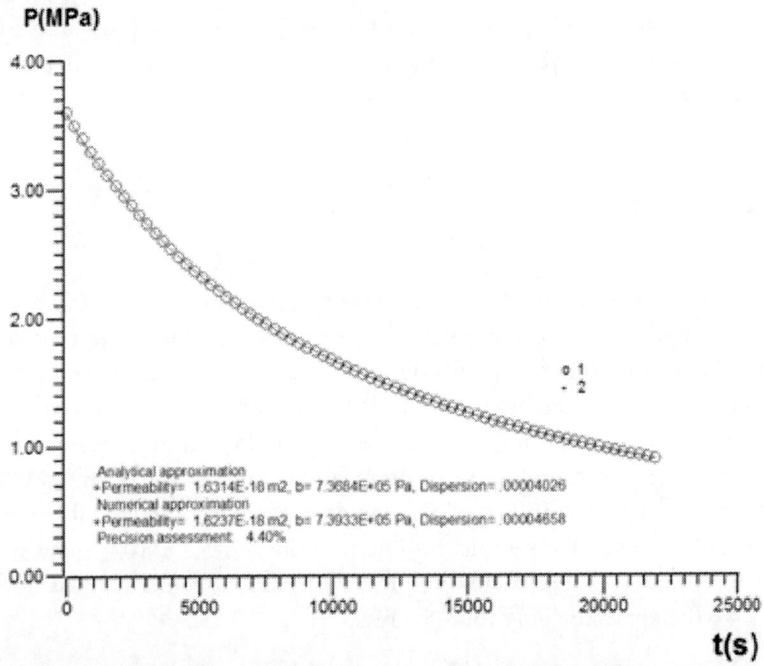

Figure 3. Results of permeability determination (granite sample t-65 (Streltsovskoye deposit, Eastern Transbaikalia, P=50 MPa, T=500 °C), (1) – experimental data, (2) – numerical solution).

The rate of convergence of the iteration procedure was also analyzed. The difference between the first approximation (analytical solution) and the result obtained by the numerical method was compared (Figure 4, table 2). Superscript (*an*) corresponds to experimental data processing, when the values of p_{in} are determined by the analytical relationship obtained from analysis of the simplified model of the filtration process of argon through the sample, which is similar to the Brace method (i.e., without taking into account the attenuating effects of gas in the pore space of the sample or the pressure dependence of the gas viscosity). Superscript (*num*) corresponds to that variant of method, when the values of p_{in} are determined by the numerical solution of the filtration problem in general formulation.

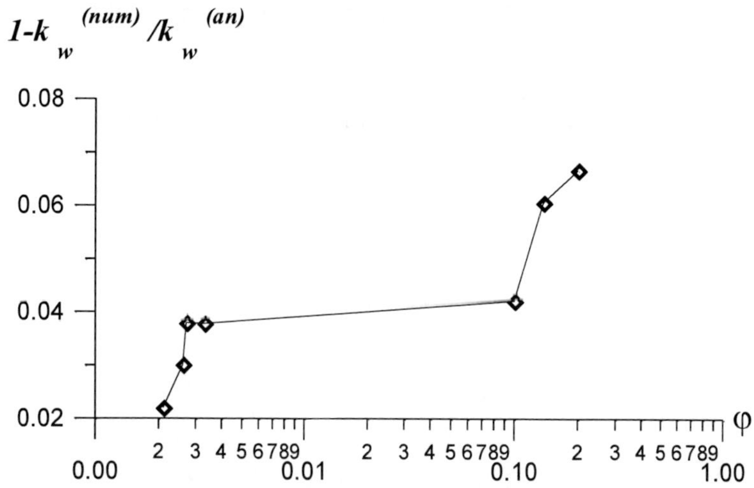

Figure 4. Dependence of the difference between numerical and analytical solutions ($k_w^{(num)}$- $k_w^{(an)}$) on porosity (φ).

Since the values of $k_w^{(num)}$ and $b^{(num)}$were obtained with a more general model of the process than $k_w^{(num)}$and $k_w^{(an)}$, they are apparently more accurate. However, it is possible to see that the differences in the values of $k_w^{(num)}$and $k_w^{(an)}$ do not exceed 7% of $k_w^{(num)}$. As noted earlier, the value of $k_w^{(an)}$is determined by analytical description of the filtration process, obtained under the assumption that the volume of the pore space of the sample is negligible and the attenuating effect of the gas contained in it can be disregarded. Hence, it follows that the differences between $k_w^{(num)}$ and $k_w^{(an)}$ should be as large as lager the value of the sample porosity - φ. A plot of the $k_w^{(num)}$- $k_w^{(an)}$ dependences on porosity (φ) drawn basing on the results obtained for all the samples studied (Figure 4) shows that $kw^{(num)}$- $kw^{(an)}$ – is actually a monotonically increasing function of φ.

In order to evaluate the reliability of the results, the value of *kw* was also determined for all the samples examined by the steady state method taking into account the Klinkenberg correction [20], i.e., according to the results of a set of measurements of permeability by the method of stationary flow, carried out with different mean gas pressures.

The values of $k_w^{(Kl)}$ obtained by the Klinkenberg method were compared with the values of $k_w^{(num)}$ determined according to the method described in the present work. It is possible to see satisfactory agreement of these values (table 2) for all investigated samples, with the sole exception of Porf-1. However, it is evident from Figure 5 that the values of *pin*(*t*) calculated from the values of $k_w^{(num)}$ и $b^{(num)}$ obtained for Porf-1 coincide with high accuracy with the experimentally measured ones, and sufficiently strict evaluation shows that the determination accuracy of sample Porf-1 is quite satisfactory (table 2).

Evidently, the relatively large disagreement between $kw^{(Kl)}$ *and* $kw^{(num)}$ for sample Porf-1 can be explained by the fact that the range of variations of the mean gas pressure in a series of experiments by the Klinkenberg method was insufficient to estimate $kw^{(Kl)}$with proper accuracy.

Thereby, testing and validation of developed technique were carried out. At the present time it is successfully used for solving important geological and ecological problems, namely, for determining the transport properties of main types of rocks of the Streltsovskoye uranium ore field (Transbaikalia, Russia) [29] and areas of significant radioactive contaminations of underground medium, related to radioactive waste disposal (South Ural and Eastern Siberia) [30, 31].

However, natural fluid flow in a hydrothermal ore forming system and contamination transport by ground water in a high level waste repository near field both take place under high temperature.

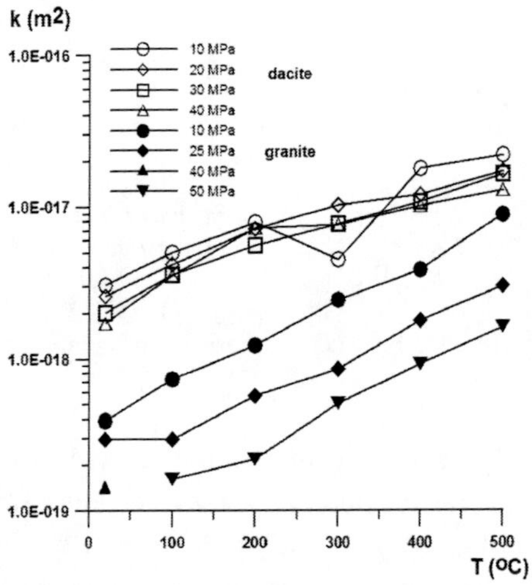

(a)

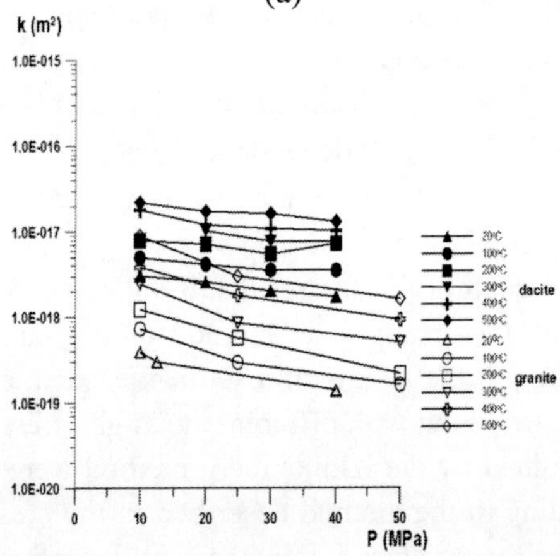

(b)

Figure 5. Permeability dependences for dacite and granite from Streltsovskoye deposit, Eastern Transbaikalia on temperature (a) and pressure (b).

So, it is very important to determine rock sample permeability under such conditions. Use of the noble gas - argon for flowing through a sample allows carrying out the experiments perfectly because it eliminates the geochemical transformations due to water-rock interaction. The new technique presented in the paper allows determining permeability in a wide range of *PT*-parameters, because it takes into account variations of argon properties under pressure and temperature. As an example we present the results obtained for the main types of rock from Streltsovskoye uranium ore field: dacite, granite and basalt. The studies under *PT*-conditions of ore formation: temperature from 20 to 500 – 600 °C and pressure from 10 to 50 MPa showed that pressure increase leads to permeability increase, in contrast, temperature increase leads to its monotonous increase (Figure 5). It is significant, that strong difference between permeability values proper for initial samples remains so under the all PTs.

CONCLUSION

A modification of the pulse decay method of permeability measurement has been developed taking into account the dependence of the properties of a flowing gas on its parameters of state, which makes it possible to simultaneously determine, according to the data of only a single experiment, the permeability values of the sample for the water and the Klinkenberg constant, which characterizes the pore space of rock. This approach made it possible to substantially increase the accuracy of measurements, reducing time and labor. The technique allows accomplishing high-accuracy determinations of permeability within the range of values of 10^{-22}–10^{-15} m^2 both under normal conditions and at elevated *PT*-parameters. The instrumentation and software for implementing the new methodologies of permeability measurement have been developed and tested thoroughly with samples of different types of rocks with various reservoir properties. The methods have been verified by comparison of the values of permeability and the parameter of anisotropy with the values obtained with the same samples using conventional methodologies, which showed good agreement.

The designed technique is successfully used for solving important geological and ecological problems, namely, for determining the filtration properties of main types of rocks of the Streltsovskoye uranium ore field and areas of significant radioactive contaminations of the underground medium, related to radioactive waste disposal.

ACKNOWLEDGMENT

This work was supported by the Russian Foundation for Basic Research (project 11-05-00778) and by the Russian Academy of Sciences: Programs 4, of the Presidium and 9 of the Department of Earth Sciences.

REFERENCES

[1] A.A. Pek, On the dynamique of juvenile solutions, *Nauka*, Moscow (1968).

[2] V.I. Malkovsky, A.A. Pek , P.L. Arseniev, and D.N. Topor, Izvestya RAN, *Fizika Zemli*, 12, 57 (1988).

[3] A.V. Tutubalin , D.V. Grichuk, and V.I. Malkovsky A complex hydrodynamic model of a convective hydrothermal system, *The 8th Int. Symp. on Water-Rock Interaction,* August 1995 (Moscow, Russia) pp. 763–766 (1995).

[4] V.I. Malkovsky, A.A. Pek, B.I. Omelyanenko, E.G. Drozhko, Izvestya RAN, *Energetika*, 3, 113 (1994).

[5] Y.D. Diadkin, S.G. Gendler, N.N. Smirnova, Geothermal physics, *Nauka*, S.-Petersburg (1994).

[6] C. Morrow, J. Byerlee, *J. Geoph. Res.,* 15, 1033 (1988).

[7] C. Morrow, D. Lockner, *J. Geoph. Res.,* 21, 2151 (1994).

[8] W.F. Brace, J.B. Walsh, W.T. Frangos, *J. Geophys. Res.,* 73,2225(1968).

[9] P.A. Hsieh, J.V. Tracy, C.E. Neuzil, J.D. Bredehoeft, S.E.Silliman, *Int. J. Rock Mech. Min. Sci.Geomech. Abstr.,* 18, 245 (1981).

[10] W. Lin, *J. Geoph. Res.,* 87, 1055 (1982).

[11] D.B. Trimmer, H.C. Bonner, A. Duba, *J. Geophys.Res.,* 85, 7059 (1980).

[12] C. Lin, G. Pirie, D.A. Trimmer, *J. Geophys. Res.,* 91, 2173 (1986).

[13] T.J. Katsube, B.S. Mudford, M.E. Best, *Geophysics,* 56, 1681 (1991).

[14] C. Le Guen, M. Deveughele, J. Billiotte, J. Brulhet, *Quarterly Journal Of Engineering Geology,* 26, 327 (1993).

[15] E.M. Zeynalyandabily, S.S. Rahman, *Measurement Science and Technology,* 6, 1519 (1995).

[16] S. Finsterle, P. Persoff, *Water Resources Res.,* 33, 1803 (1997).

[17] G.J. Fischer, M.S. Paterson, Permeability and storage capacity during deformation at elevated temperatures, in: "Fault *Mechanics and Transport Properties of Rocks"*, 187-211, Academic, San Diego (1992).

[18] S. Zhang, M.S. Paterson, S.F. Cox, J. Geophys. Res., 99, 15741, (1994).

[19] A.V. Zharikov, V.M. Vitovtova, V.M. Shmonov, Geologiya *Rudnikh Mestorojdeniy,* 32, 79 (1990).

[20] L.J. Klinkenberg, *Drilling and Production Practice,* Amer. Petroleum Inst, 200 (1941).

[21] D.S. Scott, F.A.L. Dullien, *Amer. Inst. Chem. Eng. Journal*, 15, 293 (1962).

[22] F.A.L. Dullien, *Porous media: fluid transport and pore structure.* Academic, San Diego, (1992).

[23] V.N. Balashov, G.P. Zaraisky, Experimental and theoretical investigation of decomposition process of rock at heating, in: "*Outlines of Physico-Chemical Petrology"*, pp. 69-109, Nauka, Moscow (1982).

[24] G. de Marsily, *Quantitative hydrogeology : groundwater hydrology for engineers.* Academic Press, Orlando (1986).

[25] P.J. Roache, *Computational fluid dynamics*, Hemrosa, Albuquerque (1976).

[26] V.N. Zubarev, A.D. Kozlov, V.M. Kouznetsov, L.V. Sergeyeva, G.B. Spiridonov, Thermophysical properties of gases important in techniques at high temperature and pressure, Moscow, *Energoatomizdat* (1989).

[27] V.I. Malkovsky, A.V. Zharikov, V.M. Shmonov, *Experiment in Geosciences*, 12, 32 (2004).

[28] V.I. Malkovsky, A.V. Zharikov, V.M. Shmonov, Izvestiya, *Physics of the Solid Earth*, 45, 89 (2009).

[29] A.V. Zharikov, V.M. Shmonov, A.A. Bourmisrtov, A.P. Aleshin, *Electronic Scientific Information Journal Herald of the Department of Geology, Geophysics, Geochemistry and Mining Sciences of the Russian Academy of Sciences*, 26 (2008). http://www.scgis.ru/russian/ cp1251/h_dgggms/1-2008/informbul-1_2008/geomaterial-3e.pdf.

[30] V.A. Petrov, V.V. Poluektov, A.V. Zharikov, R.M. Nasimov, N.I. Diaur, V.A. Terentiev, A.A. Burmistrov, G.I. Petrunin, V.G. Popov, V.G. Sibgatulin, E.N. Lind,

A.A. Grafchikov, V.M. Shmonov, Microstructure, filtration, elastic and thermal properties of granite rock samples: implication for HLW disposal, in: P.K. Harvey, T.S. Brewer, P.A. Pezard, V.A. Petrov, (eds) "*Petrophysical properties of crystalline rocks*". Geological Society of London. Special publications, 240, pp.237-253 (2005).

[31] V.A. Petrov, V.V. Poluektov, A.V. Zharikov, V.I. Velichkin, R.M. Nasimov, N.I. Diaur, V.A. Terentiev, V.M. Shmonov, V.M. Vitovtova, Deformation of metavolcanics in the Karachay Lake area, Southern Urals: petrophysical and mineral-chemical aspects, in: P.K. Harvey, T.S. Brewer, P.A. Pezard, V.A. Petrov, (eds) "*Petrophysical properties of crystalline rocks*". Geological Socicty of London. Special publications, 240, pp.307–321 (2005).

In: Argon: Production, Characteristics and Applications ISBN: 978-1-62618-204-2
Editors: B.N. Sismanoglu, H.S. Maciel et al.

Chapter 3

EFFECT OF ARGON ANNEALING ON CHROMIA OR ALUMINA FORMING ALLOYS OXIDIZED AT HIGH TEMPERATURES

H. Buscail[1,*]***, C. Issartel***[1]***, F. Riffard***[1] ***and A. Fleurentin***[2]
[1]Clermont Université- UdA - LVEEM EA3864, Le Puy en Velay, France
[2]CETIM Senlis, Senlis, France

ABSTRACT

In order to improve the oxide scale adherence on alloys in various corrosive environments at high temperatures, the present chapter focuses on the effect of argon annealing on sol gel coated specimens. On a chromia forming 330Cb alloy (Fe-35Ni-18Cr-1Nb-2Si), the influence of a lanthanum sol-gel coating on the oxide scale adherence as been studied at 900°C, in air. The alloy oxidation is performed in order to generate a protective chromia scale acting as a good barrier against carburization. Argon annealing of lanthanum sol-gel coatings have been performed at various temperatures in order to find the best conditions to insure the scale adherence. Kinetic results show that lanthanum sol-gel coatings lead to a lower oxidation rate compared to blank specimens. Thermal cycling tests on lanthanum the sol-gel coated specimen show that the oxide scale formed at 900°C, in air, is very adherent. The effect of lanthanum sol-gel coatings was studied in order to improve the alumina scale adherence during the model Fe-20Cr-5Al alloy oxidation, at 1100°C, in air. Various sol-gel coating procedures were applied. Argon annealing of the lanthanum sol-gel coating was tested at temperatures ranging between 600 and 1000°C. The coating crystallographic nature was characterised by X-ray diffraction (XRD) depending on the annealing temperature. The oxidation process has been determined at 1100°C by *in situ* X-ray diffraction on blank Fe-20Cr-5Al, sol-gel coated and argon annealed specimens. The present chapter shows that the coating argon annealing at 1000°C leads to the preferential formation of $LaAlO_3$ instead of La_2O_3. This coating procedure leads to an alumina scale formation showing the best adherence under thermal cycling conditions at 1100°C. It is demonstrated that argon annealing of coated alloys can largely improve the oxide scale adherence.

---

* E-mail address: henri.buscail@udamail.fr.

1. INTRODUCTION

On chromia forming alloys, conveyor units' carburization is the main reason of degradation due to a quick alloy weakening. To improve the alloys carburization resistance it is assumed that the presence of an adherent oxide scale acts as a carbon diffusion barrier. The literature indicates that a Fe-35Ni-18Cr-2Si alloy is used as resistance heating-element material and has been studied in the 1000 and 1300°C temperature range [1]. It is shown that chromia Cr_2O_3 is the protective oxide scale formed at high temperature. Stevens indicate that at temperatures higher than 1100°C the chromium oxide CrO_3 evaporation occurs. In this case chromium depletion is observed in the oxide scale and in the alloy. The concentration of elements in the protective oxide was found to be both temperature and time dependent. Chromia is stable at low temperature (< 1100°C) and was found to evaporate at higher temperatures and longer oxidation time, thereby increasing nickel and iron concentrations. On the basis of changes in the chemical composition of the oxide film, together with scaling, the failure of heating-element wire by the development of hot spots can be explained. At failure both long-term (low temperature < 1000°C) and shorter-term tests (temperature > 1000°C) were found to have oxidized to the same depth. This can be explained on the basis of a depletion of chromium, which occurred at the oxide-alloy interface, and becomes insufficient for a healing process. Other authors have shown that in the 800 - 1000°C temperature range, the 330Cb oxidation leads to a chromia scale acting as a good diffusion barrier under isothermal conditions [2]. It is expected that the manganese chromite can limit the chromia scale evaporation at temperatures higher than 1000°C [3].

Nevertheless, important scale spallation occurred after oxidation and cooling to room temperature [2]. At 900°C, spallation only appeared on the niobium and silicon rich areas. At 1000°C, scale spallation is present on the whole specimen surface indicating that this temperature is too high to obtain an adherent chromia scale on the alloy surface. The comparisons between the 46h and 160h oxidation duration have shown that at 800 and 900°C short oxidation duration (46h) is favourable to the oxide scale adherence. It then appeared that the chromia scale never shows a perfect adherence on the alloy under the tested conditions. In order to improve the chromia scale adherence previous works have proposed to introduce lanthanum as an alloying element [4] or on the alloy surface as a La_2O_3 coating [5-9]. Gleeson demonstrated the significant effects of alloy composition on long-term, cyclic-oxidation resistance. Each of the alloys exhibited scale spallation; however, the manner by which spallation occurred varied between the alloys [4]. The 230 alloy, which contains 0.02 wt. % La, exhibited partial scale spallation, thus allowing for the easier formation of a protective Cr_2O_3 -rich scale during subsequent oxidation. The HR-160 alloy exhibited complete spallation owing largely to its relatively high silicon content (2.75 wt. %). However, the silicon was also beneficial in promoting protective scale formation when the exposed alloy was subsequently oxidized. The HR-120 alloy showed the poorest cyclic-oxidation resistance, due in part to poor scale adhesion and the tendency of the iron in this alloy (33 wt. %) to eventually oxidize and result in the formation of a less-protective scale. All of the alloys underwent internal attack in the form of internal oxidation and void formation. In most cases, the extent of internal attack was significantly greater than that of metal loss.

It has been shown by Fontana that Crofer 22 APU, AL 453 and Haynes 230 alloys seem particularly suitable as interconnect in SOFC devices [5]. The beneficial effects of reactive elements on oxidation behaviour were clearly highlighted. The coatings composed of a thin reactive element oxide such as, La_2O_3, Nd_2O_3 and Y_2O_3 resulted in an important improvement in the high temperature oxidation resistance; the oxide growth rate decreased, the adhesion between alloy and oxide scale was improved. The MOCVD method showed good reproducibility and excellent efficiency to obtain nanometric layers on the interconnect surface. The objective of these coatings was to form perovskite-type oxides with the chromia growing layer because perovskite present better conductivities than chromia, especially when they are doped with iron.

It has been explained by Fernandes that sols of La_2O_3, were prepared as aqueous dispersions with nitric acid and a non-ionic surfactant, heated to 80°C for 1 h to form a sol, the RE oxide sol then sprayed on specimens of the two alloys to form a coat and finally the specimens heated to 150°C to form a 10 μm thick surface layer of RE oxide gel [6]. These specimens were isothermally oxidized at 1000°C for about 400 min in a thermogravimetric balance to obtain kinetic curves. This article presents the influence of surface additions of nanocrystalline rare earth (RE) oxide La_2O_3 and CeO_2, on the isothermal oxidation behaviour of Fe20Cr and Fe20Cr5Al at 1000°C. Thermogravimetric studies revealed parabolic kinetics in all cases and the scale thickness on specimen surfaces varied with the nature of RE oxide. The oxidation resistance of specimens coated with two RE oxides was significantly higher than those coated with either one of the two oxides. The marked increase in the oxidation resistance of the alloys coated with two RE oxides is due to optimization of RE ion radius and RE oxide grain size/shape.

Under similar conditions, Ramanathan proposed that in extended cyclic oxidation tests that were carried out from peak temperatures of 900, 1000 and 1100°C to room temperature at cooling rates of 300°C/s and 1000°C/s, the La_2O_3 coating increased cyclic oxidation resistance of the Fe20Cr alloy significantly more than the Pr_2O_3 coating [7]. The cyclic oxidation resistance of reactive element oxide coated Fe20Cr alloy was significantly higher than that of the uncoated alloy. The thickness of the chromia layer on reactive element oxide coated Fe20Cr alloy varied with the reactive element oxide. Spalling of the chromia scale occurred when mass gain due to scale formation reached a limiting value of 1.25–1.5 mg cm^{-2}. Evidence of a direct correlation between RE ion radius and oxidation resistance of reactive element oxide coated Fe20Cr alloy has been observed.

Yoon indicated that solid oxide fuel cells typically operate at temperatures of about 1000°C [8]. At these temperatures only ceramic interconnects such as $LaCrO_3$ can be employed. The development of intermediate-temperature solid oxide fuel cells (IT-SOFCs) can potentially bring about reduced manufacturing costs as it makes possible the use of an inexpensive ferritic stainless steel interconnector. However, this ferritic stainless steel suffers from Cr_2O_3 scale formation and a peeling-off phenomenon at the IT-SOFC operating temperature in an oxidizing atmosphere. Application of an oxidation protective coating is an effective means of providing oxidation resistance. In its study, he coated an oxidation protective layer on ferritic stainless steel using a precursor solution prepared from lanthanum nitrate, ethylene glycol, and nitric acid. Heating the precursor solution at 80°C yielded a spinable solution for coating. A gel film was coated on a ferritic stainless steel substrate by a dip coating technique. At the early stage of the heat-treatment, lanthanum-containing oxides such as La_2O_3 and La_2CrO_6 formed at temperatures lower than 600°C. As the heat-treatment temperature was increased over 700°C, an oxidation protective perovskite-type $LaCrO_3$ layer was produced by the reaction between the lanthanum-containing oxide and the Cr_2O_3 scale on the substrate. As the concentration of La-containing precursor solution was increased, the amount of La_2O_3 and La_2CrO_6 phases was gradually increased. The coating layer, which was prepared from a precursor solution of 0.8 M, was composed of $LaCrO_3$ and small amounts of $(Mn,Cr)_3O_4$ spinel. A relatively dense coating layer without pinholes was obtained by heating the gel coating layer at 1073K for 2 h. A La-based gel film was successfully coated on a STS444 substrate by a dip coating technique and $LaCrO_3$ perovskite phase was formed by heat-treating the gel film at 500-800°C. Transient phases such as La_2O_3 and La_2CrO_6, which were formed at lower temperature, reacted with Cr_2O_3 to produce $LaCrO_3$ perovskite. It is believed that control over the reaction between La_2O_3 or La_2CrO_6 and Cr_2O_3 by modification of the precursor solution and heat-treatment temperature is an effective means of producing a dense and thick $LaCrO_3$ coating layer on the STS substrate. Higher viscosity resulted in a thicker gel coating layer. In order to increase the film thickness, the precursor solution should be further modified. During oxidation $(Mn,Cr)_3O_4$ spinel and Cr_2O_3 phases were formed on STS444. The spinel and Cr_2O_3 might result from outward diffusion of manganese and inward diffusion of oxygen, respectively. As a consequence, large spinel grains were produced at the top surface while growth of a Cr_2O_3 layer occurred between the $LaCrO_3$ coating layer and the alloy. Compared with non-coated specimens, the oxidation resistance of $LaCrO_3$-coated STS444 was significantly improved. Zhao exposed the effects of La_2O_3 laser-clad on AISI 1115 steel substrates. The oxidation kinetic of the coatings was studied by testing the weight gain [9]. Cyclic oxidation tests were performed to determine the eventual weight loss due to spallation. The experimental results showed that Cr_2O_3 was formed on all coatings after oxidation. The coatings with La_2O_3 exhibited excellent high-temperature oxidation behaviour including low oxidation rates and high resistance to spallation. In addition, the effect of La_2O_3 was discussed. This included the change of transport mechanism in the oxide scales, the increase in the nucleation rate of the oxide, the release of stresses in the oxide scales by refining the oxide grains and the improvement of the adhesion of the oxide scales to the coatings by changing the scale growth mechanisms.

Previous works have already indicated that the argon annealing during 2 hours could improve the effect of the coating on the oxide scale adherence [10-12]. The present chapter will focus on the effect of a lanthanum sol-gel coating applied on the specimen surface before oxidation. The sol-gel coating argon annealing as been tested on the 330Cb chromia-forming alloy. Alumina forming alloys, FeCrAl, containing 20 % chromium and 5 % aluminium are often used in electrical heating elements at high temperatures. The alumina scale generally insures a good oxidation protection under isothermal conditions. Nevertheless, under thermal cycling conditions scale spallation generally occurs [13-15]. It has been shown that reactive elements, such as cerium, yttrium or lanthanum, can improve the protective character of the alumina scale and its adherence on the alloy. Most authors presented results about the effect of rare earth elements, introduced as alloying elements or by ion implantation on the alloy [16-18].

The aim of this chapter is to present another coating procedure, which consists in a lanthanum sol gel coating deposited on model FeCrAl alumina-forming alloy oxidized at 1100°C. The model FeCrAl alloy studied is free of impurities or minor elements. The effect of the lanthanum sol gel coating was compared with various argon-annealing conditions at 600, 800 and 1000°C prior to oxidation to examine the nature and efficiency of the coating. Thermal cycling tests were performed in order to compare the oxide scale adherence with the different surface treatments. Kinetic studies and *in situ* X-ray diffraction (XRD) will lead to explain the oxidation mechanism involved and the effect of the argon annealing on the alumina scale adherence.

Table 1. Alloys composition weight %

Wt. %	Fe	Ni	Cr	Si	Mn	Nb	Mo	Cu	C	Al
330Cb	Bal.	33.81	21.14	2.15	0.69	1.02	0.22	0.22	0.05	-
FeCrAl	Bal.	-	19.98	-	-	-	-	-	<0.001	5.00

2. EXPERIMENTAL

The 330Cb alloy is an austenitic stainless steel. Table 1 gives the 330Cb composition in weight %. 1.5 mm thick cylindrical specimens of around 14 mm diameter were abraded up to the 120-grit SiC paper, then degreased with ethanol and finally dried. This surface finishing was chosen in order to fit the surface state of the alloy used in industrial applications.

The FeCrAl alloy used in the present chapter is a high purity alloy provided by Prof. J. Le Coze from l'Ecole des Mines de Saint-Etienne (France). The alloy composition is given in Table 1. The samples are 1mm thick disks of 14 mm diameter. The blank specimens were polished on SiC paper up to the 600-polishing grade, then washed with ethanol and finally dried just before isothermal oxidation at 1100°C or sol gel coating. Kinetic results are obtained by thermogravimetric measurements under isothermal conditions. High temperature oxidation was performed in ambient air at the atmospheric pressure using a Setaram TGDTA 92-1600 microthermobalance. The oxide characterization was realized by *in situ* X-ray diffraction (XRD) tests. *In situ* XRD patterns were obtained by use of a high temperature MRI chamber installed in a Philips X'pert MPD diffractometer (copper radiation, λk_{α} = 0.15406 nm). The XRD conditions were 2Θ scan, step 0.05° ranging from 10 to 80°, 2.5 s counting time. Each *in situ* XRD pattern is obtained within one hour in the 10-80° 2Θ range. *In situ* XRD permit to follow the evolution of the oxide nature with time. The oxide scale surface and cross-section morphologies have been observed in a JEOL 6400 scanning electron microscope (SEM) coupled with a LINK energy dispersive X-ray spectroscopy (EDS). The EDS point analyses were performed with an electron probe focused to a 1 μm spot.

The lanthanum sol-gel coating was performed using a 0.4 mol.l^{-1} lanthanum hydroxide solution obtained from the precipitation of lanthanum hydroxide in 0.27 mol 28% ammonia. After filtration, the dissolution of the precipitate is obtained in the 69% nitric acid at 60°C [19]. The lanthanum sol-gel coatings are carried out by dipping the specimens during 2 seconds in the solution. Immediately after dipping, the specimens are dried in warm flowing air at about 40°C during 1 min. The coating thickness is close to 0.5 μm. After sol-gel coating, the argon annealing was performed in a horizontal quartz reaction tube. A heating ramp is applied at a rate of 5°C/min. until the 600, 800 or 1000°C annealing temperature is reached. The temperature is maintained during 2 h. Then, the specimen is cooled down at 5°C/min to room temperature. In order to compare the various surface treatments, 5 kinds of specimens have been tested; they are listed in Table 2. Before oxidation, the XRD pattern shows that the alloy is detected alone on the surface of non-annealed specimens. It indicates that the sol-gel coating is not sufficiently crystallised before the eating up of the specimen. With a 600°C argon annealing: before oxidation the XRD pattern shows that La_2O_3 (JCPDS 05-0602) is present on the surface.

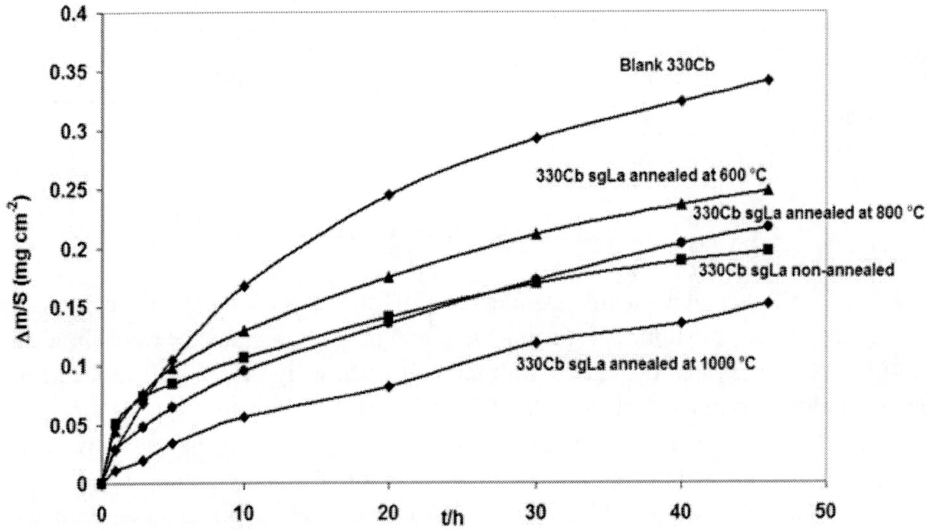

Figure 1. Mass gain curves obtained on alloy 330Cb oxidised in air at 900°C. Comparison of the kinetic curves obtained on the specimens: Blank, sol gel coated without annealing, and sol gel coated with argon annealing at 600, 800 or 1000°C.

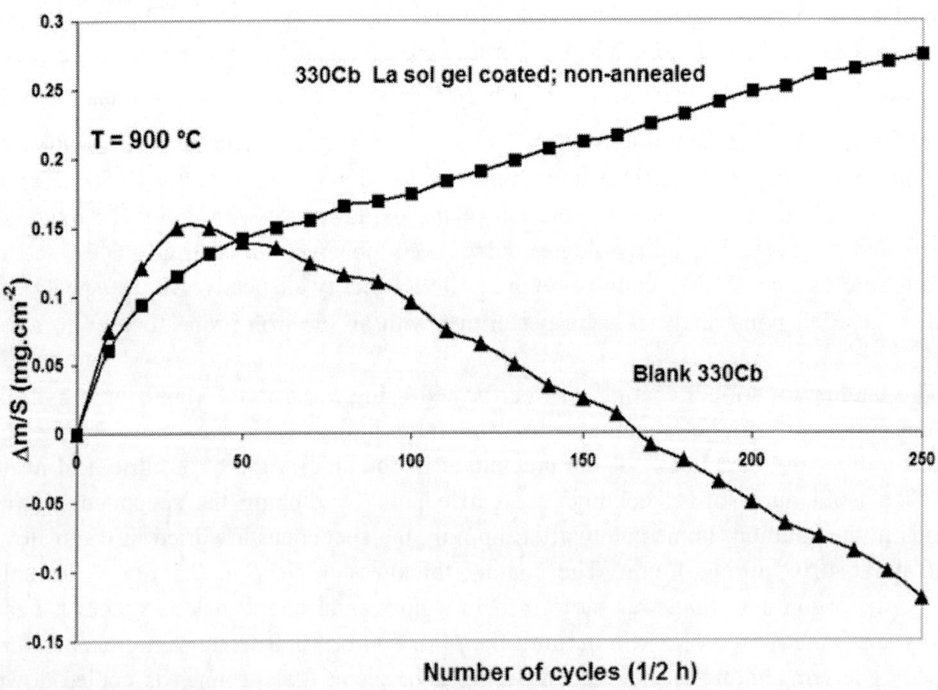

Figure 2. Mass change curves versus the number of cycles (0.5h cycles), at 900°C on blank and lanthanum sol gel coated specimens without argon annealing specimens.

Thermal cycling was performed in static laboratory air at atmospheric pressure in a classical tubular furnace. Three samples were placed at the same time into an alumina crucible to estimate the reproducibility. On the 330Cb alloy the thermal cycle consists in a 30 minutes exposure at high temperature, followed by 30 minutes at room temperature after air quenching. This frequency corresponds to industrial conditions in conveyor units. On the Fe-Cr-Al alloy the thermal cycle consists in a 20 hours exposure at high temperature, followed by 4 hours at room temperature after cooling. The weight change of the samples was determined after each cycle on a balance with an accuracy of + 0.01 mg.

Table 2. Various FeCrAl surface treatments and corresponding specimen labels

Specimen surface state	**Specimen Label**
Model FeCrAl (Blank)	FeCrAl
FeCrAl Lanthanum sol-gel coating without annealing	sg La
FeCrAl Lanthanum sol-gel coating with argon annealing at 600°C	sgLa.ann.600
FeCrAl Lanthanum sol-gel coating with argon annealing at 800°C	sgLa.ann.800
FeCrAl Lanthanum sol-gel coating with argon annealing at 1000°C	sgLa.ann.1000

3. RESULTS AND DISCUSSION

3.1. Chromia Forming Alloy Isothermal Oxidation at 900°C, in Air

On the 330Cb alloy, kinetic results show that all lanthanum sol-gel coatings lead to a lower oxidation rate compared with blank specimens Figure 1. The lowest oxidation rate is observed on the 1000°C argon annealed specimen, oxidized at 900°C in air. The other sol-gel coated specimens show similar kinetic behaviours.

Thermal Cycling Tests in Air at 900°C

Figure 2 shows the mass change curves versus the number of 30 minutes cycles at 900°C for blank specimens and lanthanum sol gel coated specimens. The comparison of the kinetic curves show that lanthanum sol gel coated largely improved the oxide scale adherence. On blank specimens spallation started after 25 cycles. On lanthanum sol gel coated specimens (non-annealed) the oxide scale is still adherent after 250 cycles.

In situ X-Ray Diffraction Analysis in Air at 900°C

In situ X-ray diffraction patterns obtained on the blank 330Cb specimen oxidized, in air at 900°C are presented on Figure 3. The oxide scale is only composed of $Mn_{1.5}Cr_{1.5}O_4$ (JCPS 33-0892) and Cr_2O_3 (JCPDS 38-1479). Chromia forms at the beginning of the oxidation process. No phase transition is detected during the 46h oxidation and during cooling to room temperature. The relatively high $Mn_{1.5}Cr_{1.5}O_4$ peaks intensities indicate that it is present at the external interface even though it corresponds to a low amount of oxide grains. The alloy peaks are still detected after 46h oxidation because the scale is thin enough to permit its detection. No iron oxides are detected in the scale during the *in situ* measurements and after cooling to room temperature.

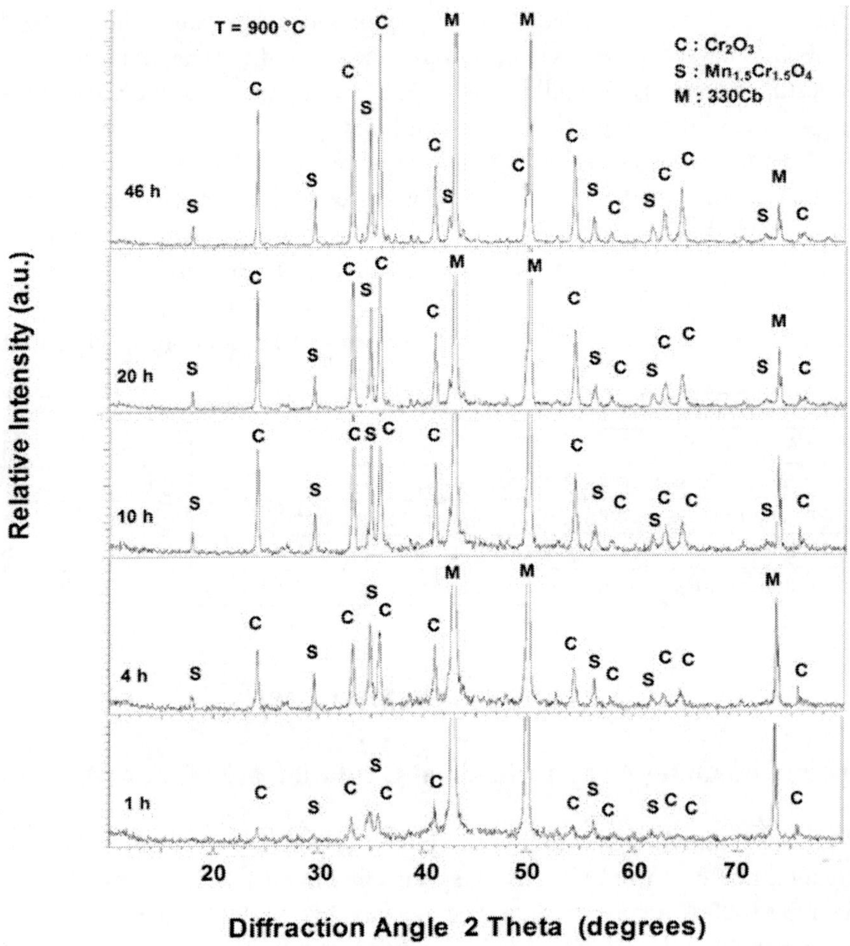

Figure 3. *In situ* XRD patterns obtained for the 46 first hours oxidation at 900°C, in air, of the blank 330Cb.

Figure 4 shows the *in situ* XRD patterns obtained during the 46 first hours oxidation at 900°C in air of the sol-gel coated 330Cb without argon annealing. Before oxidation the XRD pattern shows that the alloy is detected alone on the surface. It indicates that the sol-gel coating is not sufficiently crystallised before the eating up of the specimen. At 900°C, XRD analyses show the initial formation of $Mn_{1.5}Cr_{1.5}O_4$. Cr_2O_3 forms rapidly at the beginning of the oxidation. A lanthanum containing oxide $LaCrO_3$ (JCPDS 33-0701) is also observed from the beginning of the oxidation process. This phase is stable all along the test and is not transformed into another oxide. No phase transition is detected during cooling to room temperature.

Figure 5 shows the *in situ* XRD patterns obtained during the 46 hours oxidation at 900°C in air of the sol-gel coated 330Cb, with a 600°C argon annealing. Before oxidation the XRD pattern shows that La_2O_3 (JCPDS 05-0602) is present on the surface. During the first hour oxidation the lanthanum oxide is rapidly transformed into $LaCrO_3$. This oxide is present on the surface until the end of the oxidation test. At 900°C, XRD patterns show the growth of $Mn_{1.5}Cr_{1.5}O_4$ and Cr_2O_3.

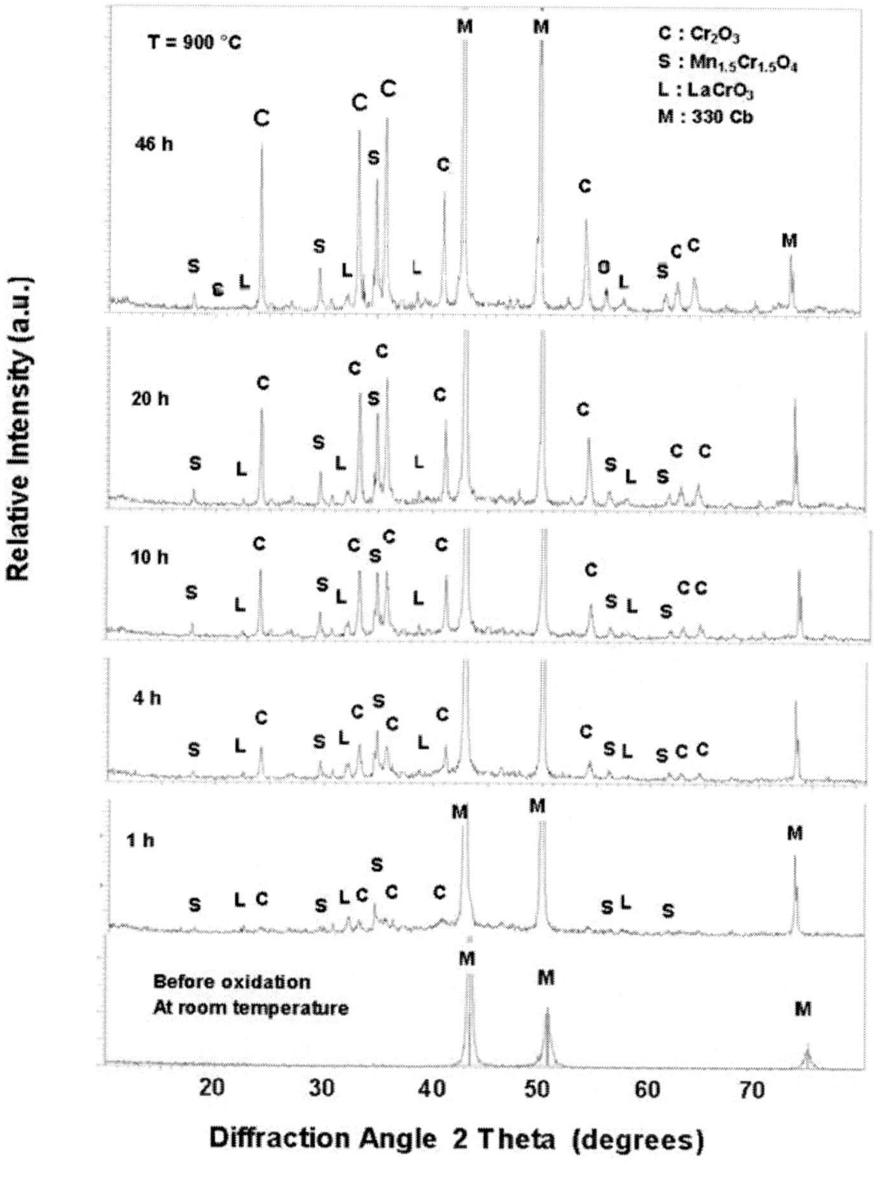

Figure 4. *In situ* XRD patterns obtained for the 46 first hours oxidation at 900°C in air of the lanthanum sol gel coated 330Cb.

Figure 6 shows the *in situ* XRD patterns obtained during 46 hours oxidation at 900°C in air of the sol-gel coated 330Cb, argon annealed at 1000°C. Similar results are obtained on 800°C argon annealed specimens. After argon annealing at 1000°C, or 800°C, La_2O_3, chromia and a small amount of $LaCrO_3$ are present on initial surface. These oxides react during the first hour oxidation to give $LaCrO_3$. Then, Cr_2O_3 and the manganese chromite grow all along the test. Lanthanum chromite $LaCrO_3$ is the major phase at the beginning of the oxidation test. It comes from the reaction between La_2O_3 and Cr_2O_3. Nevertheless, $LaCrO_3$ is less detected with time because it is incorporated inside the growing chromia scale.

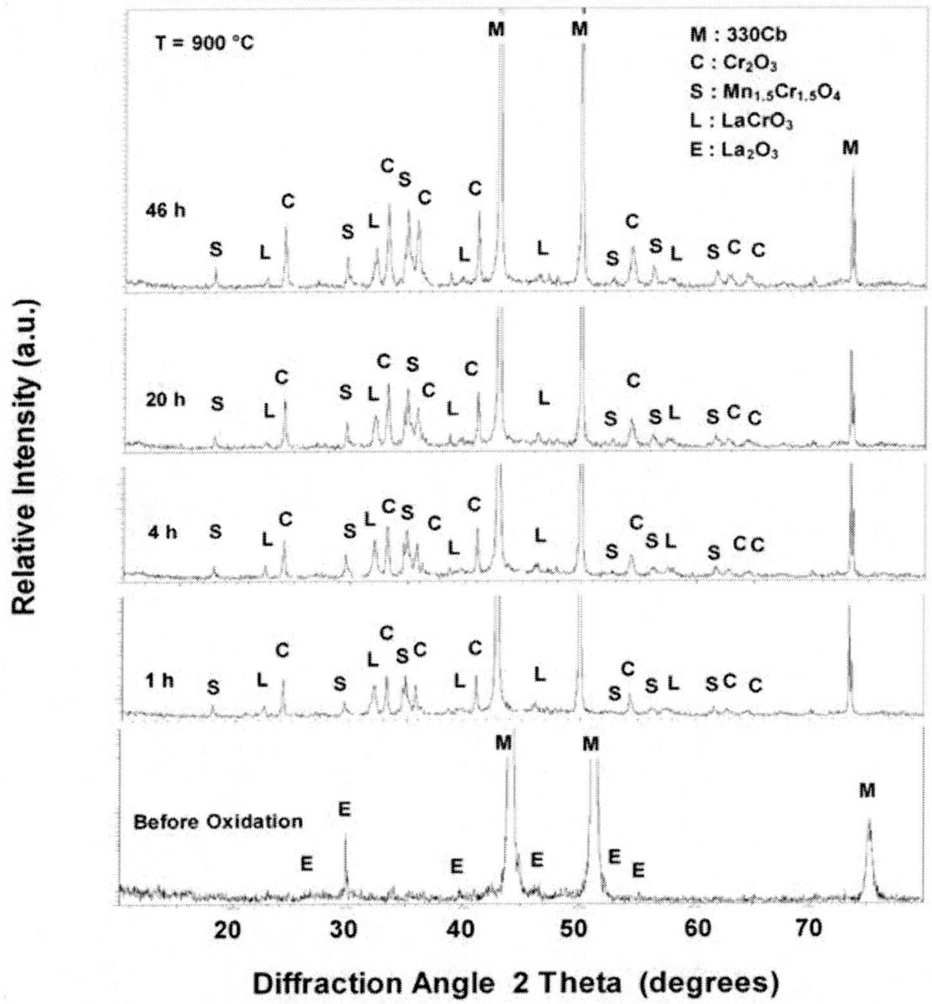

Figure 5. *In situ* XRD patterns obtained for the 46 first hours oxidation at 900°C in air of the lanthanum sol gel coated 330Cb, argon annealed at 600°C.

SEM Observations after Oxidation in Air at 900°C

The surface scale morphologies are presented on Figure 7 as BSE images and on Figure 8 with a higher magnification as SEI images (x3000). Blank specimens oxidized during 46h are covered by a chromia scale and manganese chromite grains Figure 7a and Figure 8a. Spalled areas are found on about 10% of the surface. It corresponds to the white areas on Figure 7a. Where the oxide scale spalled off, the underlying substrate surface is silicon and molybdenum rich. With sol-gel coated 330Cb specimen, no scale spallation is observed on the surface (Figure 7b and Figure 8b). The oxide scale is convoluted and the grain size is small. On Figure 8b, the biggest pyramidal grains correspond to manganese chromite. The small grains are composed of chromia and lanthanum chromite. On the BSE micrograph Figure 7b, the light grey colour corresponds to the lanthanum rich areas. Surface micrographs obtained after the 46 hours oxidation at 900°C in air of the sol-gel coated 330Cb, argon annealed at 600°C are presented on Figure 7c and Figure 8c. No spallation is observed on the specimen's surface. The convoluted scale is mainly composed of chromia. The pyramidal grains correspond to manganese chromite. Figure 7c shows the presence of lanthanum (light grey areas). Light grey areas correspond to areas where lanthanum is more easily detected from the surface. It is due to the presence of the coating closer to the external interface because less chromia has grown over it.

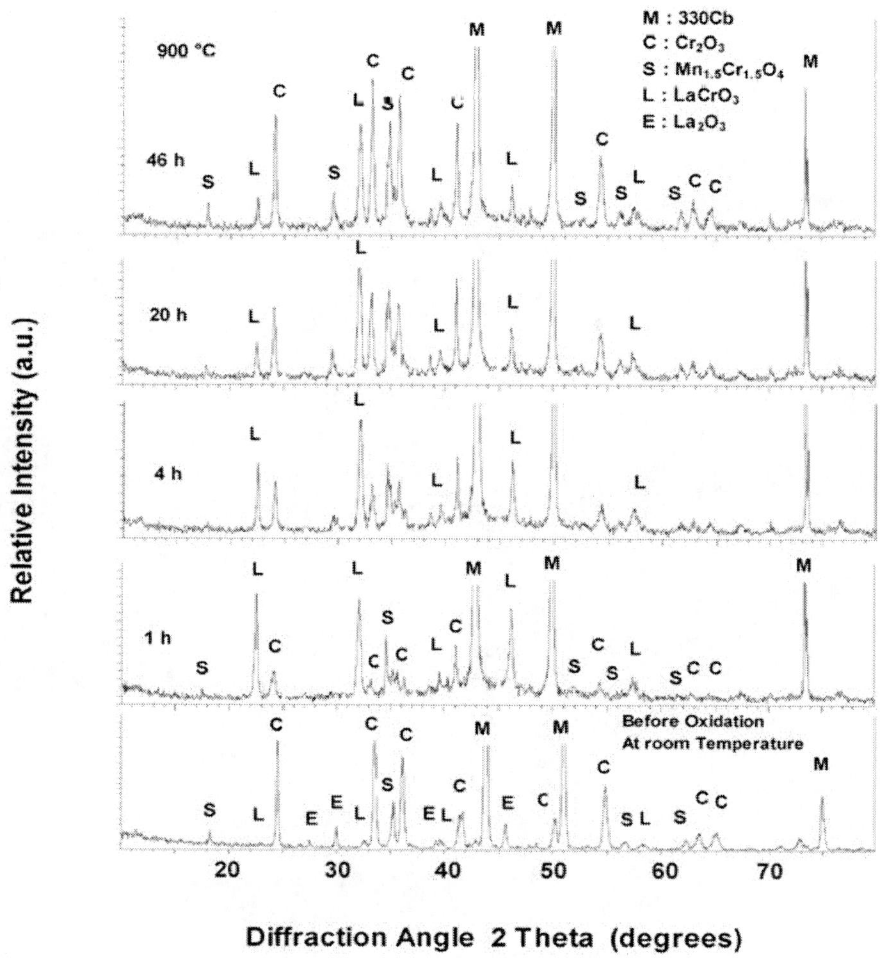

Figure 6. *In situ* XRD patterns obtained for 46 first hours oxidation at 900°C in air of the sol gel coated 330Cb, argon annealed at 1000°C. Similar results are obtained on 800°C argon annealed specimens.

Surface micrographs obtained after the 46 hours oxidation at 900°C in air of the sol-gel coated 330Cb, argon annealed at 800°C are presented on Figure 7d and Figure 8d. Spalled areas are observed on the surface and no lanthanum is found on the specimen's surface. On Figure 7d the white zones correspond to the spalled areas. Then the metallic substrate alone can be detected. It is also observed that the scale is not convoluted. The scale morphology is very similar to the one observed on blank specimens after oxidation. It is then concluded that the 800°C temperature argon annealing did not permit the incorporation of lanthanum inside the oxide scale during the oxidation process. The lanthanum coating felt down during the heating up of the specimen placed vertically in the TGA equipment. Surface micrographs obtained after the 46 hours oxidation at 900°C in air of the sol-gel coated 330Cb, argon annealed at 1000°C are presented on Figure 7e and Figure 8e. On 1000°C argon annealed specimens, no spallation is observed on the surface after oxidation at 900°C. The chromia scale is convoluted. Figure 7e shows light grey areas indicating the presence of lanthanum.

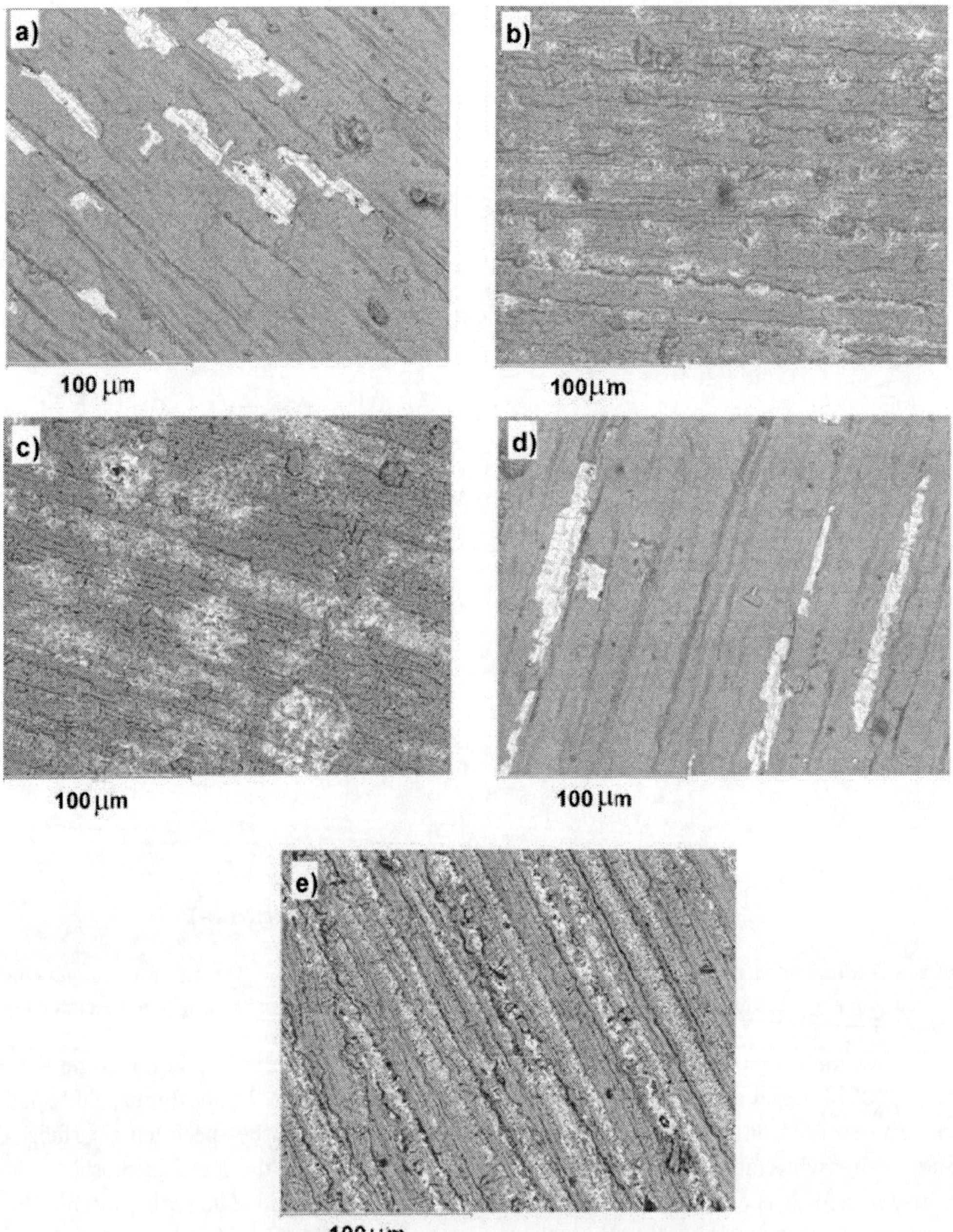

Figure 7. SEM surface micrographs of the specimens oxidized at 900°C, for 46 h, in air (BSE x 500). a) 330Cb blank, b) 330Cb sol gel coated without argon annealing, c) 330Cb sol gel coated with 600°C argon annealing, d) 330Cb sol gel coated with 800°C argon annealing, e) 330Cb sol gel coated with 1000°C argon annealing.

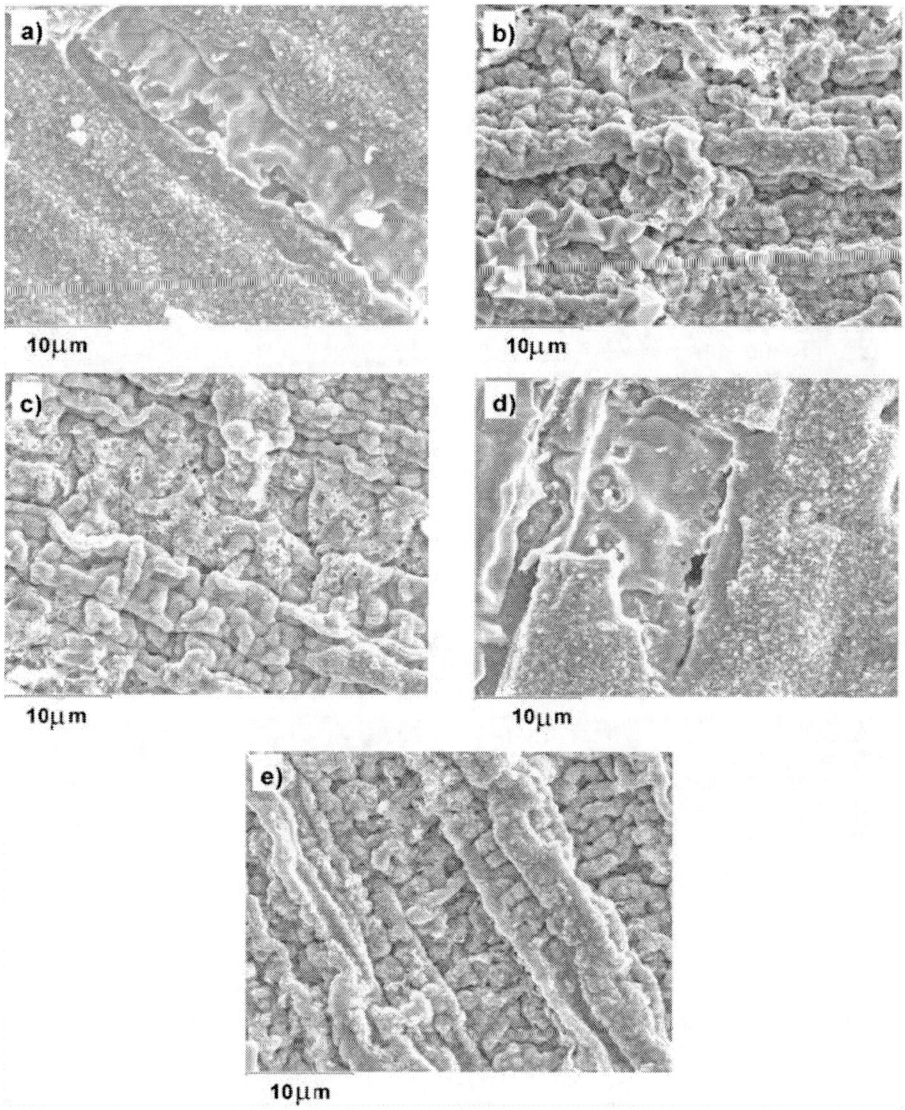

Figure 8. SEM surface micrographs of the specimens oxidized at 900°C, for 46 h, in air (SEI x 3000). a) 330Cb blank, b) 330Cb sol gel coated without argon annealing, c) 330Cb sol gel coated with 600°C argon annealing, d) 330Cb sol gel coated with 800°C argon annealing, e) 330Cb sol gel coated with 1000°C argon annealing.

Scale cross-sections are presented on Figure 9. Blank specimens, oxidized during 46h, are covered by a 3μm thick chromia scale (Figure 9a). On the sol-gel coated specimen no scale spallation is observed on the cross-section (Figure 9b). The oxide scale is 2μm thick and convoluted. $LaCrO_3$ is located at the external interface (white areas on the BSE micrograph). Cross-section micrographs obtained after the 46 hours oxidation at 900°C in air of the sol-gel coated 330Cb, argon annealed at 600°C are presented on Figure 9c. No spallation is observed on the specimen's cross-section. The very convoluted scale is mainly composed of a 2μm thick chromia scale. Figure 9c shows the presence of lanthanum (white areas). Lanthanum is present at the oxide/gas interface or incorporated inside the chromia scale. The cross-section micrograph obtained after the 46 hours oxidation at 900°C in air of the sol-gel coated, argon annealed at 800°C, is presented on Figure 8d. Spalled areas are found on the cross-section. No lanthanum is found at the external interface or inside the thin chromia scale. The 800°C temperature argon annealing did not permit the incorporation of lanthanum inside the oxide scale during the oxidation process. Cross-section micrographs obtained after the 46 hours oxidation at 900°C in air of the sol-gel coated 330Cb, argon annealed at 1000°C is shown on Figure 9e. No spallation is observed. Lanthanum is present close to the oxide/gas interface or incorporated inside the 1µm thick chromia scale.

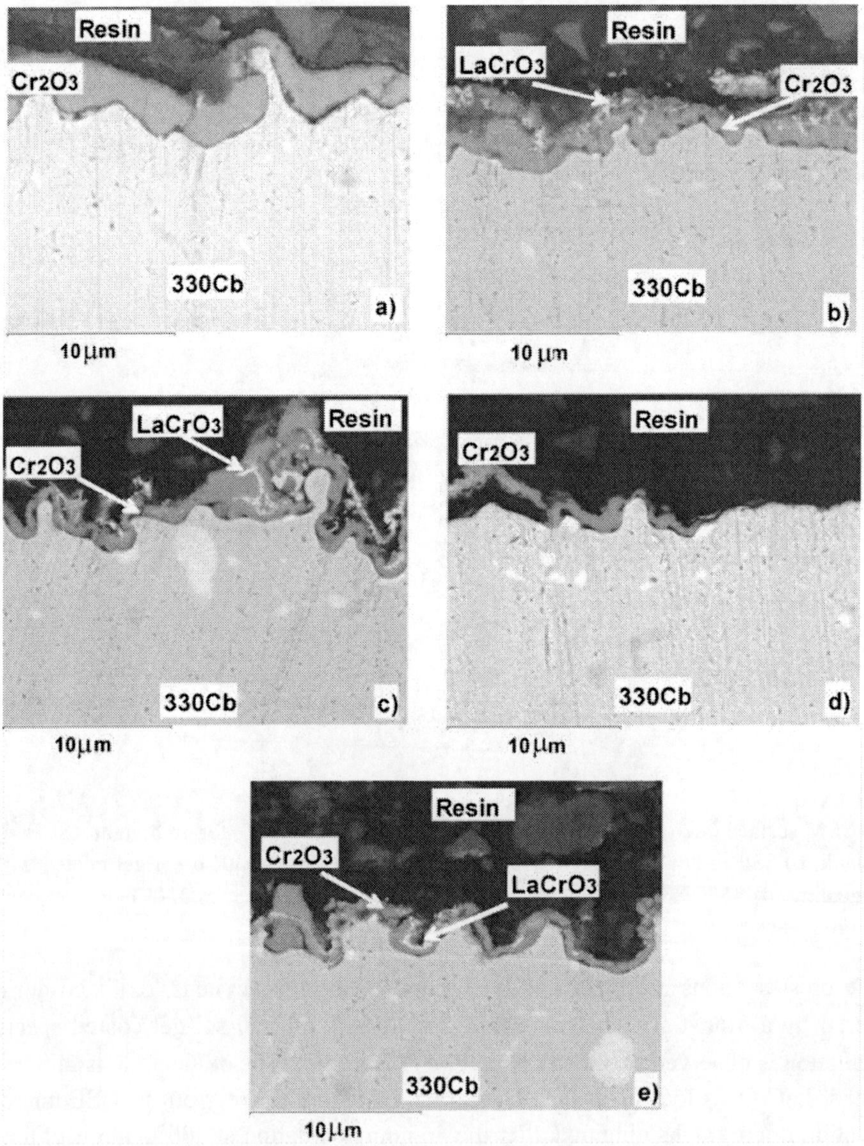

Figure 9. SEM cross-section micrographs of the specimens oxidized at 900°C, for 46 h, in air (BSE x 5000). a) 330Cb blank, b) 330Cb sol gel coated without argon annealing, c) 330Cb sol gel coated with 600°C argon annealing, d) 330Cb sol gel coated with 800°C argon annealing, e) 330Cb sol gel coated with 1000°C argon annealing.

Kinetic results show that lanthanum sol-gel coatings always lead to a lower oxidation rate compared to blank specimens. Paúl already showed this effect on austenitic steels and proposed that a lanthanum sol-gel coating lowers the parabolic rate constants during oxidation at 900°C [20, 21]. It was also demonstrated that cerium ion implantation followed by an annealing process reducing the defaults present in the alloy structure, leads to a lower oxidation rate at 900°C [22]. On the 330Cb alloy the oxide scale protective character is confirmed by *in situ* XRD analysis (Figure 3 to Figure 6), which shows the chromia scale formation instead of iron containing oxides. It also exhibits the manganese chromite growth from the beginning of the oxidation process at 900°C. SEM surface analyses confirm the presence of Cr_2O_3 and $Mn_{1.5}Cr_{1.5}O_4$. On all sol-gel coated specimens, the chromia scale growth occurs with the presence of LaCrO3 from the starting point of the XRD in situ test. This mixed oxide formation is thermodynamically explained by a reaction between Cr2O3 and La2O3 according to the Cr2O3-La2O3 phase diagram [23] and was obtained by mechanochemical synthesis by grinding constituent oxides by Zhang [24]. The LaCrO3 formation, on chromia forming alloys, has been described by other authors [9, 25-28]. The present results indicate that the reaction between Cr2O3 and La2O3 occurs at the beginning of the oxidation process on the non-annealed specimen and during the argon annealing at 600 or 1000°C. As observed by SEM, argon annealing at 600 or 1000°C improve the lanthanum incorporation inside the chromia scale during oxidation. In situ XRD shows that no structural changes occur along the test and during cooling to room temperature. It is also demonstrated that the lanthanum chromite LaCrO3 relative peaks intensity is the same before an after cooling. It indicates that LaCrO3 does not spall off when the specimen is horizontally positioned in the high temperature in situ XRD chamber. On 800°C argon annealed specimens, SEM examination shows that no more lanthanum is present on the substrate after oxidation on vertically placed specimens in the TGA rig. This annealing temperature does not permit to introduce the coating in the chromia scale because it falls down during the oxidation process at 900°C in air. SEM micrographs show that well adherent scales are convoluted. Lanthanum is mainly located close to the oxide/gas interface. Moreover, Figure 2 shows that on blank specimens spallation started after 25 cycles whereas on lanthanum sol gel coated specimens the oxide scale is still adherent after 250 cycles. As proposed by other authors a convoluted scale structure suggests that the cationic diffusion in chromia scale is lowered by the presence of lanthanum [29]. Except for the coated 330Cb argon annealed at 800°C, lanthanum leads to an adherent scale after 48h oxidation at 900°C. The oxide scale then shows a very convoluted aspect. This scale structure develops with time during the oxidation process and is not the result of a deformation during cooling to room temperature [30-34]. The convolution mechanism formation is not fully understood but it is proposed that during the oxidation process, lateral growth occurred in the chromia scale [34]. It results from the formation of new oxide inside the chromia scale [34]. It is also due to growth stresses generated inside the chromia scale leading to a plastic deformation of the scale and the underlying metallic substrate. Then, the growth stress relaxation in a thinner scale can explain the better scale adherence. Hou et Al. proposed a mechanism explaining the alumina scale convolution on a lanthanum implanted Fe-Cr-Al alloy [35]. They have shown the distribution of local stress components. It appears that the convolution results from σ_{ox} which is the compressive stress inside the oxide scale and σ_n which is the tensile stress normal to the alloy/oxide interface on the convoluted oxide scale. According to Evans et al [36], σ_n increases with the curvature radius increase. Pint indicated that during the scale growth, stresses develop inside the scale and at the oxide/alloy interface [29]. If the scale adherence is maintained, growth stresses can be high enough to plastically deform the metallic substrate an also influence the scale morphology. Interfacial voids can also play an important role on the stress development and scale morphology. With, compressive stresses in the scale, an interfacial void can lead to a stress contribution perpendicular to the alloy/oxide interface. On a hard metallic substrate, cavities can develop without scale buckling during its thickening. On the contrary, a smooth alloy can accommodate the stress increase by plastic deformation and buckling of the oxide scale. Some authors have shown that an yttrium sol-gel coating promotes a continuous silica scale formation at the internal interface on the alloy AISI 304 [37]. The silicon presence hinders the iron oxidation and the formation of non-protective iron oxides [38]. Even though alloy 330Cb contains 2.15 wt.% silicon no continuous silica scale formation was observed within the 46h oxidation at 900°C, in air. The bad scale adherence on uncoated specimens can be also explained by the presence of a too high silicon amount in the alloy [35]. The absence of iron in the oxide scale can also be due to the fact that the alloy 330Cb is a nickel-based alloy.

Table 3. Parabolic rate constants obtained after FeCrAl alloy oxidation at 1100°C, 100 h, in air

Specimen surface state	kp ($g^2.cm^{-4}.s^{-1}$)
FeCrAl	$1.13.10^{-12}$
sg La	$1.07.10^{-12}$
sgLa.ann.600	$2.06.10^{-12}$
sgLa.ann.800	$2.33.10^{-12}$
sgLa.ann.1000	$1.98.10^{-12}$

FeCrAl cyclic oxidation.

3.2. FeCrAl Alloy Isothermal Oxidation at 1100°C, in Air

Mass gain curves obtained after FeCrAl isothermal oxidation at 1100°C are shown on Figure 10. These curves allow to compare the results obtained on all the specimens: FeCrAl ; sg La ; sgLa.ann.600 ; sgLa.ann.800 and sgLa.ann1000. In all cases the kinetic curves follow a parabolic behaviour. The parabolic rate constants, kp, were calculated from the slope of the line obtained on a $\Delta m/S$ versus $t^{1/2}$ plot [39]. The calculated kp values are reported in Table 3. It appears that these values are higher for the argon-annealed specimens. The FeCrAl and "sg La" specimens show similar kp values.

The mass change versus number of cycles is presented on Figure11. The oxide scale is not always adherent to the substrate surface. FeCrAl specimens lead to a scale spallation after only two cycles. The lanthanum sol gel coating (sg La) has a slight beneficial effect on the oxide scale adherence compared to blank specimens. The best adherence is observed when the specimen argon annealing has been performed at 1000°C. In this case, very limited spallation occurred during the cycling test.

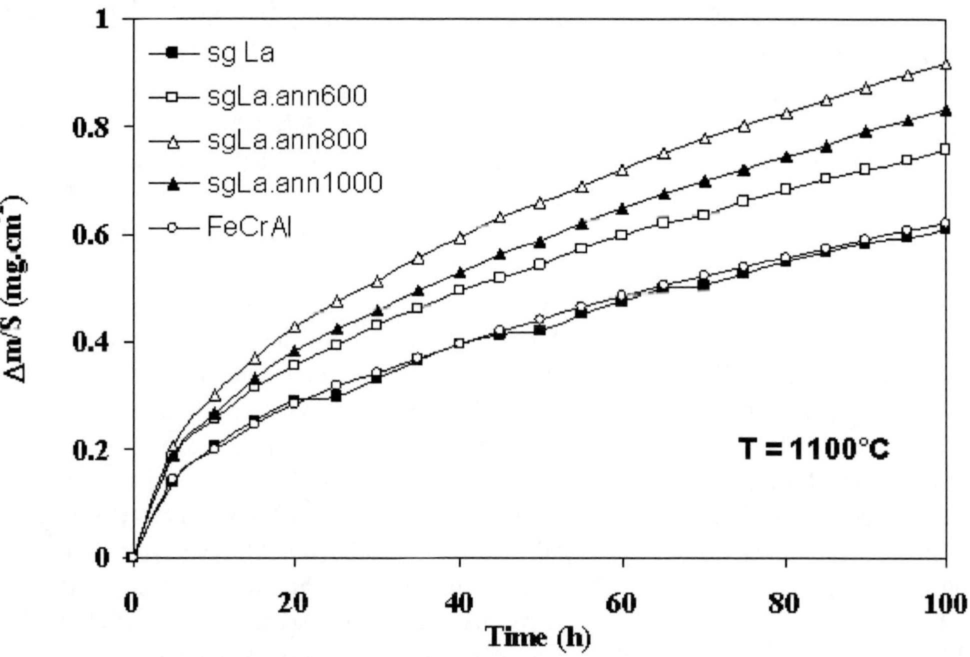

Figure 10. Mass gain curve obtained after isothermal FeCrAl oxidation at 1100°C.

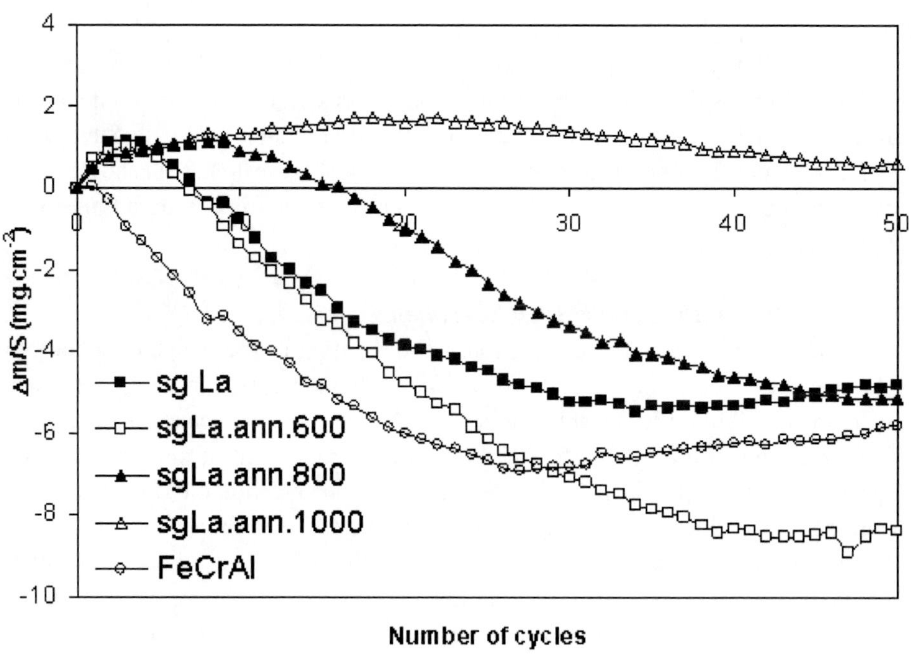

Figure 11. Specimens mass change versus the number of cycles at 1100°C (20-4 h cycles).

X-Ray Diffraction Analysis

In situ X-ray diffraction patterns obtained on the FeCrAl specimen at 1100°C are presented on Figure 12. The oxide scale is only composed of α-alumina (JCPDS 46-1212). No transition alumina is detected during the initial stage oxidation. α-alumina is detected from the first hour oxidation and grows continuously during the 30 h oxidation without showing any phase transformation. *In situ* X-ray diffraction patterns obtained on the "sg La" specimen at 1100°C are presented on Figure 13. No lanthanum containing oxides were detected before oxidation at 1100°C. This is due to the fact that the sol gel coating does not lead to sufficiently crystallized species to be detected by XRD. After heating up to 1100°C, α-alumina appears very quickly and grows with time. During the first hour oxidation, if alumina is the major phase present on the alloy, XRD patterns also show the presence of three lanthanum containing oxides: La_2O_3 (JCPDS 05-0602), $LaAlO_3$ (JCPDS 31-0022), $LaAl_{11}O_{18}$ (JCPDS 33-0699) which appear simultaneously. The La_2O_3 peaks intensity decreases with time and becomes undetectable after 2 h oxidation. It appears that La_2O_3 is transformed into $LaAlO_3$ and $LaAl_{11}O_{18}$ with time. *In situ* X-ray patterns obtained on the "sgLa.ann.600" specimen oxidized at 1100°C are presented on Figure 14.

We have noticed that before oxidation, La_2O_3 and $LaAl_{11}O_{18}$ are present on the surface. At 1100°C the La_2O_3 peaks completely disappear after the 1rst hour oxidation. $LaAl_{11}O_{18}$ peaks also become undetectable after 10 hours whereas the $LaAlO_3$ peaks grow during the 10 first hours. *In situ* X-ray patterns obtained on the "sgLa.ann.800" specimen oxidized at 1100°C are presented on Figure 15. On the initial surface, the lanthanum containing oxide La_2O_3 is present on the surface with $LaAl_{11}O_{18}$. During the oxidation at 1100°C, the La_2O_3 and $LaAl_{11}O_{18}$ peaks completely disappear after the 2nd hour of oxidation whereas the $LaAlO_3$ peaks grow with time. *In situ* X-ray patterns obtained on "sgLa.ann.1000" specimen at 1100°C are presented on Figure 16. On the initial surface, $LaAlO_3$ appears as the major lanthanum-containing phase on the surface. The small La_2O_3 peaks disappear during the first hour oxidation. All along the 30 hours oxidation test the $LaAl_{11}O_{18}$ has never been detected on the surface of the "sgLa.ann.1000" specimen and after cooling to room temperature. After 150 h oxidation, XRD patterns show that on all kinds of lanthanum sol gel coated specimens; $LaAl_{11}O_{18}$ can be detected on the surface after oxidation (Figure 17). This shows that if after 30 hours oxidation $LaAl_{11}O_{18}$ is not detected at room temperature it is found after long term testing.

Scale Cross Section of the sgLa.re1000 Specimen Oxidised at 1100°C

In order to estimate the subscale thickness and to identify the elements incorporated in the oxide scale, SEM examinations were carried out on the sgLa.re1000 specimen cross section (Figure 18). After 150 h oxidation at 1100°C, the adherent scale is 4 ± 0.5 μm thick. EDS analyses are given on Figure 19. The scale bulk is composed of alumina (spectra 1 and 2). It appears that Lanthanum is mainly located close to the air/oxide interface (spectrum 3).

According to the kinetic results presented on Figure 10, it is shown that argon annealing of the lanthanum sol gel coating always leads to a slight increase of the mass gain. This phenomenon is generally observed with lanthanum-coated specimens due to the oxygen defective nature of the lanthanum containing oxides [40]. Nevertheless, the protective character of the scale is maintained as long as the parabolic behaviour is followed and the scale acts as a diffusion barrier. Thermal cycling tests allow comparing the scale adherence depending upon the FeCrAl specimen surface treatment. Figure 11 exhibits that the lanthanum sol gel coated specimens show a better α-Al_2O_3 scale adherence during the first cycles.

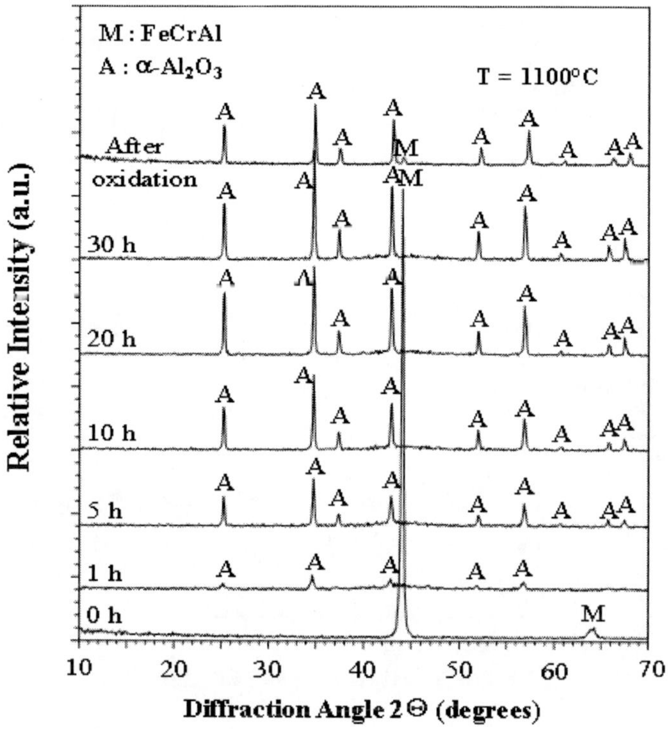

Figure 12. *In situ* X-ray diffraction patterns obtained on the FeCrAl specimen, at 1100°C.

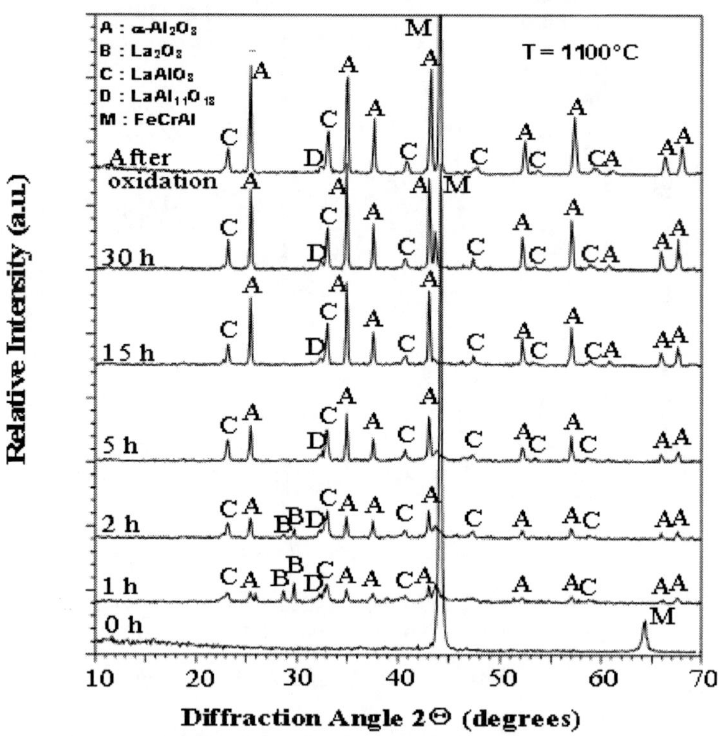

Figure 13. *In situ* X-ray diffraction patterns obtained on the "sg La specimen" at 1100°*C*.

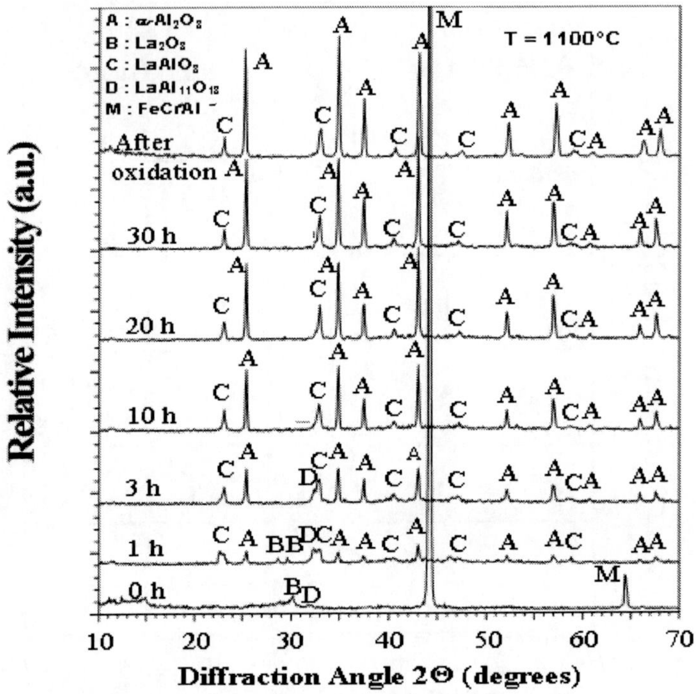

Figure 14. *In situ* X-ray diffraction patterns obtained on the "sgLa.ann.600" specimen at 1100°C.

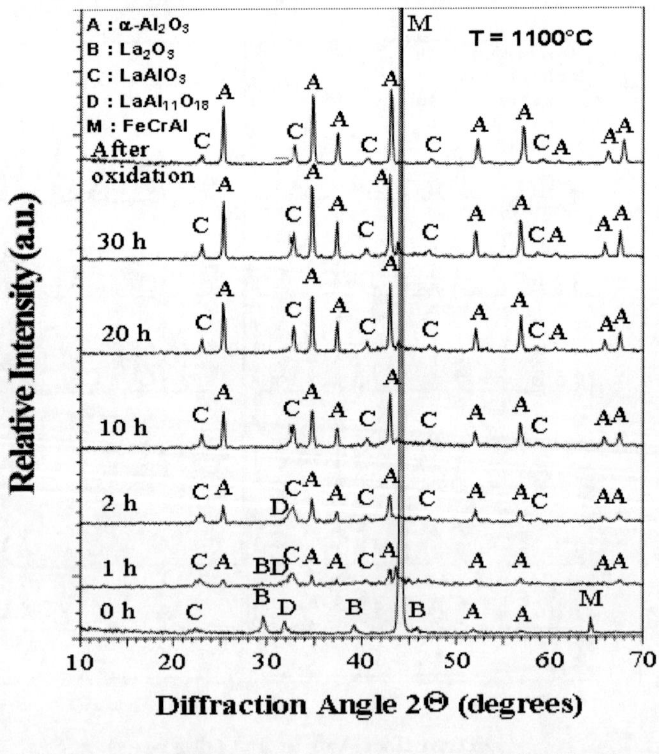

Figure 15. *In situ* X-ray diffraction patterns obtained on the "sgLa.ann.800" specimen, at 1100°C.

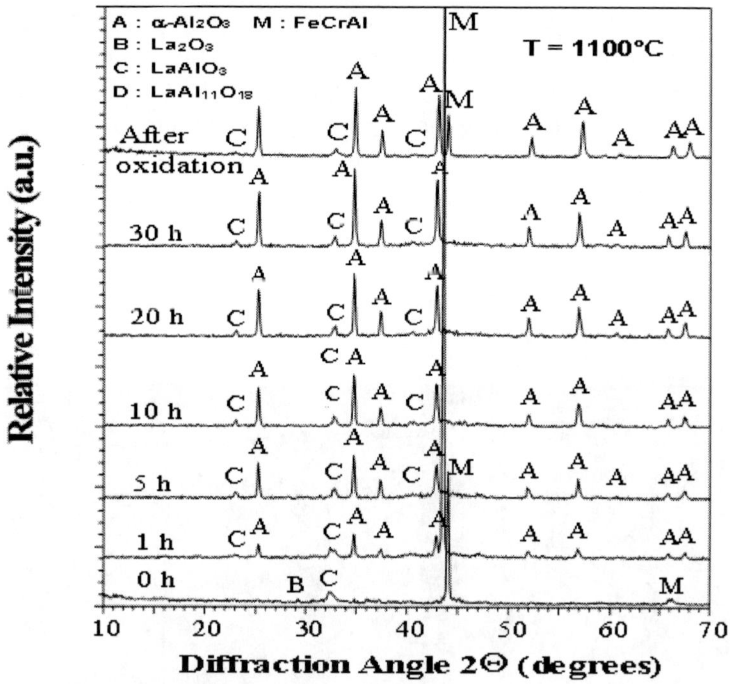

Figure 16. *In situ* X-ray diffraction patterns obtained on the "sgLa.ann.1000" specimen, at 1100°C.

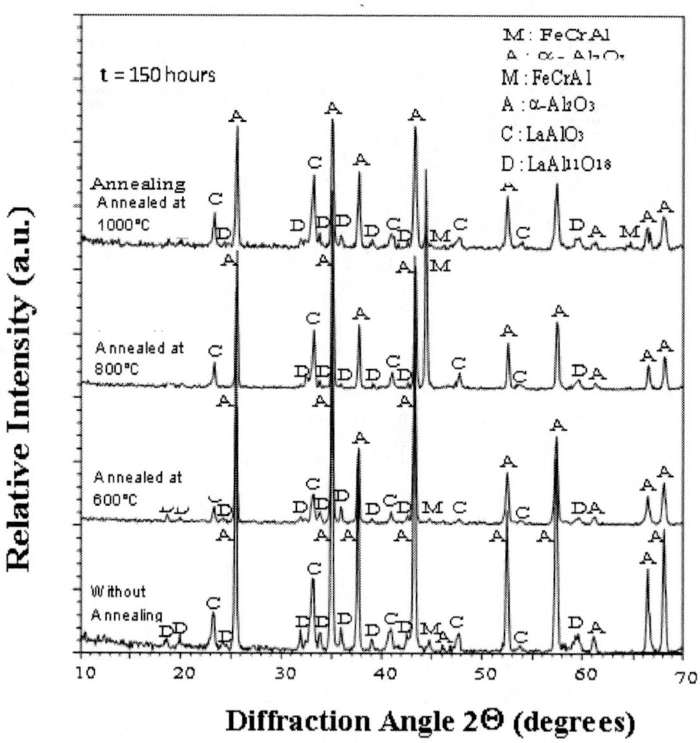

Figure 17. X-ray diffraction patterns obtained on coated specimens oxidized 150 h, at 1100°C.

A drastic improvement of the α-alumina scale adherence is observed when the argon annealing is performed at 1000°C. Thermodynamic assessments on the system La_2O_3 –Al_2O_3 show that the two mixed oxide which can be formed are $LaAlO_3$ and $LaAl_{11}O_{18}$ [41]. *In situ* XRD results (Figures 12 to 16) show that during the first 2 hours of the oxidation test, the La_2O_3 peaks intensity decreases rapidly and the $LaAlO_3$ peaks growth is observed. It means that a complete transformation of La_2O_3 into $LaAlO_3$ occurs according to reaction (1) [42, 43]:

$$La_2O_3 + Al_2O_3 \rightarrow 2LaAlO_3 \quad (1)$$

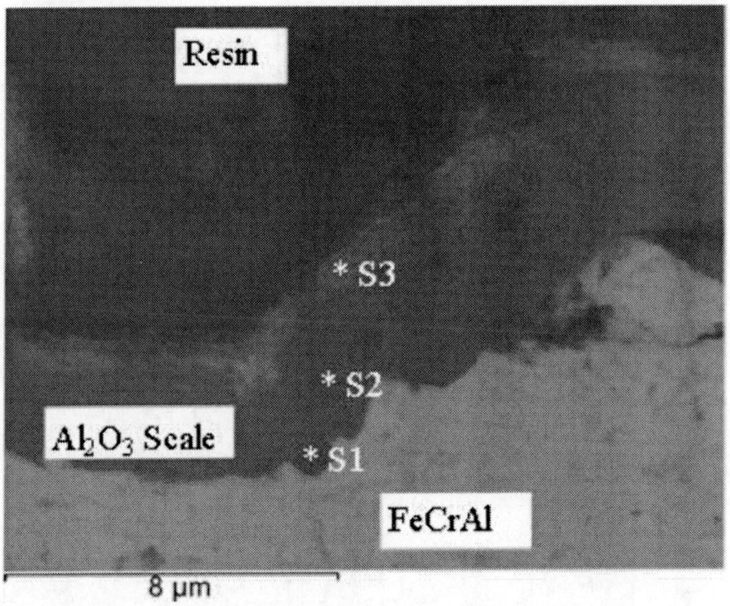

Figure 18. Scale cross section observed on the sgLa.re1000 specimen oxidised at 1100°C, in air during 150 h.

Moreover, *in situ* XRD and XRD patterns obtained after 150 h oxidation tests (Figure 17) indicate that $LaAl_{11}O_{18}$ is the main lanthanum-containing phase observed after long term testing. This means that $LaAlO_3$ is progressively transformed into $LaAl_{11}O_{18}$ all along the oxidation process according to reaction (2) [42, 43]:

$$LaAlO_3 + 5Al_2O_3 \rightarrow LaAl_{11}O_{18} \quad (2)$$

It was proposed by R. C. Ropp et al. [44] that the La_2O_3 transformation into $LaAlO_3$ is a quick phenomenon whereas the $LaAlO_3$ to LaAl11O18 transformation is a slower process. It is in accordance with our results obtained *in situ* and after 150 h oxidation. It then appears that reaction (2) has occurred after long term testing. T.S. Key et al. [45] have proposed that the La_2O_3 - Al_2O_3 oxide mixture can provide two phases: a perovskite type oxide $LaAlO_3$ and a LaAl11O18 phase. Other authors have proposed that various factors influence the preferential formation of LaAlO3 and/or LaAl11O18 from the La_2O_3 - Al_2O_3 mixture. It depends on the La_2O_3/Al_2O_3 ratio [46], the reaction duration [43], the temperature [45], the oxygen pressure [43] and the preparation mode of the oxide precursors [44]. Our results indicate that the prevalent formation of each lanthanum containing oxides is governed by: the temperature, the oxidation duration, and the ratio between La_2O_3 and Al_2O_3 on the specimen surface. Figure 13 shows that on the "sg La" specimen oxidized at 1100°C, the oxides La_2O_3, $LaAl_{11}O_{18}$, $LaAlO_3$ and Al_2O_3 initially grow at the same time. Alumina is quickly formed due to the high oxygen pressure in air. Then, the alumina proportion in the oxide scale is high enough to lead to the $LaAl_{11}O_{18}$ formation from the beginning of the oxidation test.

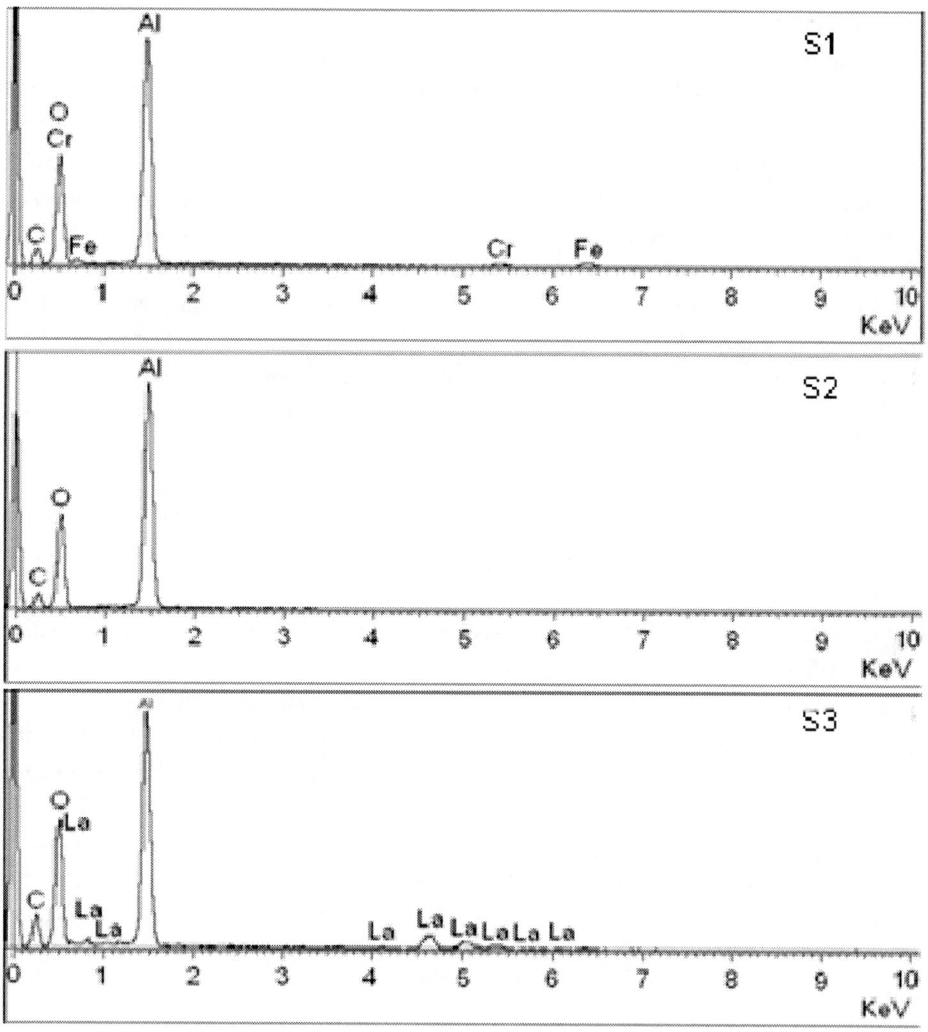

Figure 19. EDS spectra obtained on the cross section of the sgLa.re1000 specimen oxidized at 1100°C, in air, during 150 hours.

When we consider the argon-annealed specimens at 600 and 800°C, La_2O_3, $LaAl_{11}O_{18}$ and Al_2O_3 are formed under the low oxygen residual pressure present in argon (around 10 ppm). Alumina nucleates on the specimen surface and La_2O_3 and $LaAl_{11}O_{18}$ are formed by crystallisation of the initial sol gel coating and reaction with alumina. Then, we have observed that the argon annealing temperature has a great influence on the coating composition (see the 0 h patterns on Figures 14, 15 and 16). The 1000°C argon annealing temperature promotes the formation of $LaAlO_3$ due to the quick reaction (1) between La_2O_3 and Al_2O_3, which is thermally activated [42, 43]. At low argon annealing temperatures (600 and 800°C) our results show that the La_2O_3 is not completely transformed and remains present on the surface after argon annealing. The lowest annealing temperatures are not favourable to the $LaAlO_3$ formation because the reaction (1) rate is not high enough.

Our kinetic results indicate that the highest parabolic rate constants are registered on lanthanum coated-annealed specimens. XRD results show that the $LaAl_{11}O_{18}$ formation is detected after long term testing. These results are in accordance to what was proposed by Nair et al. [46] who indicate that the $LaAl_{11}O_{18}$ formation in the La_2O_3 - Al_2O_3 mixture induces the oxide grain size growth and scale porosity and an increase of the oxidation rate. According to Hou et al. [40], lanthanum on the initial surface promotes the alumina growth on the surface and a higher mass gain in air, but it also favours the scale spallation. Reactive element oxides are also known to be oxygen defective structures. The higher mass gain observed on lanthanum coated specimens can be due to a favoured oxygen transport in these oxide particles during the initial oxide growth mechanism on coated substrates [40]. Our results also show that increasing the argon annealing temperature reduces the initial amount of La_2O_3 on the substrate and improves the scale adherence. Nevertheless, our XRD results show that long-term oxidation tests always lead to the continuous increase of the Al_2O_3 amount in the scale. With time, such an alumina amount increase favours the $LaAlO_3$ transformation into $LaAl_{11}O_{18}$ mixed oxide.

CONCLUSION

Kinetic results show that lanthanum sol-gel coatings always lead to a lower oxidation rate compared to blank specimens. The smallest oxidation rate is observed on the lanthanum sol-gel coated specimens, argon annealed at 1000°C. *In situ* XRD indicate that the lanthanum coating leads to the formation of $LaCrO_3$ by a reaction between La_2O_3 and Cr_2O_3. The scale structure suggests that the cationic diffusion in a chromia scales is lowered by the presence of lanthanum. On blank 330Cb specimens, scale spallation is observed after cooling to room temperature. The chromia scale grows predominantly by outward cationic diffusion. It appears that more than 2 wt% silicon in the alloy has no beneficial effect on the chromia scale adherence. The oxide scale spallation also occurred on 800°C argon-annealed coated specimens after oxidation at 900°C. This annealing temperature is not favourable to the lanthanum-coating adherence on the substrate. On the sol-gel coated specimen without annealing and the argon annealed specimens at 600°C and 1000°C, the oxide scale formed at 900°C, in air, is adherent after 48 hours isothermal oxidation. SEM micrographs revealed that the adherent oxide scales are always convoluted. It results from the formation of new oxide inside the chromia scale due to an anionic and a cationic mixed diffusion process. Convolution can also be the result of growth stresses relaxation by plastic deformation of the scale. Thermal cycling tests performed on blank 330Cb and lanthanum sol gel coated specimens without argon annealing show that the scale adherence is improved by this coating.

On the FeCrAl alloy, thermal cycling tests indicate that the temperature of argon annealing of the lanthanum containing sol gel coating greatly improves the alumina scale adherence. X-ray diffraction (XRD) was performed in order to identify the coating crystallographic nature according to the argon annealing temperature ranging between 600 and 1000°C. In order to follow the oxidation process at 1100°C, *in situ* X-ray diffraction was performed on blank Fe-20Cr-5Al, sol gel coated and argon annealed specimens. Our results show that increasing the argon annealing temperature, on lanthanum sol gel coated specimens up to 1000°C leads to the preferential formation of $LaAlO_3$ instead of La_2O_3. The better alumina scale adherence under thermal cycling conditions is attributed to the presence of $LaAlO_3$ on the initial specimen surface before oxidation at 1100°C. With time, $LaAlO_3$ is progressively transformed into $LaAl_{11}O_{18}$ after long-term oxidation.

References

[1] Stevens R., *Oxid. Met.* 13, 353 (1979).

[2] Buscail H., Issartel C., Nguyen C. T., Fleurentin A., *Traitement Thermique* 394, 31 (2009).

[3] Holcomb G. R., Alman D. E., *Scripta Mater.* 54, 1821 (2006).

[4] Gleeson B., Harper M. A., *Oxid. Met.* 49, 373 (1998).

[5] Fontana S., Amendola R., Chevalier S., Piccardo P., Caboche G., Viviani M, Molins R., Sennour M., *J. Power Sources* 171, 652 (2007).

[6] Fernandes S. M. C., Correa O. V., Ramanathan L. V., *J. Therm. Anal. Calorim.* 106, 541 (2011).

[7] Ramanathan L. V., Pillis M. F., Fernandes S. M. C., *J. Mater. Sci.* 43, 530 (2008).

[8] Jong Seol Yoon, Jun Lee, Hae Jin Hwang, Chin Myung Whang, Ji-Woong Moon, Do-Hyeong Kim, *J. Power Sources* 181, 281 (2008).

[9] Gao-Min Zhao, Kun-Lin Wang, *Surf. Coat. Technol.* 190, 249 (2005).

[10] Karimi N., Buscail H., Riffard F., Rabaste F., Cueff R., Issartel C., Caudron E., Perrier S., *Mat. Sci. Forum* 595-598, 733 (2008).

[11] Nguyen C. T., Buscail H., Cueff R., Issartel C., Riffard F., Perrier S., *Annales de Chimie Sciences des Matériaux* 33, 33 (2008).

[12] Buscail H., Nguyen C. T., Cueff R., Issartel C., Riffard F., Perrier S., *J. Mater. Sci.* 44, 3968 (2009).

[13] Cueff R., Nguyen C. T., Buscail H., Caudron E., Issartel C., Riffard F., *Mater. Sci. Forum* 595-598, 933 (2008).

[14] Chevalier S., Tankeu A. P. D., Buscail H., Issartel C., Borchardt G., Larpin J. P., *Mater. Corros.* 55, 610 (2004).

[15] Cueff R., Buscail H., Caudron E., Riffard F., Issartel C., El Messki S., *Appl. Surf. Sci.* 229, 233 (2004).

[16] Golightly F. A., Stott F. H., Wood G. C., *Oxid. Met.* 10, 163 (1976).

[17] Quadakkers W. J., Jedlinski J., Schmidt K., Krasovec M., Borchardt G., Nickel H., *Appl. Surf. Sci.* 47, 261 (1991).

[18] Przybylski K., Garratt-Reed A. J., Pint B. A., Katz E. P., Yurek G. J., *J. Electrochem. Soc.* 134, 3207 (1987).

[19] Czerwinski F., Szpunar J. A., *J. Sol Gel Sci*. Tech. 9, 103 (1997).
[20] A. Paúl, J. A. Odriozola, *Mater. Sci. Engi*. A 300, 22 (2001).
[21] A. Paúl, S. Elmrabet, J. A. Odriozola, *J. Alloys and Compounds* 323-324, 70 (2001).
[22] A. Paúl, S. Elmrabet, F. J. Ager, J. A. Odriozola, M. A. Respaldiza, M. F. da Silva, J. C. Soares, *Oxid. Met*. 57, 33 (2002).
[23] R. Berjoan, *Rev. Int. Hautes Temp. Refract*. 13, 119 (1976).
[24] Qiwu Zhang, Jinfeng Lu, Fumio Saito, *Powder Technology* 122, 145 (2002).
[25] A. Paúl, R. Sanchez, O. M. Montes, J. A. Odriozola, *Oxid. Met*. 67, 87 (2007).
[26] Zigui Lu, Jiahong Zhu,Ye Pan, Naijuan Wu, Alex Ignatiev, *J. Power Sources* 178, 282 (2008).
[27] Wei Qu, Jian Li, Douglas G. Ivey, *J. Power Sources* 138, 162 (2004).
[28] Zhu J. H., Zhang Y, Basu A.,. Lu Z. G, Paranthaman M., Lee D. F., Payzant E. A., *Surf. Coat. Technol*. 177-178, 65 (2004).
[29] Pint B. A., *Oxid. Met*. 45, 1 (1996).
[30] Stott F. H., Wood G. C., *Mater. Sci. Engi*. 87, 267 (1987).
[31] Stott F. H., Wood G. C., Stringer J., *Oxid. Met*. 44, 113 (1995).
[32] Miner R. G., V. Nagarajan, *Oxid. Met*. 16, 313 (1981).
[33] Tolpygo V. K., Clarke D. R., *Acta. Mater*.http://www.springerlink.com/content /101598/?p=5cbd24da50e4400da98a0e76f6808176&pi=0 46, 5153 (1998).
[34] Golightly F. A., Wood G. C., Stott F. H., *Oxid. Met*. 14, 217 (1980).
[35] Hou P. Y., Stringer J., *J. Phys*. IV C9, 231 (1993).
[36] Evans A. G., Crumley G. B., Demaray R. E., *Oxid. Met*. 20, 193 (1983).
[37] Riffard F., Buscail H., Caudron E., Cueff R., Issartel C., Perrier S., *J. Mater. Sci*. 37, 3925 (2002).
[38] Pérez F. J., Cristobal M. J., Hierro M. P., Pedraza F., *Surf. Coat. Technol*. 120-121, 442 (1999).
[39] Pieraggi B., *Oxid. Met*. 27, 177 (1987).
[40] Hou P. Y., Shui Z. R., Chuang G. Y., Stringer J., *J. Electrochem. Soc*. 139, 1119 (1992).
[41] Fabrichnaya O., Zinkevich M., Aldinger F., *Int. J. Mat. Res*. 97, 1495 (2006).
[42] Barrera-Solano C., Esquivias L., Messing G. L., *J. Am. Ceram. Soc*. 82, 1318 (1999).
[43] Zhang Q., Saito F., *J. Am. Ceram. Soc*. 83, 439 (2000).
[44] Ropp R. C., Carroll B., *J. Am. Ceram. Soc*. 63, 416 (1980).
[45] Key T. S., Crist B. Jr, *J. Am. Ceram. Soc*. 88,191 (2005).
[46] Nair J., Nair P., Mizukami F., Van Ommen J. G., Doesburg G. B. M., Ross J. R. H., Burggraaf A. J., *J. Am. Ceram. Soc*. 83, 1942 (2000).

In: Argon: Production, Characteristics and Applications ISBN: 978-1-62618-204-2
Editors: B.N. Sismanoglu, H.S. Maciel et al.

Chapter 4

GENERATION AND APPLICATION OF SMALL BANDWIDTH COHERENT EXTREME ULTRAVIOLET

L. V. Dao*

Centre for Atom Optics and Ultrafast Spectroscopy and Australia Research Council Centre of Excellence for Coherent X-ray Science, Swinburne University of Technology, Melbourne, Australia

ABSTRACT

Coherent extreme ultraviolet radiation can be produced by high-order harmonic generation of energetic femtosecond laser pulses in noble gases. Using 800 nm driving pulses and argon gas, the energy conversion efficiency is about 10^{-5} and the absorption limit reaches at the photon energy of ~ 45 eV. The high-harmonic generation in semi-infinitive gas cell has been considered to obtain high efficiency, highly spatial coherence and a good beam profile. The phase-matching, propagating of the fundamental and harmonic radiation, the selection of quantum path has been discussed in terms of intensity and truncation of driving pulse. The phase-matched harmonic radiation with a high intensity and a small bandwidth in the wavelength rang of ~ 30 nm can be generated by using ~ 80 Torr argon gas. We demonstrate that this source can be used for a multiple-harmonic-order coherent diffractive imaging with spatial resolution below 100 nm.

1. INTRODUCTION

Since the first generation of high-harmonic orders in 1987 [1] laser driven high-harmonic generation sources have become an important supplyof coherent extreme-ultraviolet radiation and soft x-rays included the water window(4.4 to 2.3 nm) [2-4]. These sources are small-scale and highly versatile in comparison to synchrotrons and free electron x-ray lasers, and their characteristic output can be tailored according to the experimental requirements. The soft x-ray ultra short pulses of reasonable brightness and high coherence are used in atomic [5, 6]

* E-mail address: dvlap@groupwise.swin.edu.au.

and molecular [7-9] spectroscopy, to solid-state [10-12], plasma [13] physics andextreme ultraviolet interferometry [14, 15]. Because the HHG spectrum depends on the structure of the atom or molecule a new method for tomography of atoms and molecules has been proposed and developed [16, 17]. Due to its highly coherent degree, harmonic emission is being used for coherent-diffractive imaging based on a single harmonic [18] or even multiple harmonic orders [19].

By their nature, high-harmonic generation radiationis a laser-like beam. Harmonic emission ranges from wavelengths in the extreme-ultraviolet to the soft x-ray region. In principle, the characteristics of the laser beam are imprinted onto the harmonic beam. The phase-matched propagation of the fundamental and harmonic radiation, i.e., the coherent construction of the harmonic field in the macroscopic medium, is reflected in the observed harmonic intensity and coherent degree. Phase matching in low-order harmonics, e.g., second harmonic generation, is usually obtained by using anisotropic crystals where the signal and pump beams propagate with different polarizations and speed in the birefringent medium. To obtaina high harmonic intensity, increasing the interaction length or maintain the phase-matching condition over a long interaction length is necessary. Because of the optical properties of materials in the XUV range the medium for HHG is normally an atomic or molecular gas. The degree of phase-matching depends on the harmonic order and a number of experimental parameters including the atomic and molecular dispersion, the absorption coefficient of the target gas at the harmonic frequencies, the ionization fraction, and the gradient of the driving laser field.

The generation of phase-matched HHG in the near cut-off region by using a moveable focusing lens in a semi-infinitive gas cell has been obtained [20]. The effect of the neutral gas dispersion on the pump beam, the plasma dispersion due to ionization and the phase change along the focusing beam can be balanced under certain conditions. Phase-matching high harmonic generation at very high photon energy is difficult because a strong driving field is required that creates a large free-electron dispersion through strong ionization of the gas. Three important effects are involved in the ionization of the medium: Depletion of the medium, plasma dispersion due to the free electrons, and defocusing of the laser beam by the radial distribution of the electrons. For a 800 nm driving laser, the "critical" ionization levels above which true phase matching is not possible are ~ 3.8 % for argon; thus efficient phase-matched HHG has been demonstrated only at photon energies < 50 eV for argon [20, 21]. When the true phase matching is not possible quasi-phase matching (QPM) techniques have been employed to generate high photon energy harmonics at high ionization levels [22, 23]. In the QPM configuration the nonlinear process in the medium where the induced polarization and the harmonic field are initially out-of-phase is cancelled or their phase is changed. Previous work has demonstrated that QPM schemes can enhance HHG by modulating the driving laser intensity in modulated waveguides [24], by using interference patterns created by a sequence of counter-propagating light pulses [25], by using a sequence of gas jets [26] or by using mode beating in a multi-mode waveguide [27].

The chosen of different geometry for the harmonic generation process and the process of plasma formation and of nonlinearity lead to a complex spatially and temporally dependent induced nonlinear polarization in the interaction medium and thus strongly determine the conversion efficiency, the spatial, spectral and temporal properties, and the coherence of the harmonic emission. In particular, it has been shown that the large and time-varying refractive index due to the plasma produced when focusing the driver pulse into the nonlinear medium is the major reason why high-harmonic generation experiments only exhibit partial coherence [28]. Therefore, and because of the highly nonlinear nature of high-harmonic generation, each of the harmonic sources that are currently in operation is unique in its experimental setup and the characteristics of the resulting harmonic output.

2. THEORETICAL BACKGROUND

2.1. The Single Atom Response in High Order Harmonic Generation

The semi-classical or the so-called 'Three-step model' allows high harmonic generation to be described quantitatively and to be understood qualitatively to a certain extent. This model allows the main characteristics of high-harmonic generation to be explained. Quantum-mechanical models [29, 30] extend this model and provide corrections in both the qualitative and quantitative description. In the first step of this model, the electric field of a high-power ultra short laser pulse perturbs the Coulomb barrier binding the outer electron to the nonlinear medium. The electric field is so strong that the Coulomb barrier is suppressed and the electron can undergo tunneling through the Coulomb barrier in continuum state (Figure 1a). In the second step, the trajectory of the electron is described on the basis of a free particle in the laser electric field. The electron exhibits harmonic motion in the laser electric field and gains momentum. In the third step, when the electric field changes its sign (Figure 1b), the electron traverses its trajectory and can return to interact with the ionic parent (Figure 1c). Ifrecombination occurs, the photon with higher energy than the incident photon can be emitted.

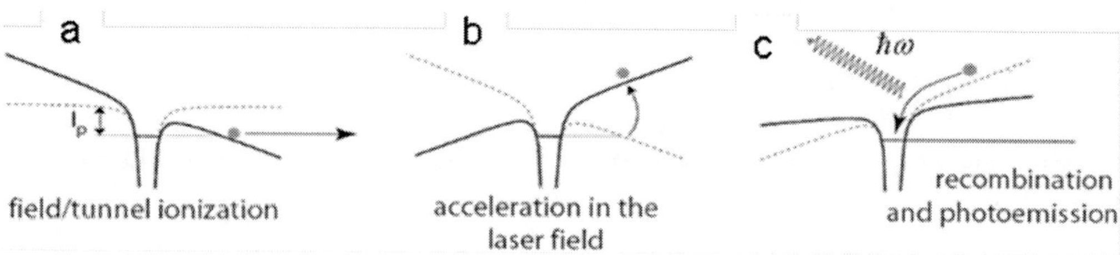

Figure 1. Three step model of the high order harmonic generation.

Only when the laser electric field is linearly polarized can the electron can return to the ionic parent. For elliptic polarization there are no classical electron trajectories and the electron cannot return to parent ion. However, for a slightly elliptical field, high-harmonic generation is not suppressed, but significantly reduced, that is, due to the finite extent of the electronic wave packet and quantum diffusion effects [29, 31]. Rescattering and recollision processes [32] play animportant role in shaping the spectrum of high-harmonic generation. To estimate the maximum harmonic order that can be generated by an intense laser beam, the intensity at which the laser field becomes comparable to the Coulomb field of an atomic nonlinear medium needs to be estimated.

Kulander et al. and Corkum suggested that the maximum of harmonic photon energy or so-called harmonic cut-off is given by [29, 33, 34]

$$E_{cut-off} = I_p + 3.17U_{pond}, \quad (1)$$

where I_p is the ionization energy of the generating medium and $U_{pond} \propto I\lambda^2$ is the ponderomotiveenergy which scales linearly with the laser field intensity I and quadratically with the wavelength of the fundamental laser beam (λ). The ponderomotive energy is the mean kinetic energy of an electron oscillating in a harmonic laser electric field. The ionization energy of common noble gases is given in table 1.

Table 1. Ionization energy in eV of the noble gases

Element	I_P
He	24.6
Ne	21.6
Ar	15.8
Kr	14.0
Xe	12.0

As a free particle in the laser electric field, the electron experiences the force

$$F = m_e\ddot{x} = -eE_0\cos(\omega_0 t + \varphi) \quad (2)$$

where m_e is the electron mass, x is the location of the electron with respect to the centre of the atom, e is the electron charge, ω_0 is the frequency of the driving laser field and t and φ are the time and the phase when the electron is injected into thevacuum. Here t = 0 is the time when the electron leaves the atom. Integration of equation (1.2) leads to

$$\dot{x} = -eE_0(\sin(\omega_0 t + \varphi) - \sin(\varphi))/m_e \quad (3)$$

Assuming there is no drift to the motion of the electron, that is, $\varphi = 0$, the mean kinetic energy is given by

$$U_p = m_e\langle\dot{x}\rangle^2/2 = e^2I/(2\omega_0^2 m_e\varepsilon_0 c) \quad (4)$$

where c is light speed and U_p is the ponderomotive energy. The maximum kinetic energy that the electron can accumulate in the laser electric field is the sum of the ionization energy I_p of the nonlinear generating medium and the ponderomotive energy that is proportional to the laser intensity I. This implies that harmonic photons of high energy can routinely be generated by means of high peak-power ultra short pulse laser systems. For example, if the laser intensity $I \sim 1.67 \times 10^{14}$ W/cm^2 and the wavelength of the driving laser pulse is 800 nm, then Up ~ 10 eV. Thus the ponderomotive potential is of the same order of magnitude as the ionization energies of the neutral noble gases.

The generation of these high-energy photons requires one to enter the regime of high-field optical science via high peak power ultra short laser pulses. Pulses with a peak power > 10^{12}W can be delivered by a table-top system. Pulses with a peak power > 10^{15} W can be obtained from even larger system and by focusing these pulses, light intensities of > 10^{20}

W/cm^2 can be achieved. However, the brightness of the harmonic radiation and the cut-off energy are determined by a complex interplay of several mechanisms and are not only determined by the laser electric field intensity. Integration of equation (3) allows one to determine the trajectory of the electron in the laser electric field. The recombination time is the time after the electron recombines with the parent ion, that is, the time when the electron wave-packet returns to the origin of location. This is determined by the following equation

$$(\cos(\omega_0 t + \varphi) - \cos(\varphi)) = -\omega_0 t \sin(\varphi)) \qquad (5)$$

The maximum kinetic energy of the electron in the laser electric field is given by

$$E_{max} = me\langle \dot{x} \rangle^2 / 2 = 2V_{pond} [\sin(\omega_0 t + \varphi) - \sin(\varphi)]^2 \qquad (6)$$

The electron acquires the maximum kinetic energy when it is released at a phase $\varphi = 17^0$. The scaling factor $2 [\sin(\omega_0 t + \varphi) - \sin(\varphi)]^2$ versus the phase φ at the time ofrelease is shown in Figure 2. The maximum photon energy after recombination with the ionic parent medium to the former ground state is therefore given by [29, 33, 34]

$$E_{cutoff} = I_p + E_{kin,max} \approx I_p + 3.17U_p. \qquad (7)$$

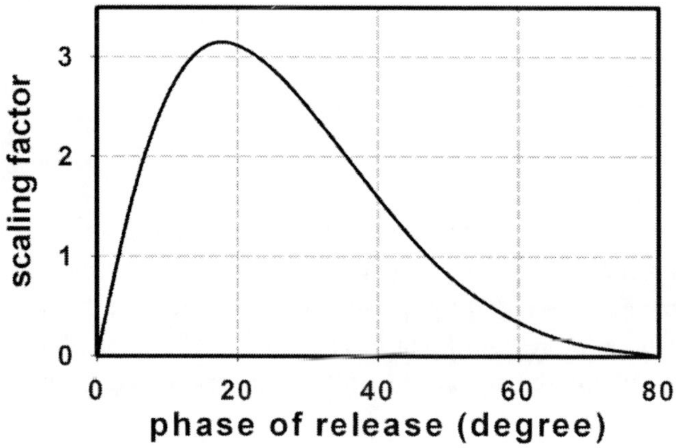

Figure 2. Scaling factor of the kinetic energy of the electron as a function of the phase of the laser electric field at the time of release.

However, different values for the scaling factor of the ponderomotive potential have been reported. Based on [35] the scaling factor is calculated to be 3, while propagation effects for very high laser intensities can possibly lead to a scaling factor of ~ 2 [36], and the harmonic spectrum can be extended to $I_p + 8U_p$ when phase-locked high-order-harmonic and sub 100 as pulses are generated from stretched molecules [37]. We note that the equation for the cut-off energy depends on the assumption that the laser intensity is a constant. This assumption does not hold when very short pulse is applied because the laser intensity can change during a single period and therefore the cut-off energy rule might not be applicable. It is interesting to note that the noble gases with high ionization energy (e.g., helium) that produce very high-order harmonics generate but the radiation is less bright due to a smaller cross-section and associated nonlinear susceptibility. In other word, in atoms of large diameter the electron wave-packet is less tightly bound to the Coulomb potential and thus undergoes a smaller spread before returning and recombining with the parent ion compared to the tightly bound electron wave-packet of smaller atoms.

2.2. Propagating and Phase Matching

Studying the HHG process in a single atom response basically provides the predicted shape of the harmonic spectrum but it cannot predict the observed harmonic yield. To obtain strong harmonic emission a long interaction medium is necessary therefore propagation effects need to be considered. The output harmonic intensity is the coherent build up of harmonic radiation from an atom ensemble. For coherent construction of the harmonic emission, the wave-fronts of the fundamental laser and the generated harmonics must be in phase. Atoms along the path of the laser produce harmonics field which is propagating with the laser field. When the generated harmonics fields are in phase, there are constructive interference and these harmonics combine to build up the total harmonic emission.The speed of the incident and the harmonic waves depend on the index of refraction and thus depend on the wavelength. The propagated distance over which the polarization and the harmonic wave-fronts slip out of phase by π, is the so-called coherence length

$$L_{coh} = \pi/\Delta k \tag{8}$$

$$\Delta k = qk_0 - k_q \tag{9}$$

where Δk is the phase-mismatch in the propagation direction and where k_0 and k_q is the wave vector of the fundamental and harmonic radiation. After traveling a distance of one coherence length, the fundamental driving laser pulse and the generated harmonic radiation of a particular harmonic order q are out of phase. The number of photons for the q^{th}harmonic N_{out} (the harmonic intensity) emitted on-axis per unit of time and area is given by [38]

$$N_{out} \sim \rho^2 d_q^2 \frac{4L_{abs}^2}{1+4\pi^2(L_{abs}^2/L_{coh}^2)}\left[1+\exp\left(-\frac{L_{med}}{L_{abs}}\right)-2\cos\left(\frac{\pi L_{med}}{L_{coh}}\right)\exp\left(-\frac{L_{med}}{L_{abs}}\right)\right] \tag{10}$$

where ρ is the gas density, L_{abs} is the absorption length, L_{med} is the interaction length and d_q is the atomic response. If the absorption of the medium is neglected, i.e., L_{abs} is much larger than L_{med}, then equation (10) reduces to

$$N_{out} \propto \rho^2 L_{med}^2 d_q^2 \frac{\sin^2\left(\frac{\Delta k_q L_{med}}{2}\right)}{\left(\frac{\Delta k_q L_{med}}{2}\right)^2} \tag{11}$$

Since L_{coh} scales linearly with $1/q$, phase matching becomes more difficult with increasing harmonic order q. This means that higher-order harmonics go in and out of phase with the fundamental radiation more often than lower-order harmonics. Two phase-matching regimes exist [39]. In the first regime phase matching is achieved early in the driver pulse. The atomic dispersion counter balances the Gouy phase shift dispersion. This regime applies to lighter elements, such as helium. In the second regime, phase-matching is achieved when the electronic dispersion counter balances the atomic dispersion. When going from lighter to heavier elements, such as argon, the first phase-matching regime is replaced by the second regime.

The amplitude of the high harmonic dipole d_q is dependent on the intensity of the driving laser field with peak intensity I_0, which varies along the focusing direction (z) and perpendicular to the focusing direction across the beam profile (r), and is also dependent on properties of the gas medium such as the ionization energy and the orientation of the atoms or molecules relative to the driving field. In the single atom response model, d_q scales as I_L^3 [40] in the intensity window of the driving pulse I to$I+dI$.

The phase mismatch Δk along the axis of the fundamental laser beam can be expressed

$$\Delta k(z) = -2\pi Pq(1-\eta)\delta n(z)/\lambda + PqN_e r_e \lambda + (\text{dipole phase term}) + (\text{geometric term}) \quad (12)$$

where λ is the laser wavelength; P, η and r_e are the gas pressure, ionization fraction and classical electron radius, respectively; N_e is the density of free electrons, $\delta n = n_{laser} - n_q$ is the difference of the refraction index of the gas per unit atmosphere at the fundamental and harmonic wavelengths, respectively.

The phase-mismatch is a separate issue from the reabsorption of harmonic emissionfrom the generating medium. Reabsorption [38, 41, 42] plays a role when the harmonic radiation is already phase-matched. The harmonic yield is limited by several phase-mismatch phenomena, such as different diffraction rates for the driving and generated radiation, the dependence of the refractive index on the wavelength of the fundamental and harmonic radiation, and the dependence of the intrinsic phase of the harmonic emission on the intensity in both the radial and longitudinal direction.

i) Material Dispersion

The phase mismatch contribution due to material dispersion for q^{th} harmonic is given by

$$\Delta k_q = \frac{2\pi Pq(1-\eta)}{\lambda}(n_{laser} - n_{harmonic}) \quad (13)$$

The refractive indices of the noble gases are close to unity and depend on the frequency. For a fundamental wavelength λ = 800 nm, $n(\lambda) \approx 1 + 2.8\times10^{-4}$ and for harmonics q> 21, $n(\lambda_q) \approx (1-10^{-4}) - (1-10^{-6})$ at 1 atm and 273K, where λ_q is the wavelength of the qth harmonic. The refractive index is proportional to the gas pressure (in atm) and when the gas medium is ionized the density of neutral atoms is also proportional to the fraction of ionization, η. For a gas pressure of 50 Torr, and assuming the ionization fraction in the medium is 1 %, the neutral dispersion phase mismatch for harmonic 25th (H25) is found to be around 45 cm^{-1}. When a laser with high intensity propagates through a medium, this also causes a change in the refraction index in proportion to the laser intensity. This change is known as the instantaneous nonlinear refraction index and given as

$$\Delta n = n_2 I(t) \tag{14}$$

where n_2 is the nonlinear refractive index.

ii) Plasma Dispersion

When harmonic radiation is generated at intensities that are high enough to ionize the generating medium, refraction due to the plasma has to be taken into account as it can severely limit the coherence length and thus the harmonic yield. The index of refraction of plasma is given by

$$n_P(\omega_0) = \sqrt{1 - \frac{\omega_P^2}{\omega_0{}^2}} \tag{15}$$

where ω_0 is the laser frequency, and ω_P is the plasma frequency. The plasma frequency is given by

$$\omega_P = e\sqrt{\frac{N_e}{\varepsilon_0 m_e}} \tag{16}$$

where N_e is the density of free electrons ε_0 is the vacuum permittivity and m_e is the electron mass. Since the laser frequency ω_0 is much larger than the plasma frequency ω_P, i.e., $\omega_P^2 << \omega_0{}^2$, equation (15) can be rewritten

$$n_P(\omega_0) \approx 1 - \frac{\omega_P^2}{2\omega_0{}^2} \tag{17}$$

From equation (17), it is clear that the refractive index for plasma is always less than one and depends on the frequency of the fundamental laser pulse. If we define $L_{coh,P}$ as the coherence length relating to the plasma dispersion, $L_{coh,P}$ can be calculated from

$$L_{coh,P}(qk_0 - k_q) = \frac{2\pi q}{\lambda_0}[n_P(\omega_0) - n_P(q\omega_0)]L_{coh,P} = \pi \tag{18}$$

Then

$$L_{coh,P} = \frac{4\pi^2 m_e}{N_e e^2 \lambda_0 \mu_0 (q - 1/q)} \tag{19}$$

where $\mu_0 = \frac{1}{c^2 \varepsilon_0}$ is the vacuum permittivity. If we assume q>>1, then equation (19) is given approximately by

$$L_{coh,P} = \frac{4\pi^2 m_e}{N_e e^2 \lambda_0 \mu_0 q} \tag{20}$$

From equation (20), we can see that the coherence length scales inversely with the harmonic order. In addition, to produce a higher order harmonic, a higher intensity of the fundamental beam needs to be applied. This leads to an increase of the ionization rate and free electron density. Therefore, the plasma phase mismatch becomes more important with increased harmonic order.

iii) Contribution of Atomic Dipole Phase

The atomic dipole phase mismatch is an intensity dependent phase mismatch and has been theoretically studied [43] as well as experimentally investigated [46]. The origin of this phase is the trajectory acquired by the electron leading to the emission of the harmonic in the continuum state. To a first approximation, the phase of the atomic dipole varies linearly with the laser intensity. The laser intensity can vary spatially in both the longitudinal and radial directions. Therefore, the atom dipole phase also varies axially as well as radially. This leads to reduced harmonic emission as well as strong spatial distortion [44]. It may be possible to find situations in which the intrinsic phase can be played against the geometrical phase mismatch in order to improve overall phase matching [43]. For a Gaussian laser beam with intensity along the direction of propagation

$$I(z) = \frac{I_0}{\left[1 + \left(\frac{2z}{b}\right)^2\right]} \tag{21}$$

the contribution of the dipole phase to the phase mismatch on axis of the qth harmonic, $\nabla \varphi_q$, is given by [45, 46]

$$\nabla \varphi_q = -\alpha_q \frac{\partial I}{\partial z} = \frac{8z}{b^2 \left[1 + \left(\frac{2z}{b}\right)^2\right]^2} \alpha_q I_0 \tag{22}$$

where α_q is a coefficient related to the electron trajectories ($\alpha_q \approx 1-5\times10^{-14}$ cm^2 / W for the short trajectory and $\alpha_q \approx 20-25\times10^{-14}$ cm^2 / W for the long trajectory) and b is the Rayleigh length.

iv) Contribution of Geometry

Geometrical phase mismatch for the focusing beam in free space is primarily due to the Gouy phase shift when focusing a laser beam as described in detail in [47]. When $W_0^2 = \lambda z_R/\pi$ is half the beam waist with respect to the centre of focus, the beam waist is $W(z)^2 = W_0^2 [1+(z/z_R)^2]$.The $R(z) = z(1 + (z_R/z)^2)$ is the radius of curvature of the wave-fronts, and $\varphi(z) = tan^{-1}(z/z_R)$ is the Gouy phase. The Rayleigh length z_R is related to the confocal parameter via $b = 2z_R$. The phase for the qth harmonic at each point where it is produced along z is $\varphi_q = q\varphi(z)$. Thus the Gouy phase shift for the harmonics is negligible. The coherence length becomes longer when the medium is away from the focus. Thus the focus of the driving laser beam can be increased relative to the gas jet to confine the interaction region to a single geometrical coherence length. The intensity of the driving laser field varies in the longitudinal and transversedirections. The dipole phase of the generating medium varies in the longitudinal and transverse directions. This can lead to on- and off-axis phase-mismatch and deterioration of the desired spatial harmonic profile and the harmonic yield [43, 44, 48]. The geometrical contribution on phase matching is dependent on the geometrical configuration, e. g., other geometrical contribution need to be considered when a hollow fibre or Bessel beam is applied.

Figure 3. Quantronix femtosecond multi-pass amplifier system.

The large and time-varying refraction index associated with free electrons has been caused the reduction of coherence of harmonic radiation. The harmonic emission at any particular wavelength can result from multiple electron trajectories. These trajectories are characterized by injection of the electron into the continuum state at different phases. For each particular harmonic order two separate electron trajectories generate the same photon energy. However, these electron trajectories involve slightly different ionization and recollision times. Thus acomplex and spatially varying multimode wave-front of the harmonic radiation can be created and can be the reason why the harmonic beam lacks the characteristics of the fundamental laser beam. In many-cycle driver pulses many optical cycles can contribute to the signal of a particular harmonic order, and thus potentially lead to a degradation of the coherence properties. Spatial coherence can easily be measured by means of Fresnel bi-mirror interferometry, and a double-pinhole or a double-slit experiment [49]. The correlation between the phase of the radiation emerging from both sources determines the visibility of the interference fringes observed. If the phase difference is constant and deterministic, and thus correlated, the visibility will be unity, whereas any coherence degradation or variation in the phase difference will result in a visibility smaller than unity.

Table 2. Specifications of the laser output of the Quantronix system

Centre wavelength	805 nm
Pulse energy	10 mJ
Pulse width ($sech^2$)	30 fs
Beam size	12 mm
Beam quality (M^2)	1.6
Instability	< 1%
Repetition rate	1 kHz
Output polarization	Horizontal

3. EXPERIMENTS FOR HIGH HARMONIC GENERATION

3.1. High Power Femtosecond Laser System

An important tool for harmonic generation is the high power laser system. The most popular table top high power femtosecond laser is a chirped pulse amplification system. The system consists of a femtosecond oscillator and a regenerative or a multi-pass power amplifiers. The femtosecond pulses with low pulse energy (sub nJ) has been generated by a Titanium: Sapphire (Ti:Sa) oscillator. The amplifier allows amplification of the pulse energy of sub nJ energy to a level of a mJ. The typical operating repetition rate is 1 kHz. The technique of chirped pulse amplification is based on stretching the short pulse in time by introducing a frequency chirp onto the pulse which increases the duration by a factor of 10^3-10^4, amplification of the long duration pulse and then compression of the pulse for obtaining short pulse at the end. The mechanism of this technique follows in steps. Firstly, the pulse is stretched by propagation of the short pulse in a special grating system which exhibits negative dispersion. The stretching of the pulse leads to a significant reduction of its intensity which removes the tendency of gain medium damage due to the high intensity of the short pulse. Secondly, the beam passes through the gain medium multiple times with (in regenerative amplifier) or without (in multi-pass amplifier) the use of a cavity. Compared to a regenerative amplifier, the number of path way in multi-pass amplifier is small that requires a higher gain per pass but that help to reduce the high order chirp in amplified pulse. Finally, the amplified pulse is compressed by using a grating pair which exhibits positive dispersion for an output pulse with ~ 30 fs duration and up to 10 mJ energy. Figure 3 shows the Quantronix multi-pass amplifier system which has been used for high harmonic generation at Centre for Atom Optics and Ultrafast Spectroscopy at Swinburne University of Technology Australia. The typical parameters are shown in table 2

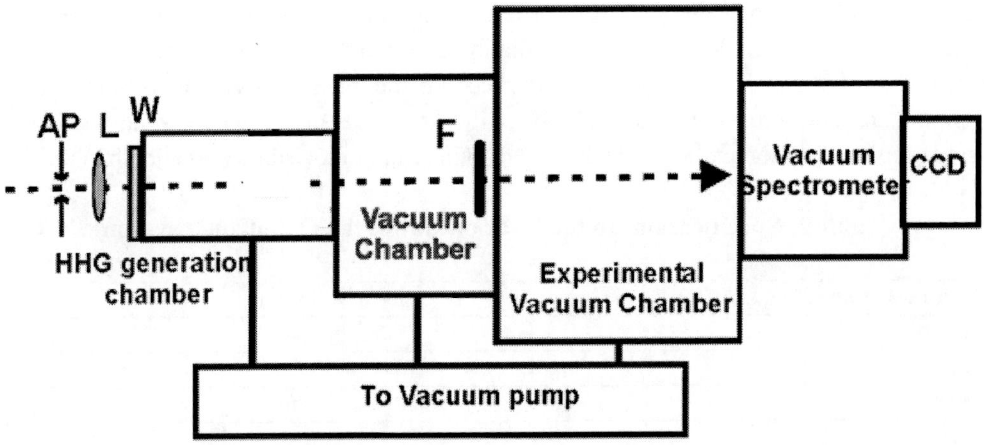

Figure 4. Schematic diagram of the fundamental experimental setup for generation and detection of the harmonic radiation.

3.2. Experimental Arrangement for High Order Harmonic Generation and Application

The principle experimental arrangement for generation and detection of the harmonic radiation is shown in Figure 4 [20]. The output of the amplifier system, known as the fundamental laser beam, is passed to the high harmonic generation apparatus. This beam is focused by a focusing lens into a 200 mm long gas cell (semi-infinite gas cell) with a glass window at the entrance and ~ 0.1 mm pinhole at the exit which is used to separate the vacuum chamber from the gas filled cell. The input window is placed close to the lens to avoid continuum generation and the output pinhole on the thin aluminum plate is drilled online by the laser to ensure the alignment of the laser beam. The effective peak intensity at the focus is approximately 10^{14}-10^{15} W/cm^{2}which can be controlled manually or automatically. The aperture and the combination of a half-wave plate and a polarizing beam splitter which are placed before the focusing lens are used to control the fundamental laser beam. Changing the laser beam by adjusting the aperture affects not only the intensity but also the beam profile of the harmonics [50]. The gases that can be used are argon, neon, helium, or molecular gases, like nitrogen, oxygen. The high harmonics generated on-axis or close to the optical axis pass through a metal filter (aluminum, zirconium or silver), which removes the fundamental beam. Depending on which XUV range will be detected, different kinds of ultra-thin metal foils having thicknesses of 100 μm to 300 μm are applied.

Using difference vacuum pump stages the pressure in the vacuum chamber is kept less than 10^{-3}-10^{-4}Torr for the last vacuum chamber section the so-called experimental chamber.

For measurement of the HHG spectrum, a grazing incidence spectrometer (GIMS #4 – Setpoint) with a grating (150, 300, 600, or 1200 grooves/mm) is inserted into the beam path. A 13 mm x 13 mm CCD chip with 13 μm pixel size (Princeton Instruments PIXIS-XO) is used as the detector. The HHG photon number N_{photon} is calculated from

$$N_{photon} = \frac{a\eta}{QE} \times \frac{3.65}{E_{ph}} \tag{23}$$

where a is the number of counts per CCD pixel, η depends on the setting of the hardware gain given by the manufacturer, QE is the quantum efficiency of the detector which is nearly constant over the range of photon energies 20–200 eV, and E_{ph} is the photon energy.

4. Characteristics of the High Harmonics Beam

In general, the high-harmonic generation power spectrum exhibits a typical behavior [51]. The intensity of the few very low (2rd, 3th) harmonic orders drops by several orders of magnitude. The power spectrum then levels off into a harmonic plateau where the intensity of adjacent harmonics varies only slightly. Then the power spectrum drops sharply in intensity over the last few orders and thus virtually comes to an end. The power spectrum depends strongly on whether the high-harmonic generation is produced by excitation pulses consisting of a few optical cycles or many optical cycles [3, 52-54]. Characteristics, such as conversion efficiency, spectral resolution and noise, spectral cut-off, observation of even and odd orders, and observation of harmonic sub-bands representing attosecond pulses, are all related to the length of the pump pulses with respect to the length of the carrier frequency.

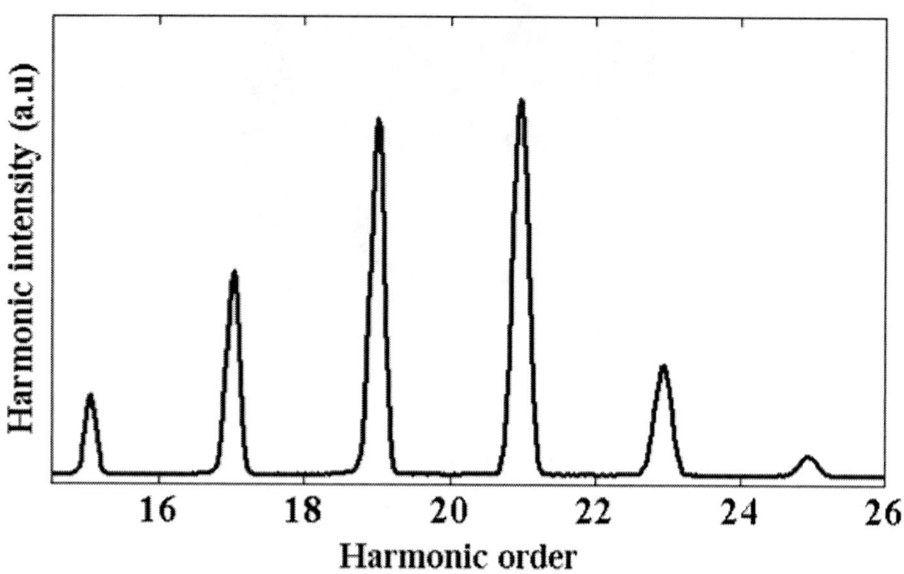

Figure 5. HHG spectrum with argon gas when the laser energy is 0.4 mJ.

shape of the harmonic spectral distribution can be observed on blue wing of spectrum. But we cannot see clearly the red wings (the dashed lines) and this is likely to be due to the significant plasma phase mismatch induced by the strong ionization rate on the trailing edge of the pulse which distorts the harmonic emission on this edge. The experimental spectral lines do not shift anymore for laser energies higher than 2.3 mJ. As explained above, this is a direct consequence of the almost saturation of the laser intensity induced by the defocusing effect and this is also seen through the saturation of the harmonic intensity for laser energies higher than 2.3 mJ in Figure 7.

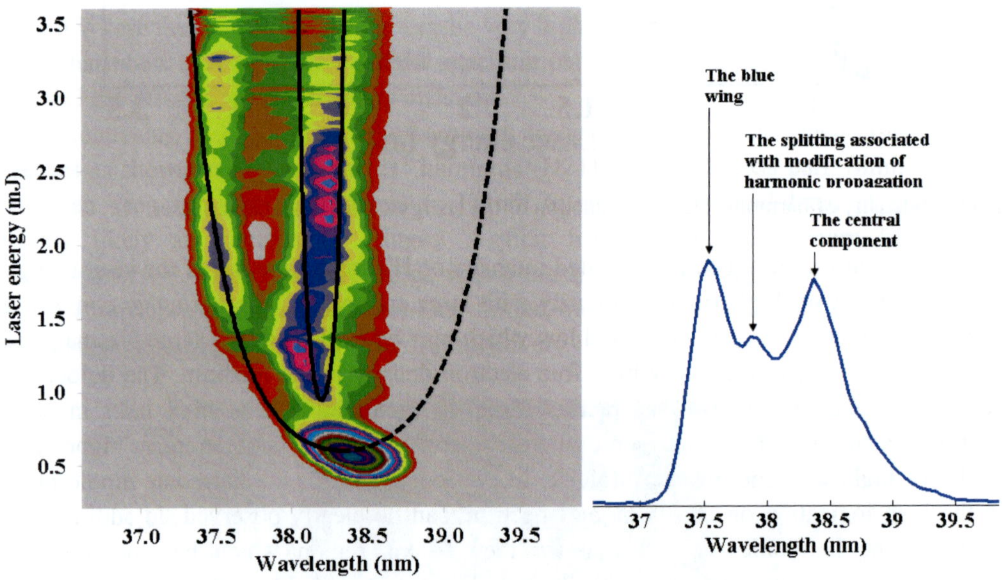

Figure 8. Measured H21 on-axis selection as a function of the laser energy. The black lines are calculated from equation 24, fitted to the experimental spectrum (a); and the harmonic spectrum for a laser energy ~ 3.3 mJ (b).

The variation of the spectral line in the on-axis region where theoretically the short trajectory predominantly contributes to the harmonic radiation is slightly different as shown in the Figure 8. For laser energies less than 0.6 mJ, as the laser energy increases, the harmonic signal increases and the spectrum still appears to be narrow. For laser energies greater than 0.6 mJ, the spectrum exhibits a blue-shifting causing the first fringe. We would observe a second fringe when the laser energy is ~ 0.9 mJ. However, because this shifting may be very small a second fringe cannot be seen. Similarly to the case of the off-axis selection, by performing a simulation and fitting to the experimental results, we find a maximum coefficient $\alpha \sim 7\times10^{14}$ cm^2/W which is in agreement with the major contribution of the short trajectory. When the laser energy is above 3 mJ, we can observe a slight splitting between the blue wing of the first fringe and the central component. The splitting becomes stronger with increase of the laser energy and may be attributed to the modification of the harmonic propagation in the ionizing medium. In order to see this splitting clearly, The inset in Figure 8 shows the measured harmonic spectrum for a laser intensity of ~ 3.3 mJ. We note that the harmonic intensity is not increased for laser intensity higher than 2.5 mJ. Because the ionization has a strong influence of harmonic spectrum and intensity we propose to use the aperture to control this parameter.

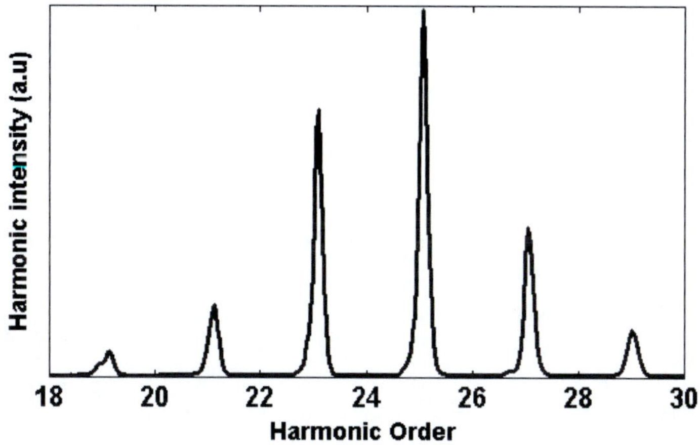

Figure 9. HHG spectrum at the laser energy of 2.1 mJ. The aperture is used for minima the phase-mismatch.

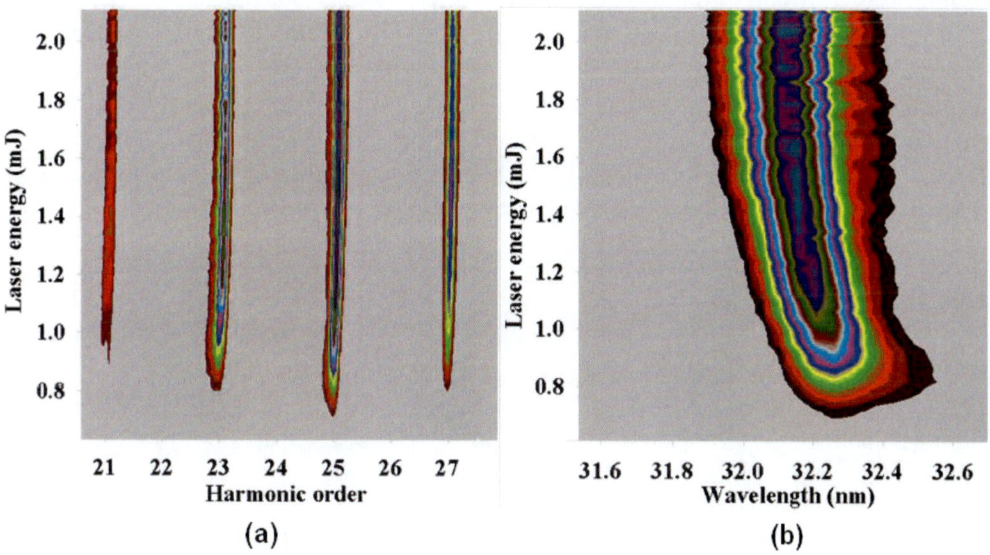

Figure 10. Spectrum of four intensive harmonics (a) and of selected 25th harmonic (b) as a function of the laser energy.

The half-wave plate and the polarizing beam splitter are positioned so that the laser energy coming after the polarizing beam splitter is maximal (~ 4 mJ). The aperture diameter is varied so that total harmonic intensity is maximum for on-axis emission and to be obtained spectrally narrow which is typical for a dominant contribution from the short trajectory emission. Under these conditions, the energy of the fundamental laser is measured to be 2.1 mJ and diameter of the aperture of the beam is ~ 6 mm. By using the aperture to control the ionization fraction at the maximum laser energy, we can ensure that during scanning of the laser energy, the ionization fraction is kept below the critical ionization level (e.g., for argon, the critical ionization fraction is less than 5 % [57]). Figure 9 shows the observed harmonics spectrum at the laser energy of 2.1 mJ.

Figure 10 shows the spectrum of the four strongest harmonics (Figure 10a) and the selected harmonic H25 (Figure 10b) as a function of the laser energy. Remarkable broadening or splitting of harmonic spectrum is not observed. The spectrum remains narrow with high intensity on-axis during scanning of the laser energy from the minimum to the maximum value. This behaviour indicates that the short trajectory whose coefficient, α_q, is relatively small contributes predominantly to the harmonic emission during the scanning of the laser energy. The influence of the ionization and the harmonic dipole phase, which vary strongly during scanning of the laser energy, on the HHG process cannot be observed by examining the harmonic spectrum in this case.

The spectrally integrated intensity of the on-axis harmonic H25 when the laser energy varies from 0.6 mJ to 2.1 mJ (the H25 harmonic appears when the laser energy is around 0.6 mJ) is displayed in Figure 11. The increasing of the harmonic signal at low laser intensity (below 1.2 mJ) can be fitted very well with the scale of I_0^{6}, where I_0 is the laser intensity. For low laser intensity, there is a minor influence of the ionization on the phase mismatching and the phase mismatch is very small (≈ 0). When the laser energy is above 1.2 mJ the ionization fraction becomes higher and, especially for the gas cell geometry, the rapid increase of ionization causes strong plasma dispersion. The laser intensity dependence of the harmonic yield in this region becomes more complicated. Indeed, in Figure 11, for $E_0 > 1.2$ mJ, it is shown that the increasing of the harmonic intensity with laser energy is slower due to the strong influence of the phase mismatch. In addition, the defocusing effect due to the high ionization fraction causes reduced focusing. Therefore, the laser intensity does not scale linearly with the laser energy. The considerable phase mismatch and the defocusing leads to almost saturation of the harmonic yield at a laser energy of ~1.9 mJ.

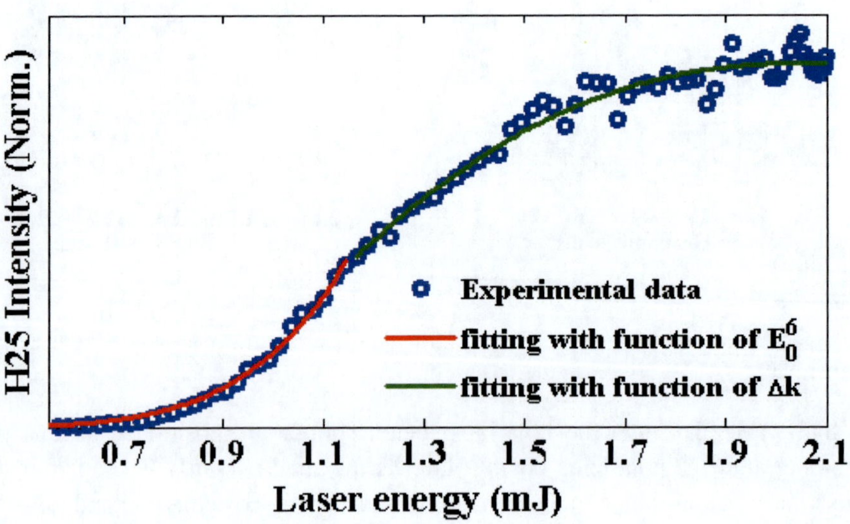

Figure 11. Intensity of the harmonic H25 for argon gas when varying the laser energy of the fundamental laser.

In the high energy regime, the development of the harmonic intensity as a function of I_0 can be fitted with a linear dependence of the phase mismatch on laser energy. The linear dependence of the phase mismatch on laser energy reflects the important contribution of the harmonic dipole phase term to the phase mismatch than that of the plasma dispersion and the neutral gas dispersion.

In Figure 10b, for laser energies less than ~ 1.8 mJ, we observe a small blue-shift ($\Delta\lambda <$ 0.1 nm) in the H25 spectrum with increasing laser energy. This may be due to the larger contribution of the electrons accelerated in the leading edge of the laser pulse to the harmonic generation process and the dominant harmonic emission with the short trajectory. However, for laser energies higher than ~ 1.8 mJ, the defocusing effect causes almost saturation of the laser intensity and therefore the shift is no longer visible.

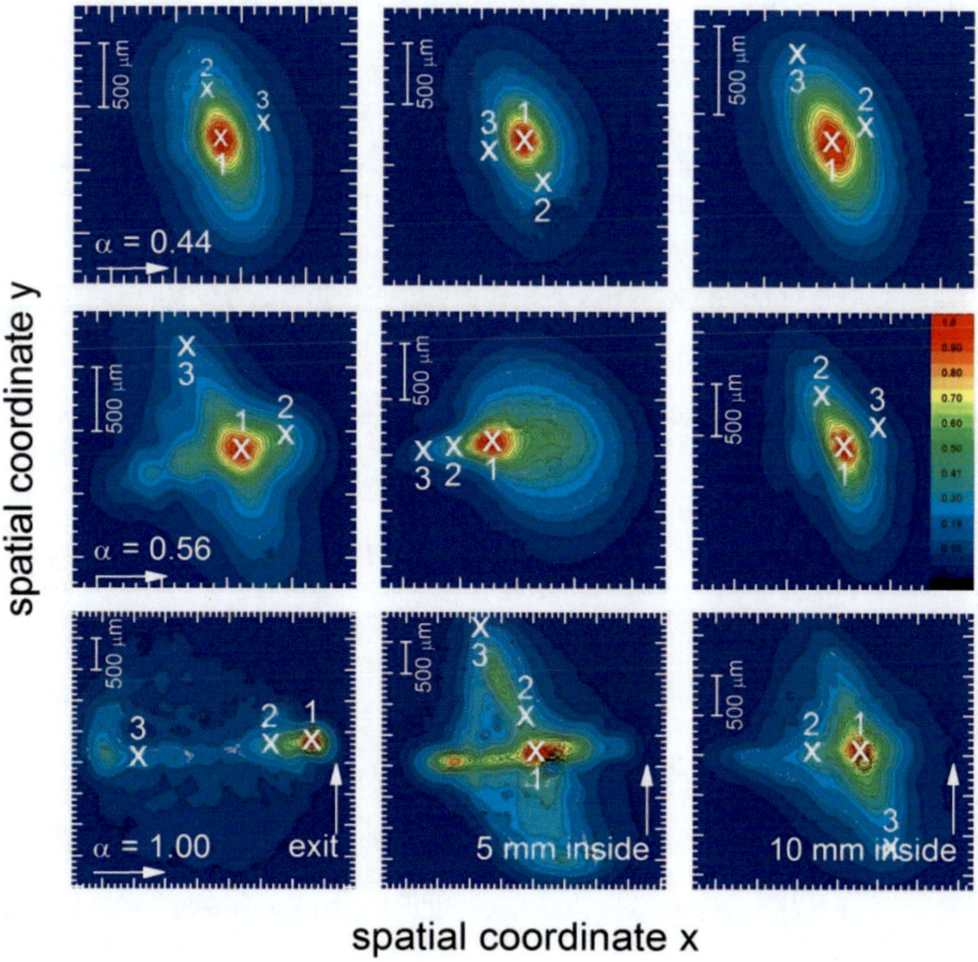

Figure 12. Profile of the harmonic beam with the laser focus position in the exit plane (left), 5 mm (middle) and 10 mm (right) inside the gas cell and for $\alpha = 0.44$ (top), 0.56 (middle) and 1 (bottom). The points indicate that the harmonic intensity is maximum (1), 41% (2) and 19% (3) of the maximum intensity, respectively.

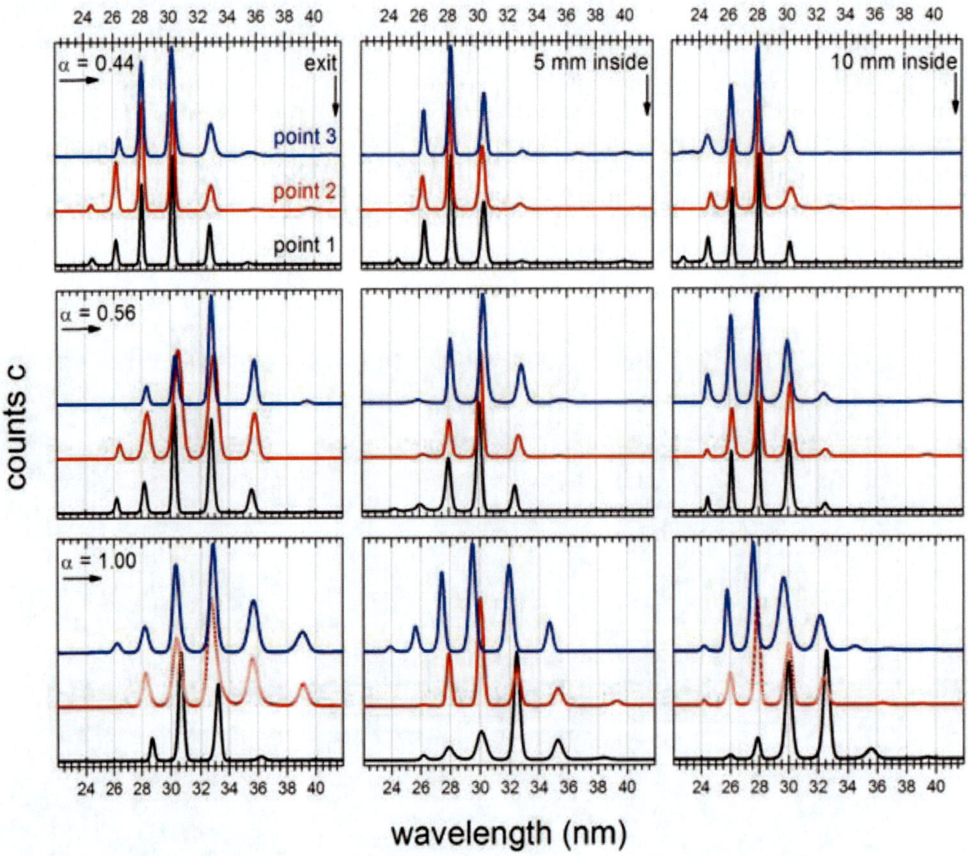

Figure 13. Spectral characteristics across the harmonic beam with the laser focus in the exit plane (left), 5 mm (middle) and 10 mm (right) inside the gas cell for, α = 0.44, 0.56 and 1. The spectrum was measured at different points (1 - black line, bottom), (2 - red line, middle) and (3 - blue line, top) which is shown in Figure 12.

4.2. Using of Aperture for Control of the Spatial and Spectral Characteristic

By properly adjusting the size of aperture which is localized in front of focusing lens we can generate harmonic emission that is confined to effectively few intense harmonic orders around ~ 30 nm in a semi-infinite argon gas cell [20, 58]. The laser intensity and also the ionization rate at focusing is decreased with decreasing of aperture diameter *a*. The focused intensity *I* will increase steadily with the truncation coefficient $\alpha = (a/w)^2$, where w is the spot size of the Gaussian beam at the plane of the aperture.

The maximum laser intensity of $I_{\max} \approx 5.0\times10^{14}$ W/cm^2 is chosen in our configuration with 80 Torr argon gas. According to the Ammosov- Delone-Krainov ionization rate model [59], the I_{max} intensity generates a plasma fraction of 0.36, which is reached approximately 8 fs after the peak of the electric field when the neutral medium is depleted by 20 %.

The beam profile of the harmonic emission is varied strongly for difference the aperture diameter and the position of the laser focus. The spectral characteristics across the harmonic beam profile with the different laser focus placed in the exit plane of the gas cell, as well as 5 and 10 mm inside the gas cell, for, $\alpha = 0.44$, 0.56 and 1 display in Figure 13. The spectra are sampled at different points 1 (centre), 2 (perimeter of 41% intensity) and 3 (perimeter of 19% intensity) across the spatial domain of the harmonic beam as denoted in Figure 12. For relative comparison the power spectra have been normalized.

For $\alpha = 0.44$ a good beam profile of elliptical beam shape is obtained as distorted parts of the fundamental laser beam are discarded due to the small aperture diameter and the effects of plasma dispersion and defocusing are very small due to the low ionization level. For this value of α the highest harmonic order - the 35th - can be detected when the laser focus is placed 10 mm inside the gas cell. The intensity associated with this particular harmonic order incorporates the ion rate $\eta \approx 0.05$, which is still small. Spectral characterization across the harmonic beam for all three interaction lengths reveals that the spectral range and the relative weighting of the harmonic orders are homogenously distributed across the harmonic beam to a good approximation.

For $\alpha = 0.56$, when distorted parts of the laser beam can pass the aperture, distortions of the harmonic beam in its spatial domain can be observed and are more severe when the laser focus is placed at the exit of the gas cell. When α is large a significant plasma formation, up to $\eta \approx 0.36$, takes place and manifests itself in an increased spectral width as well as spectral blue-shifting [29-30]. These effects are more prominent for a longer interaction length when the laser focus is placed inside the gas cell. The relative weighting of the power spectra depends strongly on the interaction length whether the maximum order can be detected in the centre or off-centre. When the laser focus is placed in the exit plane of the gas cell, the centre exhibits the highest order – 31st. However, when the focus is 10 mm inside the gas cell, the highest harmonic order – 33rd – can be measured off-centre. Due to the significant amount of generated plasma for larger α , and in particular for $\alpha \rightarrow 1$, defocusing can take place and generation of higher harmonics becomes less efficient in the centre of the beam.

When placed at the exit of the gas cell, the highest harmonic order is the 29th and detected in the centre. Off-centre the power spectrum exhibits harmonic orders up to the 29th. However, when the laser focus is placed 10 mm deep inside the gas cell, the off-centre power distribution exhibits more intense higher harmonic orders as well as the highest harmonic order. In all cases significant red- as well as blue-shifting of the harmonic orders across the beam occurs. It is also interesting to note that the off-centre spectra from the two different perimeters resemble each other but are very much different compared to spectral power distribution measured in the centre of the harmonic beam.

4.3. Spatial Coherence of High Order Harmonic Beam

In order to measure the degree of spatial coherence the different Young double-slits (YDS), which consist of two parallel slit with a same width, are used. The Young double-slits are used to extract information about the degree of spatial coherence as well as the power spectrum of the multiple-harmonic order emission as shown in [60, 61] and in our work [62]. The interference pattern of a monochromatically illuminated Young double-slit for each spatial coordinate x on the detector and each wavelength λ within the power spectrum is given by [63]

$$I(x|\lambda) \propto |\gamma_{12}|\cos[2\pi xD/Z\lambda]\ \mathrm{sinc}^2[\pi W/Z\lambda] \quad (25)$$

where W is the slit width, D is the distance between centre of two slits, Z is the distance between slit and detector, and γ_{12} is the complex degree of coherence. For a typical measurement of the coherence properties of a (quasi-) monochromatic source the fringe visibility of interference pattern is given by

$$v = (I_{max} - I_{min}) / (I_{max} + I_{min}) \quad (26)$$

where the I_{max} is the intensity of the centre and I_{min} is the intensity of the first minimum of the interferogram. The degree of spatial coherence $|\gamma_{12}|$is related to the fringe visibility v by

$$|\gamma_{12}| = v(I_1+I_2)/(4I_1I_2)^{1/2} \quad (27)$$

where I_1 and I_2 are the intensity of the field at two individual slit of Young double slit. Figure 14a shows the interferogram recorded by the CCD and the interference pattern when the Young's double slit with slit spacing of 20, 50 and 100 μm is used and the slit is illuminated by all harmonic orders. The CCD is localized 600 mm away from the YDS. For more detail study of the coherence degree, a spectrometer is used to disperse the spectrum in the different harmonic order. The YDS is perpendicular to entrance slit of spectrometer and the interference pattern will be observed along y-direction of CCD detector. The interference pattern are shown in Figure 14 b,c for YDS with slit spacing of 20 μm and 100 μm. A fringe visibility of ~ 0.95 is measured for centre part with slit spacing of 20 μm (Figure 15a). The fringe visibility decreases linearly to a value of ~0.85 and 0.70 for the 50 μm and 100 μm spaced Young double-slit pair, respectively. When all harmonics are used for illuminating the YDS, the fringe visibility decreases for large spatial coordinate especially for YDS with large slit spacing (Figure 15 a, b).

5. Multiple-Harmonic-Order Coherent Diffractive Imaging

Coherent diffractive imaging (CDI) is a new imaging technique in which no optical elements such as lenses are needed to obtain the magnified image. CDI was first demonstrated in 1999 using the light from a synchrotron by J. Miao et al. [64]. Instead of using optical elements, in CDI the sample to be investigated is illuminated with a coherent x-ray source directly. By recording the far-field coherent diffraction pattern, and using iterative phase retrieval algorithms, the image of the object can be reconstructed also in three dimensions [64-66]. Since there is no physical lens, this method is free from aberration and provides a very large depth of focus. The resolution of the CDI technique basically depends on the wavelength of the x-rays and the largest angle of scattering where diffraction patterns are recorded. Based on this technique, the structure, shape and size of finite non-periodic objects have been determined [64-71]. Using x-rays from a synchrotron, D. Shapiro et al.

have examined a frozen dried yeast cell and its internal components with about 30 nm resolution [72]. Using a soft x-ray free electron laser, H. N. Chapman et al. have successfully imaged a nanostructured non-periodic object with a spatial resolution down to 60 nm [73]. Also, three dimensional reconstruction of objects has been performed successfully at 50 nm resolution [74]. Very recently, C. G. Schroer et al. have shown that by focusing hard x-rays on a sample in the form of a small gold particle, a spatial resolution of about 5 nm can be obtained [71].

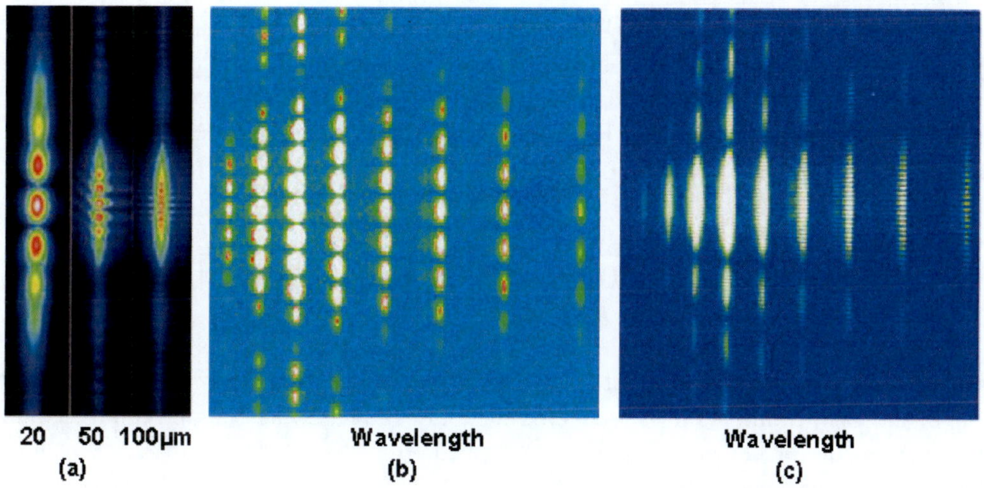

Figure 14. Interference pattern on the CCD for the 20, 50 and 100 µm spaced Young double-slits illuminated by all harmonics (a). Interference pattern of different harmonic orders measured on the focusing plane of spectrometer for the 20 µm (b), and 100 µm (c) spaced Young double-slits. The Young slit is perpendicular to entrance slit of spectrometer.

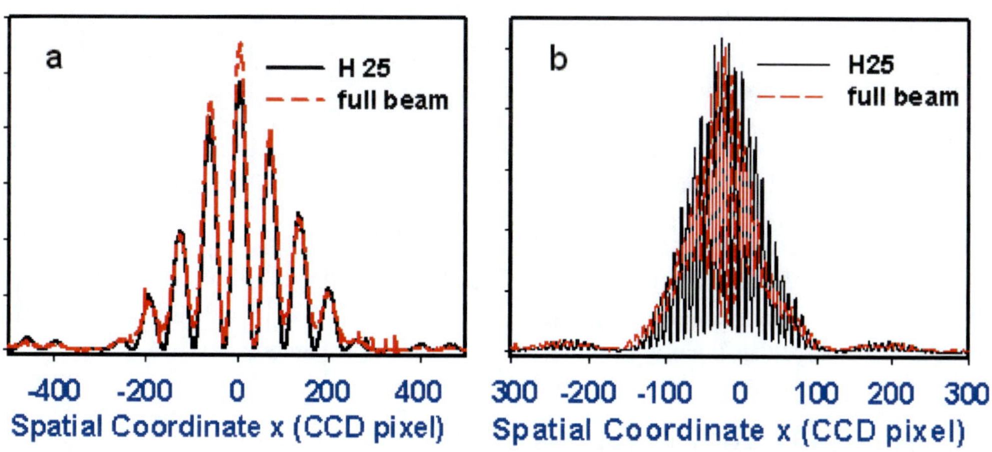

Figure 15. Cross-sections of the interference pattern for the 20, and 100 µm spaced Young double-slits. The Young slit is illuminated by all harmonic orders (full beam) or by 25th harmonic.

Besides the radiation from a synchrotron and from a free electron x-ray laser, High Harmonic Generation (HHG) can be employed as a new x-ray source in CDI. Table top HHG sources for CDI not only provide a high degree of spatial coherence but also enable small scale x-ray microscopy. In principle, HHG source is a laser-like beam of multiple longitudinal coherent modes whereas an iterative method for image reconstruction normally needs an entirely coherent monochromatic wave field. To meet this requirement, i.e., monochromaticity and coherence of the incident x-rays, CDI is usually carried out with a single harmonic order which is selected by using a monochromator or a narrow bandwidth mirrors. This can lead to a significant decrease of the photon flux of the incident beam on the sample. Alternatively, in past work, to utilize as much of the generated harmonic flux as possible in order to keep the acquisition time short and thus avoid possible long term instabilities of the harmonic source, our collaborators have conducted CDI using multiple order harmonic emission [19, 75, 76]. It has been shown that by employing HHG radiation from an argon source (wavelength ~ 30 nm) with about 5 harmonic orders and a reconstruction using a modified Gerchberg-Saxton iteration algorithm for phase retrieval and a refinement by the application of the maximum entropy method, a spatial resolution of ~ 100 nm has been successfully obtained.

5.1. Background of Coherent Diffractive Imaging

A typical CDI technique includes three basic. Firstly, the sample is illuminated with a monochromatic, coherent wave light field and the diffraction pattern is recorded. Then, the phase of the diffraction pattern is recovered by applying phase retrieval algorithms. Finally, the density of the object in real space is regained from the reciprocal space by using an inverse Fourier transform.

When an object of density f(x) is illuminated with a coherent plane wave light field, the diffracted pattern in the far field is the Fourier transform of the function f(x) [77].

$$F(k) = |F(k)| \exp(i\psi(k)) = \hat{F}(f(x)) = \int_{-\infty}^{\infty} t(x) \exp(2\pi i k x) dx \tag{28}$$

Here, x is the three-dimensional spatial coordinate in real space, k is the three-dimensional frequency coordinate in reciprocal space and $\hat{F}$ stands for the Fourier transform operator.Thus, the Fourier transform F(k) can be written as an amplitude and a phase. In the experiment, only the intensity distribution of F(k), i.e., $|F(k)|^2$, is able to be recorded by the detector and the phase of the pattern is lost. Therefore, in order to fully recover the sample f(x), we must determine the phase information of the function F(k) given by ψ(k). If both the intensity and phase of the function F(k) are known, the sample can be reconstructed by

$$f(x) = \hat{F}^{-1}\left(\hat{F}(f(x))\right) = \hat{F}^{-1}(|F(k)| \exp(i\psi(k))) \tag{29}$$

where $\hat{F}^{-1}$ denotes the inverse Fourier transform operator.

The problem of phase retrieval from the diffraction pattern has been well studied in the field of x-ray crystallography and can be solved by using an oversampling method [78-82].

Generally, an iterative phase retrieval algorithm consists of the following four basic steps [80,83] involving iterative Fourier transformation back and forth between the object and Fourier domains and application of the measured data or known constraints in each domain.

A random phase set is applied as an initial input. This phase set merging with the magnitude of the Fourier transform, which is calculated by the square root of the measured diffraction intensities, exhibits a function with both amplitude and phase for each pixel.

An inverse fast Fourier transform is applied to compute a temporary density function.

The new density function recovered in the second step is subjected to the constraints which are positive inside the finite support and zero outside the finite support.

A new set of phases can be calculated by employing a fast Fourier transform to the new density function obtained in the third step. The phase of the central pixel is set to zero and this new phase set is used for the next iteration.

The iterations continue until the computed image satisfies the object domain constraints.

In order to achieve a successful and high quality reconstructed image, some experimental requirements need to be met. The sample is illuminated with a plane wave which should be monochromatic and coherent. The spatial coherence length which is defined as

$$L_{S,coh} = \frac{a\lambda}{2\pi d_s} \tag{30}$$

needs to be larger than the transverse dimension of the sample. In equation 30, a is the distance from the source and d_s is the diameter of the source. In addition, the coherent photon flux density needs to be high enough to obtain a coherent diffractive pattern at high diffraction angles. The sample is placed far enough from the detector to ensure that a far-field diffraction pattern is recorded and the Fraunhofer criterion is satisfied.

$$z >> \frac{D^2}{\lambda} \tag{31}$$

Here D is the diameter of the sample or the diameter of the illumination area, λ is the wavelength of the illuminating radiation and z is the distance from the sample to the CCD. In a successfully reconstructed image, the size of each image pixel is given by

$$r = \alpha\frac{z\lambda}{pN} \tag{32}$$

where N is the number of pixels in an $N \times N$ CCD array, $\alpha=0.94$ (Sparrow criterion) and $\alpha=1.22$ (Rayleigh criterion). From equation (32), for a given wavelength of the illuminating radiation, the resolution is linearly proportionally to the distance between the sample and the CCD. This means that in order to enhance the resolution of the reconstructed image we can reduce this distance. On the other hand, the ratio between this distance, z, and the diameter of the sample, D, must satisfy

$$\frac{z}{D} > \frac{p}{\lambda}\sqrt{2} \tag{33}$$

Therefore, for enhancing the resolution by reducing the distance between the CCD and the sample, one needs to ensure that the diameter of the sample is also reduced so that the conditions for the oversampling ratio and the far field are maintained. Relating to the requirements of the spectral bandwidth of the illuminating beam for a desired resolution, it has been proposed that [80]

$$\frac{\lambda}{\Delta\lambda} \geq \frac{OD}{d} \tag{34}$$

Here, d is the desired resolution of the reconstructed image and $\Delta\lambda$ is the FWHM bandwidth of the incident beam.

5.2. Coherent Diffractive Imaging with Few Harmonic Source

Iterative methods [78-81] for image reconstruction inherently assume a fully coherent wave field. Thus, a single wavelength needs to be selected which leads to a significantly decreased flux of the harmonic beam. In the following, a novel approach for coherent diffractive imaging is presented that relies on illumination of the sample with a multiple-order harmonic beam. The effective spectral range and the relative weight of the single harmonic orders of the harmonic beam depend strongly on the focusing geometry and the control over the phase-matching parameters. By appropriate choice of the species of gas and its pressure, the interaction geometry, the position of the laser focus and the intensity and diameter of the laser beam, harmonic emission can be phase-matched and confined to just a few orders [20, 50]. Thus, the harmonic orders of the harmonic beam can be tailored according to the experimental requirements and all effectively available harmonic orders can be employed to keep the acquisition time for the coherent diffractive imaging short and to avoid sample damage or detrimental effects due to possible poor long-term stability of the harmonic source. However, employing a harmonic beam consisting of multiple harmonic orders for coherent diffractive imaging requires a reconstruction method that is not critically reliant on the assumption that the harmonic emission possesses essentially perfect spatial and temporal coherence. This approach extends the conventional coherent diffraction imaging algorithm to enable diffractive imaging to be achieved using polychromatic diffraction data. A detailed description of this work can be found in [19]. A schematic representation of the methodology is displayed in Figure 16 [76]

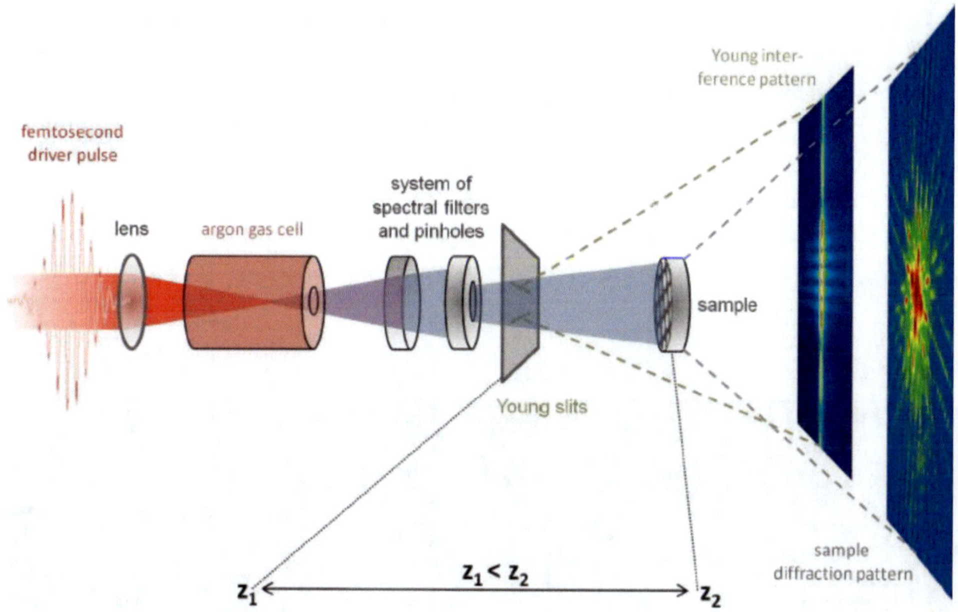

Figure 16. Schematic of the high-harmonic generation gas cell setup and setup for analysis of the harmonic spectrum and for coherent diffractive imaging.

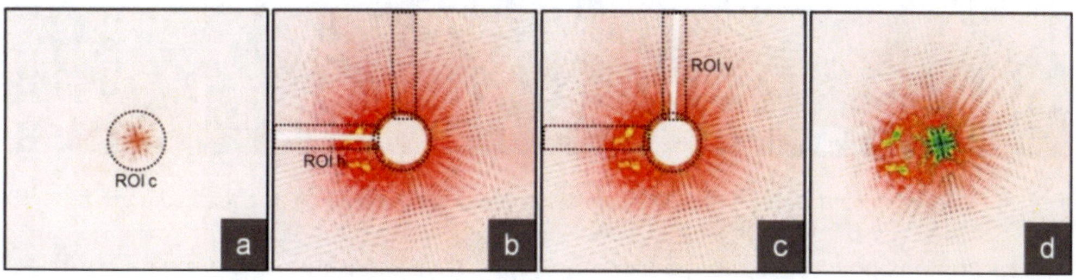

Figure 17. High-dynamic range diffraction images: An image (a) of low-angle and two images (b and c) of high-angle diffraction information are acquired and stitched together to form one single image (d).

In order to achieve a high resolution of the reconstructed object from the coherent diffractive images, images of high dynamic range need to be taken. Only then can diffractive features at high angles be captured, where the high-resolution information resides. The natural dynamic range of the charged-coupled detector is not sufficient high (~ 60000 count for 4 bit CCD), but the diffractive features at the high angles require a longer exposure time. A beam block adaptor is used to allow the acquisition of a diffraction pattern with a dynamic range that significantly exceeds the dynamical range of the charge-coupled device detector chip. Here, three different diffraction patterns of the same sample are taken with special emphasis on the diffractive features in the centre (first pattern) and in the wings of the diffraction pattern. For the latter the beam block adaptor is placed over the centre of the charge-coupled device detector chip. This is done for the x (horizontal) direction and y (vertical) direction. The combined diffractive image is the data set that the modified Gerchberg-Saxton iteration and the maximum entropy method refinement procedure are applied to in order to recover the image of the sample. In the following, the acquisition of these three coherent diffractive images and images of the interference pattern for analysis of the spectral composition of the harmonic beam is discussed. The harmonic source is optimized with respect to the harmonic intensity by adjusting the macroscopic phase-matching parameters. Here, the centre of the harmonic beam coincides with the centre of the detector chip. The Young double-slit is moved into the beam path. An interference pattern of high dynamic range from the Young double-slits is taken in order to extract the spectral information by the application of the maximum entropy method. In-situ spectral characterization allows the spectral power distribution to be matched with respect to the transmissive characteristics of the sample by adjusting the macroscopic phase-matching parameters. Then the Young double-slit is moved out of the beam path.

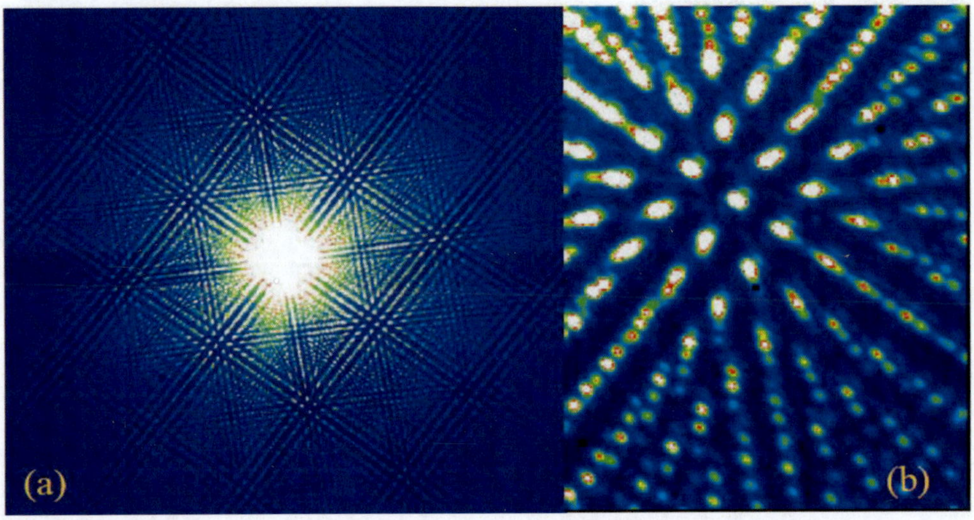

Figure 18. Diffraction pattern from the 2D pinhole sample (a) and a zoom of the marked area (b).

The sample is now positioned at the appropriate transverse and longitudinal coordinates, so that the centre of the diffraction pattern coincides with the centre of the detector chip. The acquisition of the low-angle diffraction information or the centre of the diffractive image is taken (Figure 17a). The horizontal and then the vertical beam block is moved into the beam path and placed in the centre of the detector chip then the image with the high-angle diffractive features is taken (Figure 17b and Figure 17c). The three diffractive images are stitched together as shown in Figure 17d.

The Figure 18a displays the diffraction imaging from a 2D 2μm pinholes sample, where its scanning electron microscope image is shown in the Figure 19a.

The diffraction pattern has a strong qualitative resemblance to the pattern that would be produced by an extended periodic sample. Close inspection reveals that the pattern contains some important differences, key among them being that there is significant intensity between, and there is a structure to, the diffraction "spots" as shown in Figure 18b. A second key feature is that the individual diffraction peaks themselves contain a series of peaks produced by each of the harmonics from the HHG source; so that the diffraction pattern is an incoherent superposition of the diffraction by each harmonic incident on the sample. The reconstruction, based on a modified Gerchberg-Saxton iteration is displayed in Figure 19b. The spatial resolution is ~ 100nm.

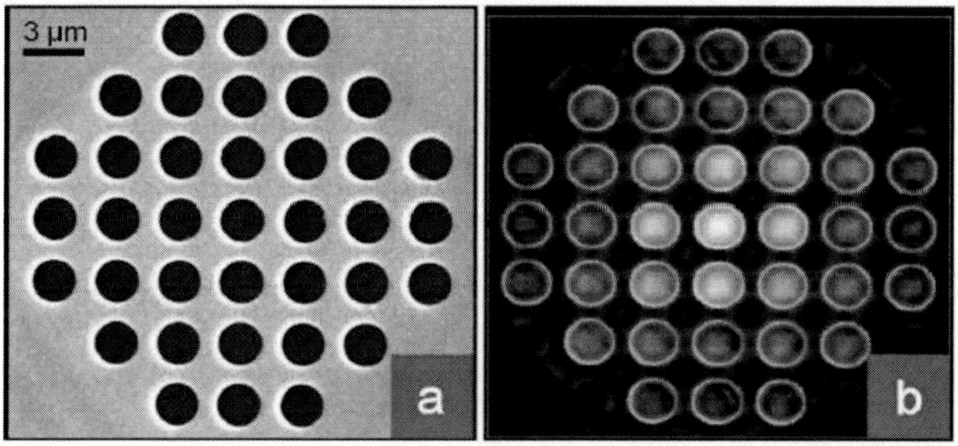

Figure 19. Scanning electron microscopy image of a binary regular sample (a), reconstruction of image from diffraction pattern (b).

CONCLUSION

Harmonic radiation with high intensity and small bandwidth in wavelength rang of ~ 30 nm can be generated by using argon gas as nonlinear medium. The efficiency up to 10^{-5} can be archived when macroscopic propagation effects or phase-matching is considered. The study of the influence of the fundamental laser intensity on the harmonic generation has been shown that without confinement of the ionization fraction also at high laser energy the harmonic conversion efficiency was low. This was found to originate mainly from a serious plasma phase mismatch. The blue-shift and splitting of harmonic spectrum resulted from the modulation of the harmonic dipole phase and transient phase matching at certain intensity during the pulse duration. When the ionization was controlled to be less than a critical level, a narrow and intense harmonic spectrum was produced with a dominant short-trajectory emission. Using an aperture to confine the ionization rate at high laser energy and to control the wavefront of the fundamental laser field we are able to generate few highly coherent, bright harmonics with sharp spectrum. The few high order harmonics generated in this way can be used directly as a coherent source of extreme ultraviolet radiation in coherent diffraction imaging.

ACKNOWLEDGMENT

The author acknowledges the financial support of the Australian Research Council. Prof. Peter Hannaford, Dr. Sven Teichmann, Mr. Khuong Ba Dinh for experiments, discussions. Prof. Keith Nugent, Assoc. Prof. Harry Quiney, Dr Ruben Dilanian and Dr. Bo Chen, Dr. Eugeniu Balaur are thanked for the guidance and contribution of Young double slit measurements and coherent diffractive imaging.

References

[1] McPherson A., Gibson G., Jara H., Johann U., Luk T. S., McIntyre I. A., Boyer K., and Rhodes C. K., *J. Opt. Soc. Am. B*4, 595–601, (1987).

[2] Chang Z., Rundquist A., Wang H., Murnane M. M., and Kapteyn H. C., *Phys. Rev. Lett.*79, 2967–2970, (1997).

[3] Spielmann Ch., Burnett N. H., Sartania S., Koppitsch R., Schnurer M., Kan C., Lenzner M., Wobrauschek P., and Krausz F., *Science* 278, 661–664, 1997.

[4] Dao L. V., Hall C., Vu H. L., Dinh K.B., Balaur E., Hannaford P., Smith T. A., *Applied Optics* 51, 4204, 2012.

[5] Larsson J., Mevel E., Zerne R., L'Huillier A., Wahlstrom C.-G., and Svanberg S., *J. Phys. B: At. Mol. Opt. Phys.*, 28, L53–L58, 1995.

[6] Gisselbrecht M., Descamps D., Lynga C., L'Huillier A., Wahlstrom C.-G., and Meyer M., *Phys. Rev. Lett.*, 82, 4607–4610, 1999.

[7] Bjorneholm O., Hjelte I., Kihlgren T., Ohrwall G., Sundin S., and Svensson S., *J. Chem. Phys.,* 112, 8038–8042, 2000.

[8] Bauer M., Lei C., Read K., Tobey R., Gland J., Murnane M. M., and Kapteyn H. C., *Phys. Rev. Lett.*, 87, 025501, 2001.

[9] Nugent-Glandorf L., Scheer M., Samuels D. A., Mulhisen A. M., Grant E. R., Yang X., Bierbaum V. M., and Leone S. R., *Phys. Rev. Lett.,* 87, 193002, 2001.

[10] Haight R., and Peale D. R., *Phys. Rev. Lett.*, 70, 3979–3982, 1993.

[11] Quere F., Guizard S., Martin Ph., Petite G., Merdji H., Carre B., Hergott J.-F., and le Deroff L., *Phys. Rev.* B, 61, 9883–9886, 2000.

[12] Sekikawa T., Ohno T., Nabekawa Y., and Watanabe S., Lumin J., 87–89, 827–829, 2000.

[13] Theobald W., Haßner R., Wulker C., and Sauerbrey R., *Phys. Rev. Lett.*, 77:298–301, 1996.

[14] Descamps D., Lynga C., Norin J., L'Huillier A., Wahlstrom C.-G., Hergott J.-F., Merdji H., Sali`eres P., Bellini M., and Hansch T. W., *Opt. Lett.,* 25, 135–137, 2000.

[15] Salieres P., le Deroff L., Auguste T., Monot P., d'Oliveira P., Campo D., Hergott J.-F., Merdji H., and Carre B., *Phys. Rev. Lett.*, 83, 5483–5486, 1999.

[16] Itatani J., Levesque J., Zeidler D., Niikura H., Pépin H., Kieffer J. C., Corkum P. B., and Villeneuve D. M., *Nature* 432, 867, 2004.

[17] Kanai T., Minemoto S., and Sakai H., *Nature* 435, 470, 2005.

[18] Sandberg R. L., Song C., Wachulak P. W., Raymondson D. A., Paul A., Amirbekian B., Lee E., Sakdinawat A. E., La-O-Vorakiat C., Marconi M. C., Menoni C. S., Murnane M. M., Rocca J. J., Kapteyn H. C., and Miao J. High numerical aperture tabletop soft x-ray diffraction microscopy with 70-nm resolution. *Proc. Natl. Acad. Sci. U. S. A,* 105:24–27, 2008.

[19] Chen B., Dilanian R. A., Teichmann S., Abbey B., Peele A. G., Williams G. J., Hannaford P., van Dao L., Quiney H. M., and Nugent K. A. Multiple wavelength diffractive imaging. *Phys. Rev. A*, 79:023809, 2009.

[20] Dao L. V., Teichmann S., Davis J., and Hannaford P., *J. Appl. Phys.* 104, 023105, 2008.

[21] Dao L. V., Teichmann S., and Hannaford P., *Phys. Lett.* A, 372, 5254, 2008.

[22] Popmintchev T., Chen M., Cohen O., Grisham M. E., Rocca J., Murnane M., Kapteyn H., *Opt. Lett.* 33, 2128, 2008.

[23] Zhang X., Lytle A. L., Popmintchev T., Zhou X., Kapteyn H. C., Murnane M. M., and Cohen O., *Nat. Phys.* 3, 270, 2007.

[24] Seres J., Yakovlev V. S., Seres E., Streli Ch., Wobrauschek P., Spielmann Ch., and Krausz F., *Nat. Phys.* 3, 878, 2007.

[25] Paul A., Bartels R. A., Tobey R., Green H., Weiman S., Christov I. P., Murnane M. M., Kapteyn H. C., and Backus S., *Nature* 421, 51, 2003.

[26] Gibson E. A., Paul A., Wagner N., Tobey R., Gaudiosi D., Backus S., Christov I. P., Aquila A., Gullikson E. M., Attwood D. T., Murnane M. M., and Kapteyn H. C., *Science* 302, 95, 2003.

[27] Lytle A. L., Zhang X., Peatross J., Murnane M. M., Kapteyn H. C., and Cohen O., *Phys. Rev. Lett.* 98, 123904, 2007.

[28] Seres J., Yakovlev V., Seres E., Streli C., Wobrauschek P., Spielmann C., and Krausz F., *Nat. Phys.* 3, 878, 2007.

[29] Dromey B., Zepf M., Landreman M., and Hooker S., *Opt. Express* 15, 7894, 2007.

[30] Bartels R. A., Paul A., Green H., Kapteyn H. C., Murnane M. M., Backus S., Christov P., Liu Y., Attwood D., and Jacobsen C., *Science,* 297, 376–378, 2002.

[31] Lewenstein M., Balcou Ph., Yu M., Ivanov, A. L'Huillier, and Corkum P. B., *Phys. Rev.* A, 49, 2117–2132, 1994.

[32] Becker W., Long S., and McIver J. K., *Phys. Rev.* A, 50, 1540–1560, 1994.

[33] Antoine P., L'Huillier A., Lewenstein M., Salieres P., and Carre B., *Phys. Rev.* A,53, 1425–1445, 1996.

[34] Protopapas M., Lappas D. G., Keitel C. H., and Knight P. L., *Phys. Rev.* A, 53, R2933–R2936, 1996.

[35] Corkum P. B., *Phys. Rev. Lett.*, 71, 1994–1997, 1993.

[36] Krause J. L., Schafer K. J., and Kulander K. C., *Phys. Rev. Lett.*, 68, 3535–3538, 1992.

[37] Gallagher T. F., *Phys. Rev. Lett.,* 61, 2304–2307, 1988.

[38] L'Huillier Anne, Lewenstein M., Salieres P., and Balcou Ph., *Phys. Rev.* A, 48, R3433–R3436, 1993.

[39] Lan P., Lu P., Cao W., Wang X., and Yang G., *Phys. Rev.* A, 74, 063411, 2006.

[40] Constant E., Garzella D., Breger P., Mével E., Dorrer C. H., Le Blanc C., Salin F., Agostini P., *Phys. Rev. Lett.*, 82, 1668, 1999.

[41] Kazamias S., Weihe F., Douillet D., Valentin C., Planchon T., Sebban S., Grillon G., Auge F., Hulin D., and Balcou Ph., *Eur. Phys. J.* D, 21, 353–359, 2002.

[42] Brichta J.-P., Wong M. C. H., Bertrand J. B., Bandulet H.-C., Rayner D. M., Bhardwaj V. R., *Phys. Rev. A* 79, 033404, 2009.

[43] Schnurer M., Cheng Z., Hentschel M., Tempea G., K´alm´an P., Brabec T., and Krausz F., *Phys. Rev. Lett.*, 83722–725, 1999.

[44] Kazamias S., Douillet D., Weihe F., Valentin C., Rousse A., Sebban S., Grillon G., Aug´e, D F. Hulin, and Balcou Ph. Global optimization of high harmonic generation. Phys. Rev. Lett., 90, 193901, 2003.

[45] Peatross J., Fedorov M. V., and Kulander K. C., *Journal of the Optical Society of America* B 12, 863, 1995.

[46] Salières P., L'Huillier A., and Lewenstein M., *Physical Review Letters* 74, 3776, 1995.

[47] Gaarde M. B., Salin F., Constant E., Balcou Ph., Schafer K. J., Kulander K. C., and L'Huillier A., *Physical Review* A 59, 1367, 1999.

[48] Auguste T., Carré B., and Salières P., *Physical Review* A 76, 011802, 2007.

[49] Miloni P. W., and Eberly J. H. *Lasers*. John Wiley & Sons, Inc., New York, Wiley Series in Pure and Applied Optics edition, 1988.

[50] Peatross J., and Meyerhofer D. D., *Phys. Rev. A,* 523976–3987, 1995.

[51] Jaegle P. *Coherent sources of XUV radiation*. Springer, Berlin, first edition, 2006.

[52] Teichmann S., Hannaford P., and van Dao L., *Appl. Phys. Lett.*, 94, 171111, 2009.

[53] L'Huillier A., Schafer K. J., and Kulander K. C., *J. Phys. B: At. Mol. Opt. Phys.,* 24, 3315–3341, 1991.

[54] Kan C., Burnett N. H., Capjack C. E., and Rankin R., *Phys. Rev. Lett.,* 79, 2971–2974, 1997.

[55] Christov P., Murnane M. M., and Kapteyn H. C., *Phys. Rev. Lett.*, 78, 1251–1254, 1997.

[56] Zhou J., Peatross J., Murnane M. M., Kapteyn H. C., and I. P. Christov, Phys. Rev. Lett., 76, 752–755, 1996.

[57] Rae S. C., and Burnett K., *Physical Review* A 46, 1084 , 1992.

[58] Hey C. M., Güdde J., Höfer U., and L'Huillier A., *Physical Review Letters,* 107, 033903, 2011.

[59] Popmintcheva T., Chena M. C., Bahabada A., Gerritya M., Sidorenko P., Cohen O., Christov I. P., Murnanea M. M., and Kapteyna H. C., *Proceedings of the National Academy of Sciences of the United States* of *America* 106, 10516, 2009.

[60] Teichmann S., Hannaford P., and van Dao L., *J. Phys. D: Appl. Phys.,* 42, 135108, 2009.

[61] Ammosov M. V. and Delone N. B., and Krainov V. P., *Sov. Phys. JETP,* 64, 1191–1194, 1986.

[62] Zhang X., Libertun A. R., Paul A., Gagnon E., Backus S., Christov I. P., Murnane M. M., Kapteyn H. C., Bartels R. A., Liu Y., and Attwood D. T., *Opt. Lett.*, 29, 1357–1359, 2004.

[63] Liu Y., Wang Y., Larotonda M. A., Luther B. M., Rocca J. J., and Attwood D. T., *Opt. Expr.*, 14, 12872–12879, 2006.

[64] Dilanian R. A., Chen B., Teichmann S., van Dao L., Quiney H. M., and Nugent K., *Opt. Lett.*, 33, 2341–2343, 2008.

[65] Jenkins F. A. and White H. E., *Fundamentals of Optics*, 1967, McGraw Hill, New York.

[66] Miao J., Charalambous P., Kirz J., and Sayre D., *Nature* 400, 342, 1999.

[67] Robinson I. K., Vartanyants I. A., Williams G. J., Pfeifer M. A., and Pitney J. A., Physical Review Letters 87, 195505, 2001.

[68] Williams G. J., Quiney H. M., Peele A. G., and Nugent K. A., *Physical Review* B 75, 104102, 2007.

[69] Pfeifer M. A., Williams G. J., Vartanyants I. A., Harder R., and Robinson I. K., Nature 442, 63, 2006.

[70] Williams G. J., Quiney H. M., Dhal B. B., Tran C. Q., Nugent K. A., Peele A. G., Paterson D., and de Jonge M. D., *Physical Review Letters* 97, 025506, 2006.

[71] Marchesini S., He H., Chapman H. N., Hau-Riege S. P., Noy A., Howells M. R., Weierstall U., and Spence J. C. H., *Physical Review* B 68, 140101, 2003.

[72] Miao J., Ishikawa T., Johnson B., Anderson E. H., Lai B., and Hodgson K. O., *Physical Review Letters* 89, 088303, 2002.

[73] Schroer C. G., Boye P., Feldkamp J. M., Patommel J., Schropp A., Schwab A., Stephan S., Burghammer M., Schöder S., and Riekel C., *Physical Review Letters* 101, 090801, 2008.

[74] Shapiro D., Thibault P., Beetz T., Elser V., Howells M., Jacobsen C., Kirz J., Lima E., Miao H., Neiman A. M., and Sayre D., *Proceedings of the National Academy of Sciences of the United States of America* 102, 15343, 2005.

[75] Chapman H. N., Barty A., Bogan M. J., Boutet S., Frank M., Hau-Riege S. P., Marchesini S., Woods B. W., Bajt S., Benner W. H., London R. A., Plonjes E., Kuhlmann M., Treusch R., Dusterer S., Tschentscher T., Schneider J. R., Spiller E., Moller T., Bostedt C., Hoener M., Shapiro D. A., Hodgson K. O., van der Spoel D., Burmeister F., Bergh M., Caleman C., Huldt G., Seibert M. M., Maia F. R. N. C., Lee R. W., Szoke A., Timneanu N., andHajdu J., *Nature Physics* 2, 839, 2006.

[76] Chapman H. N., Barty A., Marchesini S., Noy A., Hau-Riege S. P., Cui C., Howells M. R., Rosen R., He H., Spence J. C. H., Weierstall U., Beetz T., Jacobsen C., and Shapiro D., *Journal of the Optical Society of America* A, 1179, 2006.

[77] Dilanian R. A., Chen B., Williams G. J., Quiney H. M., Nugent K. A., Teichmann S., Hannaford P., Dao L. V., and Peele A. G., *Journal of Applied Physics* 106, 023110, 2009.

[78] Teichmann S., Chen B., Dilanian R. A., Hannaford P., and Dao L. V., *Journal of Applied Physics* 108, 023106, 2010.

[79] Miao J. and Sayre D., *Acta Crystallographica* A, 56, 2000.

[80] Miao J., Ishikawa T., Anderson E. H., and Hodgson K. O., *Physical Review* B 67, 174104, 2003.

[81] Sayre D., *Acta Crystallographica* 5, 843, 1952.

[82] Miao J., Sayre D. and Chapman H. N., *Journal of the Optical Society of America* A 15, 1662, 1998.

[83] Fienup J. R., *Applied Optics* 21, 2758, 1982.

In: Argon: Production, Characteristics and Applications
Editors: B.N. Sismanoglu, H.S. Maciel et al.
ISBN: 978-1-62618-204-2

Chapter 5

SOME APPLICATIONS OF ARGON IN MICROWAVE PLASMAS SUSTAINED AT ATMOSPHERIC PRESSURE

M. D. Calzada*, J. Muñoz, M. Jiménez, R. Rincón and M. Sáez

Laboratory of Innovation in Plasmas (LIP),
Edificio Einstein, Campus de Rabanales,
Universidad de Córdoba, Spain

ABSTRACT

In natural state, argon is an inert gas unable to induce chemical reactions with other reactants. However, in plasma state, argon becomes active due to the different and highly energetic species produced in its ionization such as electrons, ions, excited neutral particles, and photons. All these species play an important role in the kinetics of plasmas generated with this gas and determine plasma capabilities when interacting with other substances. Among the numerous plasma devices developed nowadays, some special features of microwave sustained plasmas generated at atmospheric pressure make them particularly appealing both for scientific research and technological applications. These features include their reproducibility, the wide range of conditions under which they can be operated, and their ease of implementation. When used in fundamental research, argon plasma-emitted spectral lines can be used to develop methods to measure density and energy (temperatures) of plasma particles (electrons, ions, molecular species). In the technological field, argon and argon-based plasmas are used in applications covering medical instrument sterilization, hydrogen production, synthesis of carbon nanostructures, analytical chemistry, etc. All this makes argon one of the most widely-used gases in scientific research.

* E-mail address: md.calzada@uco.es.

1. INTRODUCTION

The earliest experiments with electrical discharges in gases date back as far as the 17th century, but it was in the second half of the 20th century that the study of plasmas was considered a discipline in itself. From there on, this field of physics that studies electrical discharges has evolved astonishingly, breaking new grounds in relevant industrial applications such as energy generation or development of new materials for the broadest imaginable range of industrial applications.

Formally, plasma can be defined as a quasineutral gas where neutral and charged particles (electrons and ions) coexist [1]. However, plasmas' most important feature hides behind a well-known statistical fact: for neutral and charged particles to coexist, it is also required in almost every case to have a plethora of excited particles, not necessarily ionized, but still with significant energies to allow them to take part into reactions where this energy can be transferred. Furthermore, if the characteristic energies into the discharge are high enough to excite and ionize a fraction of the particles within it, it will also be typically possible to break molecular bonds –in case a molecular gas is introduced in either as a majority component or as an additive– giving place to molecular fragments that otherwise could not exist, and that can be used as precursors of new substances different to those used to feed the discharge.

Among the gases that can be used for plasma feeding and operation, argon is one of the most extended, since being a rare gas, it keeps relatively simple plasma chemistry while interacting with other substances introduced in the discharge, and enables fundamental studies aimed at characterizing the discharge or exploring the potential capabilities of the considered plasmas. Besides, it offers good compromise between availability, cost, and ease of ionization due to its relatively low ionization energy when compared to other rare gases. In plasma state, argon becomes active due to the different and highly energetic species produced in its ionization –electrons, ions, excited neutral particles, and photons. All these species play an important role in the kinetics of plasmas generated with this gas, and determine plasma capabilities when interacting with other substances.

However, creating and sustaining such an environment demands an energy supply. Some techniques were developed to couple an external energy source to plasmas by means of nonconventional energy sources such as lasers [2], but the most extended and developed ones are those using an external electromagnetic field. These devices are usually classified according to their working frequency (direct current, radiofrequency, microwave...), and the way energy is coupled to the discharge. Particularly, microwave plasmas sustained at atmospheric pressure show some special features that make them particularly appealing both for scientific research and technological applications, such as high reproducibility, the wide range of conditions under which they can be operated, electrode-less operation, and ease of implementation.

In this chapter we present a practical classification of microwave plasmas sustained at atmospheric pressure, followed by the experimental techniques commonly used to characterize the most relevant parameters of an argon plasma –ionized and excited particle densities, and temperatures. Finally, some of the most relevant and state-of-the-art applications of argon microwave discharges sustained at atmospheric pressure are presented.

2. MICROWAVE DISCHARGES SUSTAINED AT ATMOSPHERIC PRESSURE

2.1. A Practical Classification

While a rigorous classification of microwave discharges strictly according to physical criteria is possible, their broad applications have extended their use among specialists in very different fields such as chemistry, biology, or materials engineering. Consequently, a more handy classification attending to functional criteria and based on easily identifiable features is frequently used. Among the most relevant microwave coupling devices, we find:

- Resonant cavities.
- Surface-wave discharges.
- Torches.
- Microplasmas.

The oldest microwave plasma sources made use of the resonant modes of cavities and the maximums that electromagnetic fields excited within them [3] for plasma sustain. Resonant cavities were primarily used in chemical analysis applications, showing good tolerance to sample entrance, capability of working at both high and low pressures and under a wide range of power conditions (50-1000 W) [4, 5], and very good detection limits (e.g.: from 15 Li ng/ml to 290 Cr ng/ml in water) [6], though in the later years, the number of publications dealing with this kind of discharges has decreased significantly.

Plasmas sustained by surface waves are a special type of microwave plasmas where an electromagnetic wave is excited at a certain point of a dielectric tube containing the discharge and propagates along the plasma-dielectric interface. The first of them, the surfatron, was introduced in 1974 [7], and its design was later improved to allow higher powers (up to 2000 W) –giving place to the surfaguide [8] and guide-surfatron [9]– and a wider range of operating frequencies (1-1000 MHz)– robox [10]. Though a relatively newcomer to the microwave plasma family, surface wave discharges and, particularly, argon and argon-based surface-wave plasmas have been extensively studied, characterized [11-13], and used during the last three decades in diverse applications such as chemical analysis [14], sterilization and inactivation of biological samples [15], and thin film deposition [16].

One of the limitations of resonant cavities and surface wave discharges is that, since plasma is contained totally or partially in a dielectric tube, when power increases above a few hundred watts or plasma extends towards the walls of the tube, it can be eroded and, therefore, some external cooling –with either air or a dielectric liquid– is required [17]. Plasma torches can overcome these limitations by sustaining a discharge at atmospheric pressure even in contact with air. The use of these systems begun in 1991 [18] with the MPT (microwave plasma torch), and its design was later improved in the TIA (*torche à injection axiale*, axial injection torch) [19]. Plasmas generated with these devices are generally smaller and thus power is dissipated in smaller volumes. Consequently, these energy couplers generate plasmas with higher electron densities and temperatures, with the only shortcoming of being highly contaminated due to their exposure to ambient air, though the atmosphere surrounding the discharge can be controlled, and thus also the kinetics of the reactions taking place in it. Apart from chemical analysis [20], these devices have also been successfully applied to pollution abatement [21] and material synthesis and processing [22].

Table 1. Plasma parameters and operating conditions of microwave plasmas sustained in argon and argon-based plasmas

Device	Electron density (10^{20} m^{-3})	Electron temperature (K)	Gas temperature (K)	Power (W)
Resonant Cavities	1 - 10	5000 - 20000	1000 - 3000	1000 - 3000
Surface-wave	1 - 10	4500 - 6000	1200 - 1500	80 - 250
Torch	5 - 30	12000 - 22000	1200 - 3500	100 - 1000
Microplasma	0.1 - 1	1800 - 3000	500 - 700	5 - 50

The development of microstrip technology –microwave transmission lines formed by two planar electrodes separated by a thin dielectric film: one of the electrodes remains grounded while the other is connected to a microwave power source, usually by means of a coaxial connector– offered new possibilities for plasma generation. Beginning in 2000 [23], this technology was successfully applied to sustain very small (milimetric and submillimetric) plasmas. Discharges are sustained in a channel perforated along the dielectric film, parallel to the electrodes, and involve the advantage of requiring very low microwave power (5-50 W) and gas flow (as low as 50 ml/min), together with the small scale of the equipments used to sustain them. These features are being exploited and the possibility of building low-cost and portable chemical analysis devices has been proposed. A number of significant successes were reached in chemical analysis, allowing carbon and sulfur detection [24], and reaching the trace level detection limits on mercury [25]. Besides, recent research [26] proved that these discharges can also be used to inactivate (sterilize) microorganisms, opening a way towards the design of point-of-care biomedical applications.

Table 1 summarizes the typical features of the aforementioned devices operated with argon and argon-based plasmas.

2.2. Characterization Techniques

As explained above, understanding the behavior of the most relevant plasma parameters is crucial for optimizing their applications. Since in microwave discharges electrons are the particles responsible for absorbing the energy from the electromagnetic field and redistributing it among the remaining particles [27], electron density (n_e) is a fundamental parameter in plasmas, as well as the electron temperature (T_e), which is an statistical mean of the energy in the electrons in the discharge.

Besides, in most atmospheric-pressure microwave-sustained plasmas, the lack of effective thermal coupling between electrons and heavy particles leads to the existence of different temperatures characterizing the motion of electrons and atoms and molecules. The later temperature is also known as gas temperature (T_g), and is a measure of average energy in the heavy particles in the discharge. This temperature has a large influence on the atomization reactions of the molecules in the discharge, along with the dependence of the speed of these reactions (reaction rate) on this parameter, and, consequently, is of great importance for applications involving the addition of molecular species to Ar plasmas.

All this parameters, together with the density of active and excited species, can be measured using active or passive techniques, the later being the preferred ones, since they do not perturb plasma during measurement. Thus, while active methods such as Thomson scattering [28] or titration [29] allow accurate electron temperature calculation and active species density determination, respectively, these methods are either expensive and difficult to implement or induce perturbations in plasma, which make them impractical for online monitoring of plasma parameters at the same time the discharge is being used on an application.

Since the atoms and molecules in a discharge are in excited states with limited lifetimes, when they deexcite, the energy difference between the transition upper and lower level is emitted as a photon (i.e., as light). Given that energy levels in atoms and molecules are quantized, this photon emission gives place to characteristic spectra once the emitted light is decomposed. This is the base of emission spectroscopy, a set of passive optical techniques that make use of the intensity and shape of the radiation spontaneously escaping the discharge to determine the parameters characterizing the discharge.

2.2.1. Atomic and Molecular Emission Spectroscopy

When discharges are generated using pure argon there are few interactions between atoms leading to the formation of molecules or molecular fragments except for those coming from impurities such as water or nitrogen, or some particular excited states such as argon molecular ions Ar_2^+ are formed in the discharge. In those cases the study of the discharge can be restricted to atomic emission spectroscopy (AES), and the spectra arising from the deexcitation of excited atoms are studied. The diverse techniques developed using AES allow determining the relative and absolute densities of the atoms in the discharge, as well as other important parameters such as excitation temperatures and charged particle densities, allowing understanding the microscopic kinetic processes that take place in the discharge.

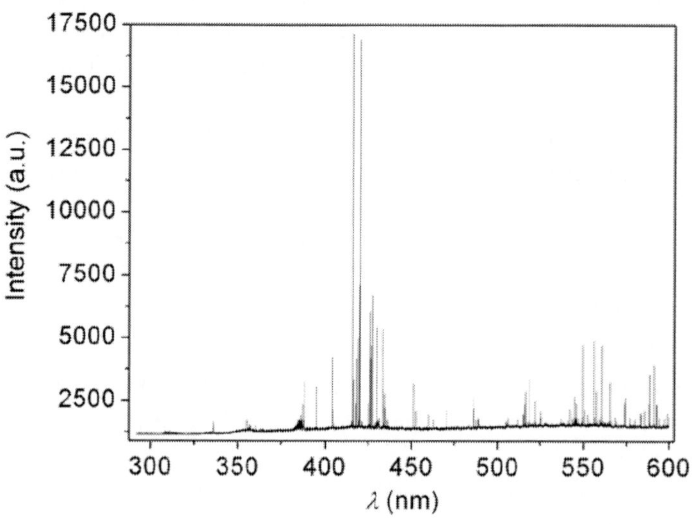

Figure 1. Typical emission spectrum of an atmospheric pressure Ar plasma.

These transitions between excited atomic states give place to the lines appearing in atomic spectra. Figure 1 shows a characteristic spectrum of an atmospheric pressure microwave Ar plasma.

Since the intensity of emitted spectral lines is proportional to the concentration in the upper level of the transition, the relative or absolute density of the excited particles in the discharge can be estimated from the aforementioned parameter. However, an experimentally-observed spectral line never appears as perfectly monochromatic but as an intensity distribution around a central wavelength or frequency, also called line profile. This can be due to several processes taking place in the discharge [30]:

Natural broadening due to Heisenberg's uncertainty principle (Lorentzian profile).

Doppler broadening due to thermal agitation of emitting particles (Gaussian profile).

Stark broadening due to the interaction of emitting particles with surrounding charged particles (Loretzian profile).

Van der Waals broadening due to the interaction of emitting particles with surrounding neutral particles in fundamental state (Loretzian profile).

Instrumental broadening due to the features of the optic device used for radiation measurement. While the exact shape of these profiles depends on the optic device, it is typically considered as Gaussian in many studies.

Consequently, at atmospheric pressure the shape of a line arises from the convolution of the Gaussian and Lorentzian, giving place to a Voigt profile, whose analysis can be a source of information on electron density and gas temperature, as it will be seen in the following sections.

On the other hand, in those cases where molecular species are present, the emission spectra become more complex. We can take a diatomic molecule as an example. The nuclei of the two atoms can be in relative motion one respect to the other without implying necessarily a mass-center translation, giving place to rotation and vibration movements associated to rotational and vibrational energies, respectively. Besides, the formation of a molecule leads to the creation of a molecular orbital where one or more electrons are shared. These electrons can be in different states, having an associated electronic energy. Finally, the molecule can have a translational energy associated to the displacement of its mass center. The energy of the molecule will be the sum of all translational, rotational, vibrational, and electronic energies.

Each of the energies mentioned above, except translation energy, is quantized and, consequently, transitions between the different excited states could happen, giving place to molecular spectra. The shape of these depends, among other factors, on the nature of the observed transition, the most usual situation being simultaneous rotational, vibrational and electronic transition.

Finally, leaving aside the particular techniques for the determination of specific plasma parameters that will be examined in the following sections, the mere registration and analysis of spectra is a simple yet powerful technique that allows fast identification of both atomic and molecular excited species and, once understood, can be easily implemented for online application monitoring.

2.2.2. Electron Density

Since electrons in microwave discharges are the particles responsible for absorbing the energy from the electromagnetic field and redistributing it among the remaining particles, electron density (n_e) is a fundamental parameter in plasmas. At atmospheric pressure, the most frequently applied method for electron density diagnostics relies on the Stark broadening of the discharge-emitted spectral lines, particularly from the Balmer series of hydrogen. In non-hydrogen plasmas, hydrogen atoms can also be present as traces, coming either from gaseous hydrogen or water dissociation.

The first theory establishing a relationship between the Stark broadening of hydrogen lines and electron density was proposed by Kepple and Griem [31], and consisted on assuming that ions are static particles as compared to electrons due to their comparatively larger mass. According to this quasi-static approximation, the full width at half the maximum of hydrogen lines caused by Stark broadening, $\Delta\lambda_S$ is proportional to $n_e^{2/3}$, and its value can be obtained from the following relation

$$n_e = C_H(n_e, T_e)\Delta\lambda_S^{3/2} \qquad (1)$$

C_H being a constant that depends on electron density and electron temperature.

Several reasons justify the use of this method. First, it requires no absolute measures of line intensities. Second, the intensities of hydrogen lines are typically large enough to allow their detection, even if present at very low concentrations. Consequently, there is no need to add a supplementary hydrogen input to the discharge. Last, the theory used by this method holds valid regardless of the particular equilibrium conditions of the discharge.

Griem [32] calculated the relation between electron density and Stark broadening for several hydrogen lines considering large electron density ($10^{19} - 10^{25}\ m^{-3}$) and temperature ($5000 - 40000\ K$) ranges that encompass the typical values of these parameters in atmospheric pressure microwave plasmas (see Table 1 above). In this relation

$$\Delta\lambda_S = 2.50 \times 10^{19}\alpha_{1/2}n_e^{2/3} \qquad (2)$$

n_e being electron density in cm^{-3}, and $\alpha_{1/2}$ a parameter called fractional width or reduced Stark width at half height, expressed in Å. This parameter is tabulated [32] and differs for every wavelength. However, these data do not take the effect of ion dynamics into account, that was later considered in the studies by Gigosos *et al.* [33].

Particularly, the most widely-used line to diagnose electron density is the H_β line, at 486.13 nm. Apart from the aforementioned advantages, its features make this line particularly suitable from an experimental point of view. On one hand, C_H data are availed by a large number of studies, typically reaching 5% accuracy [34-36]. Furthermore, the H_β line is placed in a free range of the argon spectra, is sufficiently intense, and has a large broadening. Finally, its Stark broadening is dominant as compared to other broadening effects [30], is barely affected by ion dynamics and self-absorption, and depends very weakly on electron temperature. The use of the Stark broadening of other atomic lines emitted by non-hydrogen atoms can also be used for electron density determination. In this case, the relation between Stark broadening of the atomic line and electron density can be expressed as [32]

$$\Delta\lambda_S = 2\omega_e[1 + 1.75\alpha(1 - 0.75r)] \tag{3}$$

ω_e being the so-called electron collision impact parameter expressed in nm, α the ion broadening, and r the ratio between the average distance between ions and the Debye length. In general, the main contribution to collisional broadening for these lines is due to electrons and, consequently, α is usually relatively small. In the particular case of argon, several studies [37-38] successfully used the Stark broadening of its spectral lines for electron temperature diagnosis.

2.2.3. Excited Atom Densities, Excitation and Electron Temperature

In microwave discharges sustained at atmospheric pressure, and particularly in Ar plasmas, electrons are the particles controlling the internal kinetics of the discharge. The electron excitation/deexcitation equilibrium of atoms in level p can be expressed as

$$Ar(p) + e^- \leftrightarrow Ar(q) + e^- \tag{4}$$

q being the label of an upper energy level ($p < q$). In these processes, the used energy can be written as kT_{exc}, k being the Boltzmann constant, and T_{exc} the excitation temperature.

The intensity of a spectral line corresponding to an spontaneous deexcitation from level q to level p is given by expression

$$I(\upsilon) = n_q A_{qp} \frac{h\upsilon}{4\pi} \tag{5}$$

n_q being the density of excited particles in level q, υ the frequency of the emitted radiation, A_{qp} the Einstein transition probability, and h the Planck constant. On the other hand, if Boltzmann equilibrium is verified, the relation between the density of atoms in level q and the density of atoms in ground state, n_0, is given by

$$n_q = \frac{n_0}{Q(T_{exc})} g_q e^{-E_{exc}/kT_{exc}} \tag{6}$$

g_q and E_{exc} being the level's statistical weight and excitation energy, respectively, and $Q(T_{exc})$ the partition function of the considered species. Substituting in the previous equation and considering that $Q(T_{exc})$ is the same for the different levels –which holds as long as excitation temperature is the same for every level considered– the following relation can be obtained

$$\log\left(\frac{I\lambda}{g_q A_{qp}}\right) = -\frac{\log e}{kT_{exc}} E_q + \log C \tag{7}$$

Thus, the representation of $log(I\lambda/g_q A_{qp})$ against E_{exc} for different lines of the spectrum results in a straight line from whose slope the excitation temperature can be derived as long as the Boltzmann balance is verified (Figure 3).

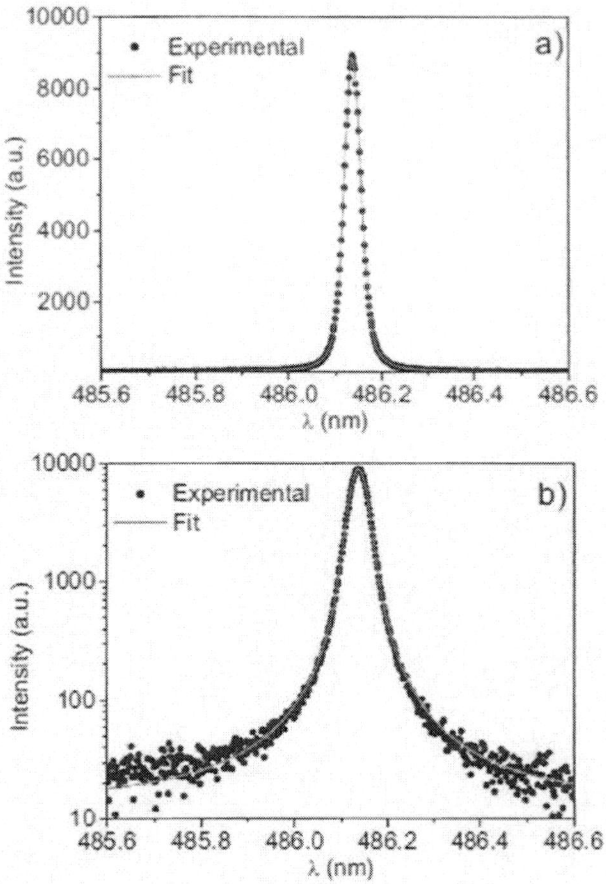

Figure 2. Experimental profile of the Balmer series H_β line (486.13 nm) used for n_e determination (a) and logarithmic detail (b) on the fit to a Voigt profile.

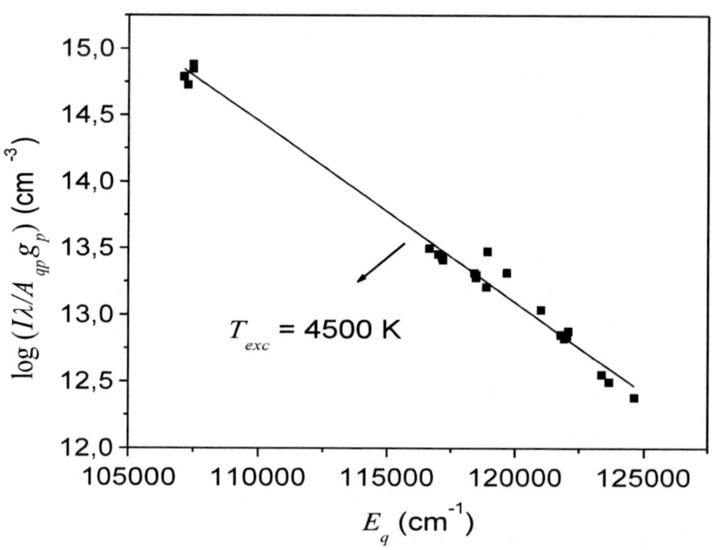

Figure 3. Boltzmann-plot of Ar atomic lines. The excitation temperature is obtained from equation (7) using data from Table 2.

Table 2. Spectral line list and features of the ArI used in the Boltzmann-plot

λ (nm)	E_q (eV)	A_{qp} (10^8 s^{-1})	g_q	Transition
425.1	14.464	0.00113	3	5p-4s
425.9	14.738	0.0415	1	5p-4s
426.6	14.529	0.00333	5	5p-4s
427.2	14.525	0.0084	3	5p-4s
430.0	14.506	0.00394	5	5p-4s
433.3	14.688	0.0060	5	5p-4s
433.5	14.687	0.00387	3	5p-4s
434.5	14.680	0.00313	3	5p-4s
518.8	15.296	0.0138	5	5d-4p
522.1	15.450	0.0092	9	7d-4p
549.6	15.331	0.0176	9	6d-4p
555.8	15.137	0.0148	5	5d-4p
560.6	15.118	0.0229	3	5d-4p
565.1	15.123	0.0333	1	5d-4p
591.2	15.003	0.0105	3	4d'-4p
603.2	15.130	0.0246	9	5d-4p
641.6	14.839	0.0121	5	6s-4p
675.3	14.742	0.0201	5	4d-4p
696.5	13.328	0.067	3	4p-4s
706.7	13.302	0.0395	5	4p-4s
714.7	13.283	0.0065	3	4p'-4s
727.3	13.328	0.0200	3	4p'-4s

This representation is known as *Boltzmann-plot*. Besides, if the absolute response of the optical system is known, equation (5) also allows the calculation of the absolute density values in excited levels, providing valuable information on the atomic energy distribution function.

2.2.4. Electron Temperature and Excitation Temperature

Electron temperature determination is probably the most difficult task in atmospheric pressure plasmas, particularly in those far from thermodynamic equilibrium. The $T_{exc} \approx T_e$ equality holds only when the levels used to calculate the excitation temperature are in equilibrium with electrons. However, this assumption usually needs to be verified by a theoretical model or by comparison with other method that does not assume the existence of such equilibrium.

There are several other spectroscopic methods for electron temperature measurement, but their applicability is usually limited by either the equilibrium degree or their difficult and/or expensive implementation. Firstly, the emission of the continuum radiation can be related to electron density and temperature [39]. Given that the emission of the continuum is strongly influenced by electron density, another method –line-to-continuum ratio– was proposed to circumvent this problem, but again this required the upper level of the transition to be in thermodynamic equilibrium. Later this method was proved to give place to important errors and some corrections were proposed [40]: ionization temperature –relating the density of ions in ground state with the density of neutral in ground state– is required to equal electron temperature, which can be again a source of errors in electron temperature determination.

Some recent methods using the Stark broadening of hydrogen Balmer series lines have been proposed [41]. However, while this method requires no assumption on the equilibrium state of the discharge and simultaneously provides a value for electron density and temperature in a relatively simple way, uncertainty in the determination of the later involves a typical error of 40% [42].

Finally, it is also worth mentioning scattering Thomson, based on the dispersion of light by the free electrons in the discharge [43-45]. This method allows simultaneous electron temperature and density determination. The main problem of this technique is its difficulty of implementation for applications –the external light source must be very precisely aligned– and the need of using a high power laser, which makes this technique much more expensive.

While in general there is no affordable and easy way to implement a method for electron temperature measurement, some qualitative predictions can be done on the knowledge of the behavior of the excited atom distribution function and excitation temperature, which is in some cases regarded as a lower limit or underestimation of the electron temperature value.

2.2.5. Gas Temperature

Gas temperature is a measure of the kinetic energy of heavy particles in the discharge such as atoms, ions and molecule fragments. For its diagnosis it is frequent to use the rovibrational spectra of molecular species such as OH [37], N_2^+ [46], CN [47] or C_2 [48], existing in the discharge as traces of impurities in the plasmogen gas. Due to the fast and efficient collisional energy interchange between the kinetic energy and the rovibrational states of molecules at atmospheric pressure, rotational temperature (T_{rot}) can be assumed to be equal to gas temperature (T_g). Thus, if the heavy particles are capable of imposing their energy in the rotational equilibrium, then $T_{rot} \approx T_g$ equality holds true.

The intensity of a rovibrational line is proportional to the population in the upper level of the transition. If Boltzmann equilibrium is verified, this intensity can be expressed as

$$I = \frac{SDc^4}{\lambda} e^{-\frac{E_J}{kT_{rot}}} \tag{8}$$

D being a coefficient including the rotational partition function, λ the transition wavelength, and S and E_J the oscillator strength and the energy in the upper level of the transition, respectively. Taking logarithms in the previous equation and grouping constants, expression (8) can be rewritten as

$$\log\left(\frac{I\lambda}{A}\right) = C - \frac{\log e}{kT_{rot}} E_J \tag{9}$$

and a linear expression whose slope is related to the rotational temperature is thus obtained. As an example, Figure 4 shows the spectrum of the OH radical, corresponding to the $A^2\Sigma^+ - X^2\Pi\ (0-0, \Delta v = 0)$ transition located within the 306-312 nm range. Correspondingly, Table 3 contains the spectroscopic data required for using this for temperature determination.

However, it is important to take into account the possible limitations of the technique. For the previous example of the OH radical, some authors claimed a lack of sensibility of this specie for gas temperatures above 1800 K [49].

Table 3. Rotational transitions of the 0-0 vibrational band of the $A^2\Sigma^+$- $X^2\Pi$ electronic transition of the OH radical typically used for T_g diagnosis

λ(nm)	A (10^8s)	E(cm^{-1})
308.3	33.7	32779
308.5	42.2	32948
308.7	50.6	33150
309.2	67.5	33652
309.5	75.8	33952
309.8	84.1	34283

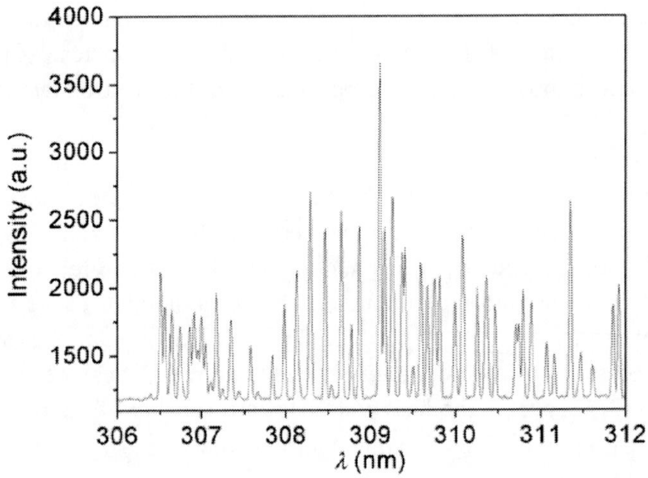

Figure 4. Spectrum corresponding to the OH $A^2\Sigma^+ - X^2\Pi\ (0-0, \Delta\nu = 0)$ transition.

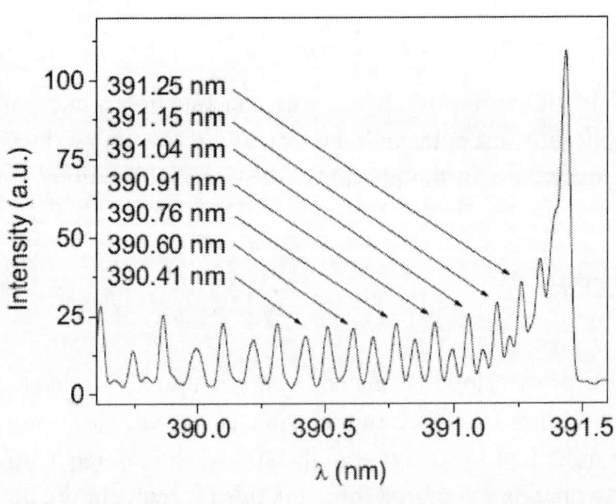

Figure 5. Rovibrational band corresponding to the first negative system ($B^2\Sigma_u^+$- $X^2\ \Sigma_g^+$) of the N_2^+ (0-0 rotational band; 391.4 nm) molecular ion. Lines used for temperature determination are indicated.

In those cases the 0-0 vibrational band of the $B^2\Sigma_u^+$- $X^2\ \Sigma_g^+$ electronic transition of the N_2^+ radical, whose bandhead is placed at 391.4 nm (Figure 5), can be used. While the intensity of this band is typically low in atmospheric pressure pure argon plasmas sustained in controlled atmospheres, it becomes particularly handy when the discharge is in direct contact with the surrounding air. For this species, according to [50], equation (10) can be substituted by the following one:

$$\log\left(\frac{I}{A}\right) = -\frac{1.298}{T_{rot}} R \tag{10}$$

where A and B are constants for each rovibrational line in Table 4.

Another way to evaluate T_g is using the van der Waals broadening [38] of the plasma-emitted lines.

Table 4. Rotational transitions of the 0-0 vibrational band of the $B^2\Sigma_u^+$- $X^2\ \Sigma_g^+$ electronic transition of the N_2^+ radical typically used for T_g diagnosis

λ(nm)	A	B
391.25	44	462
391.15	48	552
391.04	52	650
390.91	56	756
390.76	60	870
390.60	64	992
390.41	68	1122

The collisions of the emitting atoms with the surrounding neutral particles induces a broadening, $\Delta\lambda_W$, given by

$$\Delta\lambda_W = 10^{-9}\frac{\lambda^2}{2\pi c}2\gamma \tag{11}$$

λ being the wavelength in nm, c the speed of light in m/s, and γ the half width at half the maximum of the line expressed in frequency terms. This expression can be related to gas temperature using the Lindholm-Foley impact approximation [51-52]. Considering a Lennard-Jones potential for the interaction and averaging over particles' speed distribution [53-54], the following expression is obtained.

$$\Delta\lambda_W = 8.18 \times 10^{-26}\lambda^2(\alpha\langle\bar{R}^2\rangle)^{2/5}\left(\frac{T_g}{\mu}\right)^{3/10} N \tag{12}$$

α and N being the polarizability and density of perturbing (neutral) atoms, μ the reduced mass of the interacting atoms, and $\langle\bar{R}^2\rangle = \langle\bar{R}_U^2\rangle - \langle\bar{R}_L^2\rangle$ the difference of the averaged radii of the atom in the upper (U) and lower (L) levels of the considered transition. A detailed calculation of this expression can be found in [38].

Table 5. $C_W(\lambda)$ coefficients for the T_g determination in a pure argon discharge from the van der Waals broadening

λ (nm)	$C_W(\lambda)$
425.9	1.479
522.1	5.895
549.6	4.923
603.2	4.217

Though apparently complex, once the proper atomic parameters are applied and considering the ideal gas law, which holds valid for the case of argon at relatively high temperatures, equation (12) can be simplified to the following form for a given line

$$\Delta\lambda_W = \frac{C_W(\lambda)}{T_g^{7/10}} \tag{13}$$

with $C_W(\lambda)$ a coefficient containing all the dependencies of the considered transition and the nature of emitters and perturbers.

Table 5 contains some $C_W(\lambda)$ coefficients for pure argon discharges, though the method has been successfully extended to the case of Ar-He [55] and Ar-Ne [56] gas mixtures, its main limitation being the need of a good-resolution detection system, particularly in the case of high gas temperatures (>2000 K for the case of pure Ar).

3. Surface Wave Discharges (SWDs)

Surface wave discharges are a special type of microwave plasmas in which the electric field necessary to ionize the gas is imposed by an electromagnetic surface wave propagating along the discharge-containing dielectric tube [57]. Thus, no external wave-guiding structure is needed because the interface between the plasma and the surrounding dielectric behaves as such. Consequently, discharges created by a surface wave are characterized by larger dimensions than the wavelength of the electromagnetic field used to sustain them and extending outside the microwave coupling device with the applied microwave power. Besides, as a result of their physical properties, electron density decreases linearly away from the launcher up to critical electron density. Below this electron density value the wave cannot propagate, and the column ends abruptly; thus, the plasma shows a specific length (Figure 6).

An essential feature of surface wave discharges is that plasma column length increases with plasma-transferred power. Thus, when the power is increased, the preexisting plasma column is "displaced", without any modification, away from the launcher, and an additional plasma length is created. The electron density plot of this additional plasma length shows that the newly created column is the continuation of the previous plasma column. As a result of this behavior, the axial position along the plasma column is better referenced from the end of the column rather than from the surfatron gap region [58]. Consequently, surface-wave sustained plasma columns are considered equivalent measures from their ends (Figure 7).

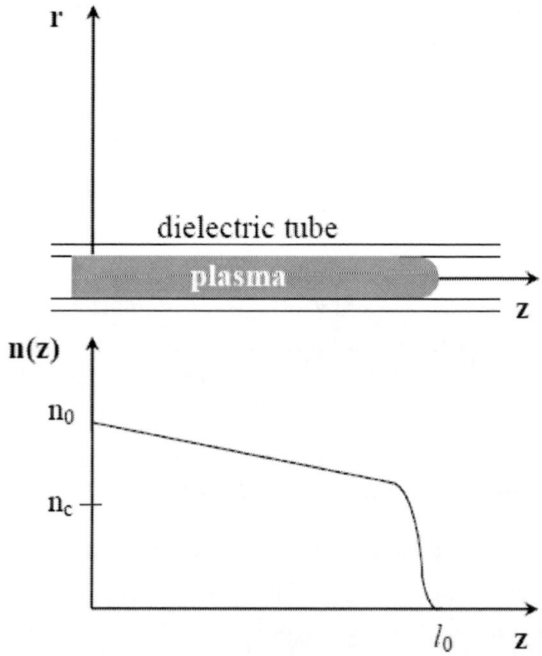

Figure 6. Plasma column and axial profile of electron density represented for a l_0-length column. n_c is the critical electron density value.

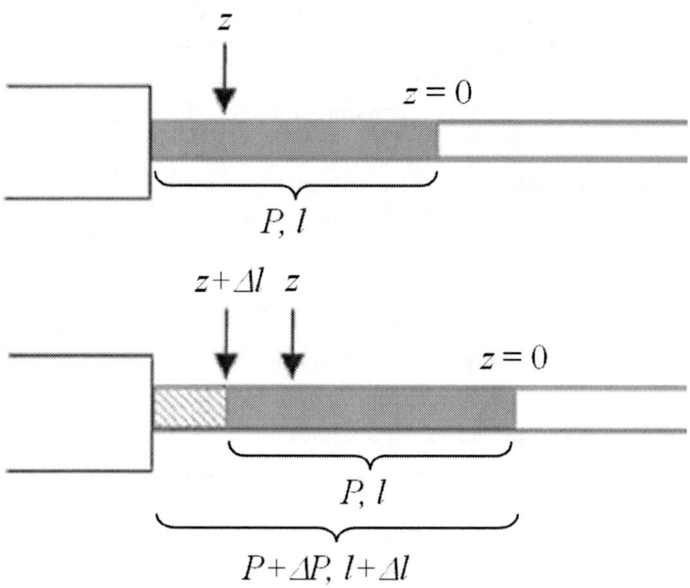

Figure 7. Schematic view of the equivalent columns in a SWD.

From an experimental point of view, surface-wave discharges have several useful properties for lab experiments such as the wide range of frequencies that can be used for plasma generation, the use of different (atomic and molecular) gases and their mixtures, and the absence of fluctuations and instabilities joined to very good reproducibility.

4. Applications of Argon Surface Wave Discharges at Atmospheric Pressure

Surface wave discharges can efficiently yield a wide variety of neutral, excited and ionized atoms and molecules, which allows developing applications in various scientific and technological fields. In this chapter, we focus on two of them: their use as radiation sources for atomic emission spectroscopy (AES), and their application for hydrogen generation.

4.1. Source for Atomic Emission Spectroscopy (AES)

The use of plasma excitation sources in atomic emission spectroscopy (AES) has grown significantly in recent decades as a result of the great amount of emission lines it provides, which enable inter-element selectivity and multi-element detection. In the past three decades, inductively coupled plasmas (ICPs) were the most extensively-applied excitation source for AES. Progress in the use of ICPs as an analytical tool was completed showing not only their usefulness as excitation sources, but also their advantages in sample vaporization, as they can evaporate samples prior to atomization. Thus, high gas temperature in ICP allows direct introduction of sample aerosols without extinguishing the discharge.

However, surface wave discharges involve some advantages as excitation sources in plasma emission spectrometry relative to ICPs. These advantages can be summarized as follows: 1) SWDs are the most flexible microwave discharges; using various tube shapes and dimensions enables optimizing discharge parameters for the best analytical performance under various practical conditions; 2) impedance matching of SWDs can be made very little sensitive to discharge conditions so as to ensure operation stability and restarting reproducibility; and 3) the possibility of adjusting plasma column length, and hence samples' residence time is a useful feature of SWDs.

Research on halogen detection (Cl, Br and I) with Ar-SWD proved this plasma type's excellent capability. A simple continuous flow generation of volatile bromine is described for the determination of low concentrations of this element [59]. Bromine was continuously produced by mixing bromine with sulfuric acid and hypochlorite solutions. Bromine, solvent in the resulting solution, went into a gas-liquid continuous separator where a mild argon flow (plasma gas) carried volatile Br out of the aqueous solution. Bromine volatile species and argon plasma gas were desiccated by passing them through sulfuric acid before entering the plasma formed in the dielectric tube. Optimum flows of sample and reagents were also investigated with selected reagent concentrations so as to obtain maximum intensity of bromine emission at 827.2 nm. This line was selected because it produced the most intense emission and the highest *Signal to Background Ratio* (SBR) for bromine relative to other published analytical lines such as 889.9 and 888.9 nm.

To test the feasibility of faster bromine separation/volatilization, and thus to attain quicker signal stabilization, the effect of bromine separation temperature was studied. Results showed that temperatures over 70° cause serious plasma instability and lead to its extinction. Room temperature was considered for bromine generation in routine operation because improvements in bromine generation with higher temperatures were not dramatic.

Table 6. Optimum operating conditions for bromine determination with the 827.2 nm line

Plasma	
Forward power	100 W
Reflected power	< 5 W
Generator frequency	2450 MHz
Inner diameter (plasma tube)	1 mm
Argon flow rate	0.5 l/min
Monochromator	
Photomultiplier voltage	1200 V
Integration time	500 msec
Slits Widths (entrance and exit)	50 μm
Sequential scanning (bandpass)	< 0.03 nm
Bromine generation	
H_2SO_4 concentration	3 *M*
NaClO concentration	4.32×10^{-3} *M*
H_2SO_4 solution flow	1.23 ml/min
NaClO solution flow	1.23 ml/min
Sample solution flow	3.81 ml/min

Using the optimum conditions summarized in Table 6 and the proposed general procedure, the attained detection limit was 2 μg/l, and precision was ± 0.7% (at 80 μg/l). The proposed determination was very selective if oxidizing/reducing agents were absent. Besides, the procedure was tested in two commercial drug separations, obtaining good agreement between experimental results and the certified values of these drugs.

Furthermore, an analytical method combining on-line gas generation and an argon-SWD for sample introduction and analysis is reported in [60]. The main objective of this work was simultaneous and continual generation of chlorine, bromine, and iodine from aqueous solutions by means of a lead (IV) oxide minicolumn connected to the plasma excitation source via a gas-liquid separator.

This enabled the introduction of the vaporized sample directly into the plasma by means of the support gas flow as analyte carrier gas. Argon plasma parameters such as microwave power, gas flow rate, and excitation temperature were studied to optimize experimental conditions.

A schematic diagram of the sample introduction system is shown in Figure 8. Chlorine, bromine, and iodine were instantaneously generated from the aqueous samples by oxidation with lead (IV) oxide column, thermostated at 60 °C. This column was made by packing a glass capillary with lead dioxide powder. Small glass wool beads on each end prevented material losses. A mild stream of argon, plasma gas at flow rate of 0.5 l/min, carried the volatile species while the solution was drained to waste. The gas phase was desiccated by bubbling through concentrated sulfuric acid before insertion into the discharge tube.

The atomic emission lines of chlorine, bromine, and iodine selected for condition optimization and calibration curve obtaining were 837.6, 827.2 and 804.4 nm, respectively. Optimum instrumental conditions are summarized in Table 7.

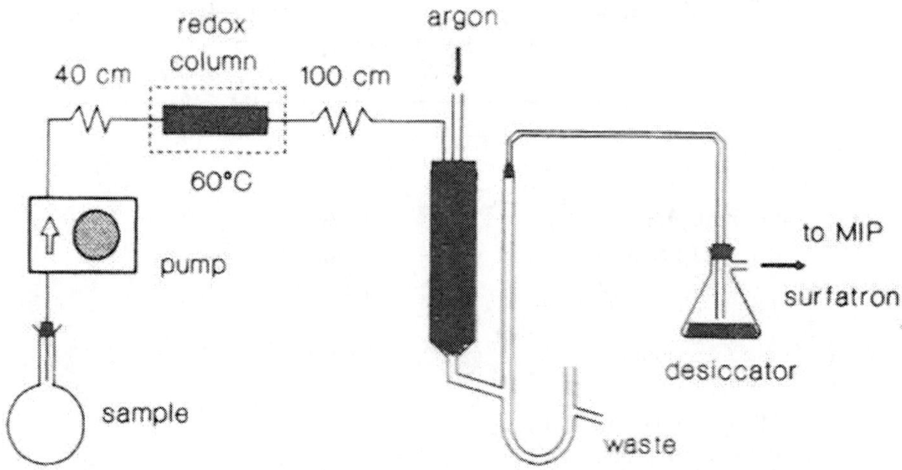

Figure 8. Flow diagram of the sample introduction system.

Table 7. Optimum instrumental conditions for Cl, Br, and I halide determination

Plasma	
Forward power	100 W
Reflected power	< 10 W
Generator frequency	2450 MHz
Inner diameter (plasma tube)	1 mm
Argon flow rate	0.5 l/min
Monochromator	
Photomultiplier voltage	1200 V
Integration time	500 msec
Slits Widths (entrance and exit)	50 μm
Sequential scanning (bandpass)	< 0.03 nm

By using this system under the optimum conditions stated above, calibration graphs for chlorine, bromine, and iodine were built, obtaining 50, 50, and 200 ng/ml detection limits for these three elements, respectively. Lower sensitivity for iodine was due to the emission line selected (803.37 nm). This line was chosen to accomplish the simultaneous determination of the three halogens with the same optical detector. Method applicability was tested by resolving various synthetic mixtures of halides. All assayed mixtures provided excellent results.

Concerning plasma parameters, argon gas flow of 0.5 l/min was chosen. The analytical signal for the three halogens increased as the argon flow rate decreased. However, flows below 0.5 l/min did not allow halogens to be swept to the plasma. SBR for lines increased with increasing microwave power up to 100 W. Powers over 100 W resulted in significantly-increased background emission, so a microwave of 100 W was considered for halogen determination in routine work. Excitation temperature –a characteristic parameter of plasmas used as excitation sources– was also measured [61] using a different technique. Similar values for excitation temperatures (≈ 6000 K) calculated for Ar and Cl species indicate that plasma can be satisfactorily used as a halogen excitation source. Besides, this temperature is scarcely affected by the variation of the sample amount introduced into the plasma, which is also an indicator of the excellent behavior of plasma as an excitation source for analytical halogen determination in vaporized aqueous samples.

However, as previously explained, while using surface-wave discharges as excitation source in analytical chemistry, the sample must be evaporated prior to its introduction into the plasma. This drawback can be overcome by working with either gas or vapor samples [59-60, 62-63]. The production of volatile species from solutions proved an effective means, and analyte preconcentration combined with sample introduction by covalent hydride generation (HG) is the most extensively-used gas-phase sample-introduction technique for the determination of elements capable of forming volatile covalent hydrides such as arsenic, bismuth, germanium, lead, cadmium, tin, antimony, etc. The main advantage of hydride generation for AES arises from the very high transport efficiency of analytes, which were injected into the excitation source along with other reaction by-products (typically H_2, H_2O and CO_2). For HG-plasmas in analytical chemistry, the carrier gas containing the volatile products is always mixed with plasma gas before the coupling device. Specifically, in surface wave discharge, the presence of by-product molecules in plasma causes instability or even discharge extinction. A new design, Lateral Sample Introduction (LSI), has been proposed [64]. LSI consists of a dielectric tube with one or more side entrances for gaseous products in which the carrier gas is not used to create plasma (Figure 9). With this arrangement, the by-products of the HG reactions reach the plasma away from the area where power is coupled to the discharge, thus not compromising discharge stability.

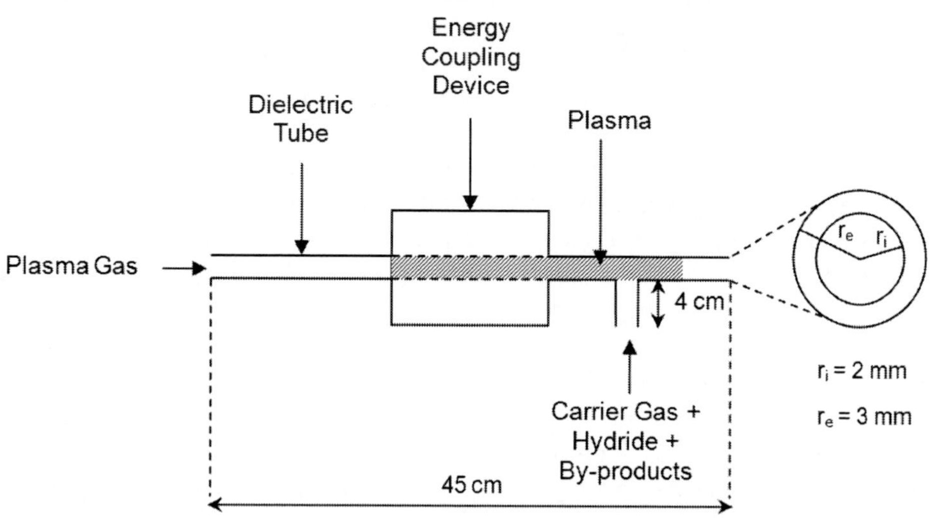

Figure 9. Diagram of the proposed device for introducing HG reaction products into the plasma source.

This way, the discharge can support a large amount of by-products requiring no additional intermediate steps between the HG system and the plasma source for their removal, which is the main advantage of the proposed system. A more detailed description of this device can be found in [64]. The analytical quality of LSI was verified through As determination in aqueous solution after hydride generation. In this study, Ar was used as plasma and carrier gas.

The analytical features as SBR and RSDB of As determination were 24 ng/ml and 3%, respectively; these values are close to those found by other authors in this field [65-66].

4.2. Hydrogen Production

Nowadays, due to the current decrease in natural reserves of fossil fuels, together with the economical problems derived from their extraction, and the emission of pollutant gases from their use a lot of research is focused on the development of new energy sources and new fuels. Among new fuels, hydrogen is becoming important because its combustion liberates a great quantity of energy with no CO_2 emissions. Besides, hydrogen is easily converted into electricity by fell cells. So, the development of procedures for its production is continuously under study.

The amount of free molecular hydrogen in the atmosphere is very small. Then, hydrogen must be obtained from other substances containing it. Different processes are used to extract hydrogen from diverse substances. Plasma features such as high temperature, energy density, degree of ionization, fast response in time, etc. have led several researchers to use plasmas such as dielectric barriers, microwave, and corona discharges as reactive media, in which the extraction reactions can happen for H_2 production [67-70]. Among different plasma types, surface wave discharges have not been researched enough for this purpose yet. Yanguas-Gil et al. [71] and Wang et al. [72] used this discharge for ethanol decomposition in reduced pressure. Only Jiménez et al. [73] have studied the capability of surface wave discharges generated at atmospheric pressure for hydrogen production from molecules of different alcohols such as methanol, ethanol, propanol, and buthanol. Plasmas were created inside a quartz tube open at one end (1.5 and 4 mm inner and outer diameters, respectively), using a total gas flow of 0.5 slm at a purity of 99.999% [73]. Microwave power of 200 W and 2.45 GHz frequency was used for plasma generation. Power was supplied by a *surfatron* as surface wave launcher [7]. Two possible configurations for surface wave plasma were considered: flame or column [11,74], but the former (Figure 10a) was chosen in this work for experiments due to its great robustness withstanding the introduction of alcohol opposite to the plasma column (Figure 10b). A method called *bubbling* was used to introduce alcohols (methanol, ethanol, propanol, and butanol) into the plasma at room temperature so as to involve no additional energy in the process. This way, total gas flow was divided into two parts: one (20% of total gas flow), used as the carrier gas, was bubbled into the alcohol-containing recipient to drag its molecules. The resulting mixture (gas and alcohol) was then joined to the remaining argon flow (80% of total gas flow) before introducing it all into the discharge-containing quartz tube. Two types of analysis were completed. On one hand, plasma capability to break down the bonds of alcohol molecules was analyzed through plasma-emitted radiation.

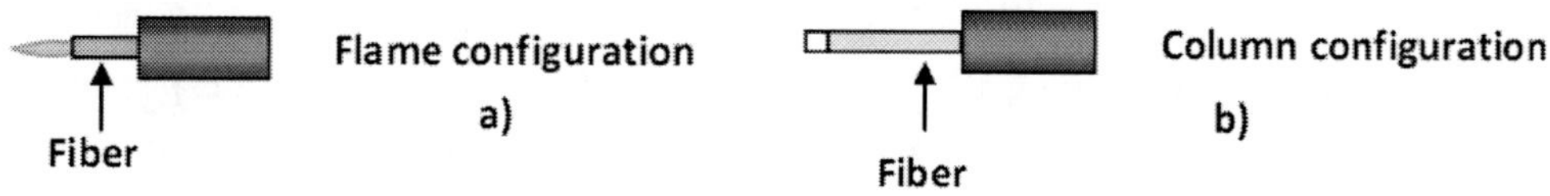

Figure 10. Two configurations for surface wave plasma: (a) Flame configuration; (b) Column configuration.

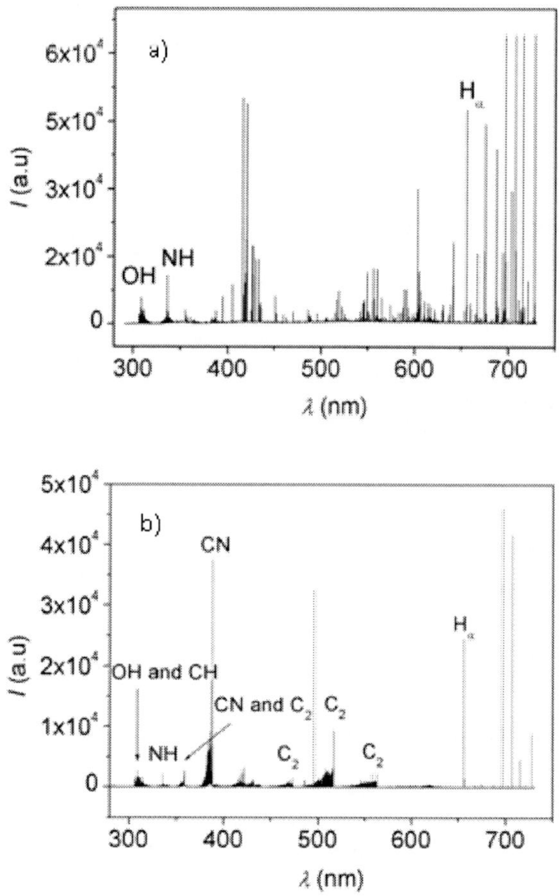

Figure 11. Plasma-emitted spectrum (a) without alcohols, and (b) when alcohols have been introduced into the plasma.

This radiation was collected by an optical fiber coupled to the entrance slit of a Jobin-Yvon Horiba monochromator, type Czerny-Turner, with a focal distance of 1 m, a holographic grating of 2400 lines/mm, and a computer-controlled CCD camera was used for spectrum registration.

To detect the molecular hydrogen produced by the proposed method, a mass spectrometer (PT M63 112, mod. Omnistar) was placed at plasma exit. A detailed description of the plasma source and devices used in plasma radiation analysis and molecular hydrogen detection can be found in [73].

Figure 11(a) shows the emission of pure Ar plasma, and Figure 11(b) shows the species obtained from the decomposition of the methanol molecules introduced into the discharge.

Comparing both spectra, one observes emissions from molecular species such as CN, C_2 and a minority contribution of CH together with increased OH band when alcohol is present in the discharge. The observation of these bands indicate the existence of dissociation of alcohol molecules by plasma. Similar spectra were obtained from the remaining considered alcohols.

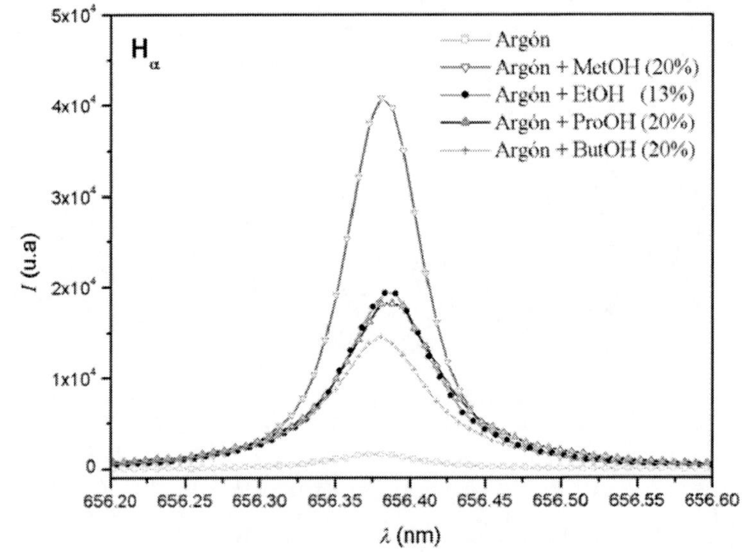

Figure 12. Pure-Ar plasma-emitted H_α line when the four alcohols were introduced into it.

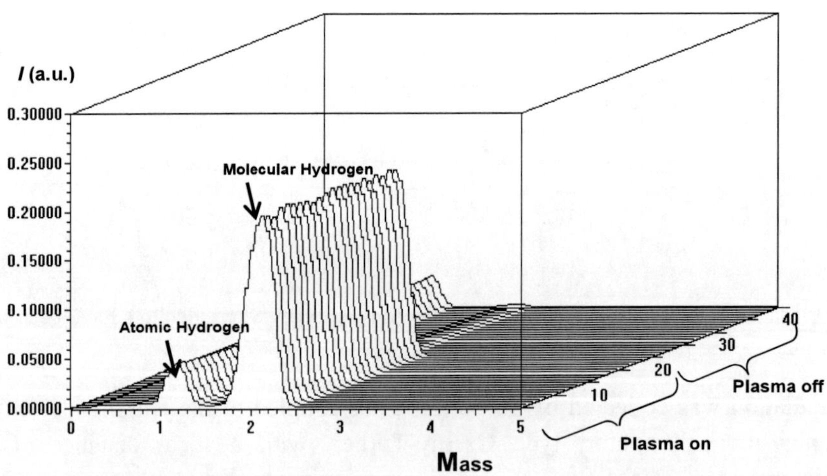

Figure 13. Mass spectra: molecular hydrogen production.

In Figure 12, the intensity of the H_α line, corresponding to atomic hydrogen, increases when alcohols are introduced into the plasma, which implies that hydrogen production takes place through the decomposition of alcohol molecules. In this figure, methanol seems the most appropriate alcohol to obtain hydrogen.

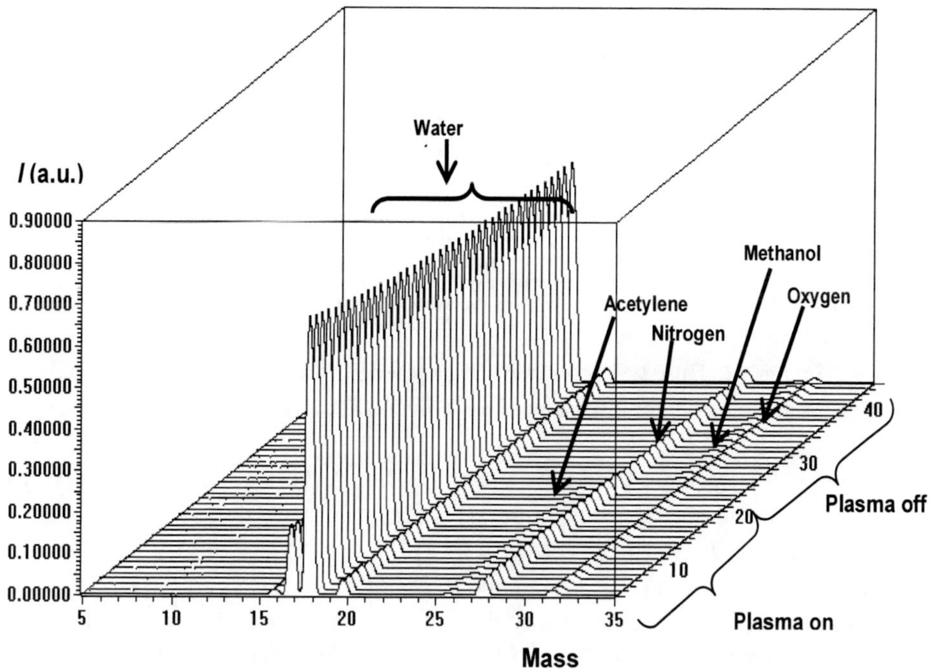

Figure 14. Mass spectra: acetylene production.

However, from an environmental point of view, ethanol can be considered the most interesting one because it can be obtained through starch, sugar or crop-waste fermentation.

To detect molecular hydrogen formation at plasma exit, a mass spectrometer was used and methanol was used in this case. Forty spectra were recorded: the first twenty ones when plasma was switched on, and the next twenty ones when plasma was switched off. Figure 13 shows hydrogen molecular production from methanol decomposition by plasma in a significant amount. Spectra corresponding to Figure 14 show the formation of acetylene (26 a.m.u.), which is obtained from the association of CH radicals. It results in the possibility of obtaining other sub-products from alcohol decomposition.

All these results show the good capability of the SWD to break down alcohol molecules and liberate their hydrogen content.

Acknowledgments

These works have been partially subsidized by the Spanish National Ministry of Science and Innovation under contract ENE2008-01015/FTN and FEDER Funds (European Community).

References

[1] F. Chen, Introduction to Plasma Physics and Controlled Fusion (2nd Edition) New York: Plenum (1984).

[2] W. B. Lee, J. Y. Wu, Y. I. Lee, J. Sneddon Recent applications of laser-induced breakdown spectrometry: A review of material approaches, *Appl. Spectros. Rev.* 39, 27 (2006).

[3] L. Bovey, A tuneable microwave cavity for excitation at 2425 MC-S of an optical spectroscopic source, *Spectrochimica Acta* 10 432 (1958).

[4] K. Wagatsuma, Recent advances in microwave induced plasma atomic emission spectrometry with Okamoto-cavity, *Appl. Spectros. Rev.* 40, 229 (2005).

[5] H. Yamada, Y. Okamoto, Characteristics of a high-power microwave-induced helium plasma at atmospheric pressure for the determination of nonmetals in aqueous solution, *Appl. Spectros.* 55, 114 (2001).

[6] C. Kin Ng and P. Bucay, Preliminary characterization of a low-powered microwave induced flame plasma for direct organic solvent nebulization, *Rev. Sci. Instrum.*, 82, 125101 (2011).

[7] M. Moisan, P. Leprince, C. Beaudry and E. Bloyet, Perfectionnements apportés aux dispositifs d'excitation par des ondes HF d'une colonne de gaz enfermée dans un enveloppe, *Brevet France* (1974).

[8] M. Moisan, Z. Zakrzewski, R. Pantel and P. Leprince, A Waveguide-Based Launcher to Sustain Long Plasma Columns Through the Propagation of an Electromagnetic Surface Wave, *IEEE Transactions on Plasma Science* 12, 203 (1984).

[9] M. Moisan, M. Chaker, Z. Zakrzewski, J. Paraszczak, The Wave-guide surfatron – A high power surface-wave launcher to sustain large-diameter dense-plasma columns, *J. Phys D: Sci. Instrum.* 20 (1987) 1356.

[10] M. Moisan and Z. Zakrzewski, New surface wave launchers for sustaining plasma columns at submicrowave frequencies (1–300 MHz), *Rev. Sci. Instrum.* 58, 1895 (1987).

[11] M.D. Calzada, M. Moisan, A. Gamero and A. Sola, Experimental investigation and characterization of the departure from local thermodynamic equilibrium along a surface-wave-sustained discharge at atmospheric pressure, *J. Appl. Phys.* 80 46 (1996).

[12] E. Castaños Martinez, Y. Kabouzi, K. Makasheva and M. Moisan, Modeling of microwave-sustained plasmas at atmospheric pressure with application to discharge contraction, *Phys. Rev. E* 70 066405 (2004).

[13] J. Muñoz and M.D. Calzada, Thermodynamic equilibrium in atmospheric pressure Ar–He surface wave discharges, *Spectrochim. Acta B: Atom. Spectros.* 65 1014 (2010).

[14] M. Jiménez, J. Muñoz and M. D. Calzada, A novel device for gas-phase sample introduction into microwave plasma: Lateral Sample Introduction, *J. Anal. At. Spectrom.*, 26, 1863 (2011).

[15] M. K. Boudam, M. Moisan, B. Saoudi, C. Popovici, N. Gherardi and F. Massines, Bacterial spore inactivation by atmospheric-pressure plasmas in the presence or absence of UV photons as obtained with the same gas mixture. *J. Phys. D: Appl. Phys.*, 39 3494 (2006).

[16] A. Walkiewicz-Pietrzykowska, J. Cotrino, A.R. Gonzalez-Elipe, Deposition of thin films of SiOxCyH in a surfatron microwave plasma reactor with hexamethyldisiloxane as precursor, *Chem. Vap. Dep.* 11 317 (2005).

[17] J. Muñoz and M.D. Calzada, Experimental study on equilibrium deviations in atmospheric pressure argon/helium surface wave discharges, *Spectrochimica Acta B: Atom. Spectros.* 65, 1014 (2010).

[18] Q. Jin, C. Zhu, M.W. Borer and G.M. Hieftje, A microwave plasma torch assembly for atomic emission spectrometry; *Spectrochim. Acta B: Atom. Spectros.* 46, 417 (1991).

[19] M Moisan, G Sauve, Z Zakrzewski and J Hubert, An atmospheric pressure waveguide-fed microwave plasma torch: the TIA design, *Plasma Sources Sci. Technol.* 3 584 (1994).

[20] L.W. Zhao, M.J. Li, X.H. Xie, D. Song, Y. Tian, L. Zhang, D. Jin, H. Zhang and Q. Jin, A simple flow-injection on-line clean-up system for microwave plasma-torch atomic emission spectrometry, *Fresen. J. Anal. Chem.* 370, 1061 (2001).

[21] M. Jasinski, J. Mizeraczyk, Z. Zakrzewski, T. Ohkubo and J. Chang, CFC-11 Destruction by microwave torch generated atmospheric-pressure nitrogen discharge, *J. Phys. D: Appl. Phys.* 25, 2274 (2002).

[22] Y.C. Hong, J.H. Kim, C.U. Bang and H.S. Uhm, Gas-phase synthesis of nitrogen-doped TiO2 nanorods by microwave plasma torch at atmospheric pressure *Phys. Plas.* 12, 114501 (2005).

[23] A. M. Bilgic, U. Engel, E. Voges, M. Kuckelheim., J. A. C. Broekaert, A new low-power microwave plasma source using microstrip technology for atomic emission spectrometry, *Plasma Sources Sci. Technol.* 9 1 (2000).

[24] J. A. C. Broekaert, V. Siemens, N. H. Bings, Microstrip microwave induced plasma on a chip for atomic emission spectral analysis, *IEEE Transactions on Plasma Science*, 33 560 (2005).

[25] S. Schermer, H.B. Bings, A.M. Bilgic, R. Stonies, E. Voges, J. A. C. Broekaert, An improved microstrip plasma for optical emission spectrometry of gaseous species, *Spectrochimica Acta B: Atom. Spectros.* 58 1585 (2003).

[26] S-J. Park, J. Choi, G. Y. Park, S-K. Lee, Y. Cho, J. I. Yun, S. Jeon, K. T. Kim, J. K. Lee and J-Y. Sim, Inactivation of S. mutans Using an Atmospheric Plasma Driven by a Palm-Size-Integrated Microwave Power Module, *IEEE Transactions on Plasma Science* 38, 1956 (2010).

[27] M.D. Calzada, M.C. García, J.M. Luque and I. Santiago, Influence of the thermodynamic equilibrium state in the excitation of samples by a plasma at atmospheric pressure, *J. Appl. Phys* 92 2269 (2002).

[28] A. Scheeline and M. J. Zoellner, Thomson scattering as a diagnostic of atmospheric-pressure discharges, *Appl. Spectros.* 38 (1984) 245 – 258.

[29] A. Ricard, J. E. Osegura-Pena, L. Falk, H. Michel and M. Gantois, Active Species in Microwave Postdischarge for Steel-Surface Nitriding, *IEEE Transactions on Plasma Science* 18 940 (1990).

[30] J.M. Luque, M.D. Calzada and M Saez, Experimental research into the influence of ion dynamics measuring the electron density from the Stark broadening of the H_α and H_β lines, *J. Phys. B*: 36 1573 (2003).

[31] P. Kepple and H. R. Griem, Improved Stark profile calculations for hydrogen lines Halpha Hbeta Hgamma and Hdelta, *Phys. Rev.* 173 317 (1968).

[32] H.R. Griem, Plasma Spectroscopy, New York: McGraw-Hill (1964).

[33] M. A. Gigosos and V. Cardeñoso, New plasma diagnosis tables of hydrogen Stark broadening including ion dynamics, *J. Phy. B* 29 (1996) 4795 – 4838.

[34] W.L. Wiese, Plasma Diagnosis Techniques, New York: Academic Press (1965).

[35] A. Scheeline and M. J. Zoellner, Thomson scattering as a diagnostic of atmospheric-pressure discharges, *Appl. Spectros.* 38 (1984) 245 – 258.

[36] H. Huang and G. M. Hieftje, Thomson scattering from an ICP, *Spectrochimica Acta B: Atom. Spectros.* 40 (1985) 1387 – 1400.

[37] M. Christova, E. Castaños-Martínez, M.D. Calzada, Y. Kabouzi, J.M. Luque and M. Moisan, Electron density and gas temperature from line broadening in an argon surface-wave-sustained discharge at atmospheric pressure, *Appl. Spectros.* 58 1032 (2004).

[38] C. Yubero, M.S. Dimitrijević, M.C. García and M.D. Calzada, Using the van der Waals broadening of the spectral atomic lines to measure the gas temperature of an argon microwave plasma at atmospheric pressure, *Spectrochimica Acta B: Atom. Spectros.* 62 169 (2007).

[39] G. J. Bastiaans and R. A. Mangold, The calculation of electron-density and temperature in Ar spectroscopic plasmas from continuum and line spectra, *Spectrochimica Acta B: Atom. Spectros.* 40 885 (1985).

[40] A. Sola, M. D. Calzada and A. Gamero, On the use of the line-to-continuum intensiry ratio for determining the electron-temperature in a high-pressure argon surface-microwave discharge, *J. Phys. D: Appl. Phys.* 28 1099 (1995).

[41] J. Torres, J. Jonkers, M. J. van de Sande, J. J. A. M. van der Mullen, A. Gamero and A. Sola, An easy way to determine simultaneously the electron density and temperature in high-pressure plasmas by using Stark broadening, *J. Phys. D: Appl. Phys.* 36 L55 (2003).

[42] C. Yubero, M. D. Calzada and M. C. García, Using the stark broadening of the H alpha, H beta and H gamma lines for the measurement of electron density and temperature in a plasma at atmospheric pressure, *J. Phys. Soc. Jap.* 74 2249 (2005).

[43] K. Muraoka, K. Uchino and M. D. Bowden, Diagnostics of low-density glow discharge plasmas using Thomson scattering, *Plasma Phys. Controll. Fus.* 40 1221 (1998).

[44] H. Kempkens and J. Uhlenbusch, Scattering diagnostics of low-temperature plasmas (Rayleigh scattering, Thomson scattering, CARS), *Plasma Sources Sci. Technol.* 9 492 (2000).

[45] M.J. van de Sande and J.J.A.M. van der Mullen, Thomson scattering on a low-pressure, inductively-coupled gas discharge lamp, *J. Phys. D: Appl. Phys.* 35 1381 (2002).

[46] M.C. García, C. Yubero, M.D. Calzada and M.P. Martínez-Jiménez, Spectroscopic characterization of two different microwave (2.45 GHz) induced argon plasmas at atmospheric pressure, *Appl. Spectros.* 59 519 (2005).

[47] B.A. Cruden, M.V.V.S. Rao, S.P. Sharma, M. Meyyappan, Neutral gas tyemperature estimates in an inductively coupled CF4 plasma by fitting diatomic emission spectra, *J. Appl. Phys.* 91 8955 (2002).

[48] G. Lombardi, F. Bénédic, F. Mohasseb, K. Hassouni and A. Gicquel, Determination of gas temperature and C-2 absolute density in Ar/H-2/CH4 microwave discharges used for nanocrystalline diamond deposition from the C-2 Mulliken system, *Plasma Sources Sci.Technol.* 13 375 (2004).

[49] A. Ricard, L. St-Onge, H. Malvos, A. Gicquel, J. Hubert and M. Moisan, Microwave-excited plasma torch – 2 complementary configurations, *Journal de Physique III France* 5 1269 (1995).

[50] J. Mermet, Inductively Coupled Plasma Emission Spectrometry. PartII: Applications and Fundamentals, New York: Wiley-Interscience (1987).

[51] W.R. Hindmarsh, A.D. Petford and G. Smith, Lectures at the Winter College on Atoms, Molecules an Lasers, Trieste Vienna: IAEA (1973).

[52] M. Findeisen, T. Grycuk, A. Brielski and J. Szudy, On the role of maxwelian averaging on the interpretations of pressure broadening and shift of spectral lines, *J. Phys. B* 20 5997 (1987).

[53] G. Peach, Theory of the pressure broadening and shift of spectral lines, *Advances in Physics* 30 367 (1981).

[54] B.N.I. Al-Saqabi and G. Peach, Unified theories of the pressure broadening and shift of spectral lines. II. Van der Waals interactions, *J. Phys. B* 20 1175 (1987).

[55] J. Muñoz, M.S. Dimitrijević, C. Yubero and M.D. Calzada, Using the van der Waals broadening of spectral lines to measure the gas temperatura of an argon-helium microwave plasma at atmospheric pressure, *Spectrochim. Acta B: Atom. Spectros.*, 64 167 (2009).

[56] J. Muñoz, C. Yubero, M.S. Dimitrijević, M. D. Calzada, Gas temperature determination in atmospheric pressure surface wave discharges from atomic line broadening, 36th EPS Conference on Plasma Physics (Sofia, Bulgaria) ECA 33E O-2.017 (2009).

[57] M. Moisan, J. Pelletier, Plasma Technology, Vol. 4- Microwave Excited Plasmas, Elsevier Science B.V. (1992).

[58] M. Moisan and Z. Zakrzewski, Microwave discharges: Fundamentals and Applications, ser. NATO ASI, Series B. Vol.302, C.M. Ferreira and M. Moisan, Eds. New York: Plenum (1993).

[59] M.D. Calzada, M.C. Quintero, A. Gamero, J. Cotrino, J.E. Sánchez-Uría and A. Sanz-Medel, Determination of bromide by low power surfatron microwave induced plasma after bromine continuous generation, *Talanta* 39 341 (1992).

[60] M.D. Calzada, M.C. Quintero, A. Gamero and M. Gallego, Chemical generation of Chlorine, Bromine and Iodine for simple introduction into a surfatron-generated argon Microwave-Induced Plasma, *Anal. Chem.* 64 1374 (1992).

[61] J. Mermet, Inductively Coupled Plasma Emission Spectroscopy, Part II, Chapter X, Wiley: New York (1987).

[62] R. Baena, M. Gallego, M. Valcárcel, J. Leenaers and F.C. Adams, Comparison of three coupled gas chromatographic detectors (MS, MIP-AES, ICP-TOFMS) for organolead speciation analysis, *Anal. Chem.* 73 3927 (2001).

[63] J.A.C. Broekaert and V. Siemens, Recent trends in atomic spectrometry with microwave-induced plasmas, *Spectrochim. Acta B: Atom. Spectros.* 59 1823 (2004).

[64] M. Jiménez, J. Muñoz and M.D. Calzada, Patent N P201031621, Spain (2010).

[65] E. Bulska, J.A.C. Broekaert, P. Tschöpel and G. Tölg, Comparative study of argon and helium plasmas in a TM_{010} cavity and surfatron and their use for hydride generation microwave-induced plasma atomic emission spectrometry, *Anal. Chim. Acta* 276, 377 (1993).

[66] R. Pereiro, M. Wu, J.A.C. Broekaert and G.M. Hieftje, Direct coupling of continuous hydride generation with microwave plasma torch atomic emission spectrometry for the determination of arsenic, antimony and tin, *Spectrochim. Acta B: Atom. Spectros.* 49 59 (1194).

[67] L. Bromberg, D.R. Cohn, A. Rabinovich, N. Alexeev, A. Samokhin, R. Ramprasad and S. Tamhankar, System Optimization and cost analysis of plasma catalytic reforming of natural gas, *Int. J. Hydrogen Ener.* 25 1157 (2000).

[68] L. Bromberg, D.R. Cohn, A. Rabinovich and N. Alexeev, Plasma catalytic reforming of methane, *Int. J. Hydrogen Ener.* 24 1131 (1999).

[69] M. Deminsky, V. Jivotov, B. Potapkin and V. Rusanov, Plasma-assisted production of hydrogen from hydrocarbons, *Pure Appl. Chem.* 74 413 (2002).

[70] B. Sarmiento, J.J. Brey, I.G, Vera, A.R. González-Elipe, J. Cotrino and V.J. Rico, Hydrocarbon production by reforming of hydrocarbons and alcohols in a dielectric barrier discharge, *J. Power Sources* 169 140 (2007).

[71] A. Yanguas-Gil, J.L. Hueso, J. Cotrino, A. Caballero and A.R. González-Elipe, Reforming of etanol in a microwave surface-wave plasma discharge, *Appl. Phys. Letters* 85 4004 (2004).

[72] Z. Wang, P.J. Hong, C.C. Zhang and S.S. Dai, The microwave-induced plasma chemical reaction of the lower alcohol, *Chem. J. Chinese- Universities* 16 1298 (1995).

[73] M. Jiménez, C. Yubero and M.D. Calzada, Study of the reforming of alcohols ina surface wave discharge (SWD) at atmospheric pressure, *J. Phys. D: Appl. Phys.* 41 175201 (2008).

[74] M.D. Calzada, M. Sáez and M.C. García, Characterization and study of the thermodynamic equilibrium departure of ar argon plasma flame produced by a surface-wave sustained discharge, *J. Appl. Phys.* 88 34 (2000).

In: Argon: Production, Characteristics and Applications ISBN: 978-1-62618-204-2
Editors: B.N. Sismanoglu, H.S. Maciel et al.

Chapter 6

AR CLUSTER ION: ITS PRODUCTION AND APPLICATION TO NANOTECHNOLOGIES

*K. Mochiji**
Graduate School of Engineering, University of Hyogo,
Shosha, Himeji, Hyogo, Japan

ABSTRACT

A gas cluster ion consists of gaseous atoms such as argon (Ar) aggregated by Van der Waals forces, and the cluster typically contains hundreds to thousands of atoms or molecules. The average level of kinetic energy per atom forming the cluster ion is equal to the value of the energy obtained by accelerating the cluster ion divided by the number of constituent atoms of the cluster ion (called cluster size). For example, when a cluster of 1,000 atoms is accelerated to 10 keV, the average kinetic energy per atom of the cluster ion is as low as 10 eV. Furthermore, when the cluster ion collides with surfaces, chemical reactions, such as sputtering or etching, intensively occur within a very thin layer as a result from the multiple collisions among the numerous atoms constituting the cluster ion. The nature of the bombardment by the gas cluster ions has a potential to drastically change ion-beam technology. In this chapter, an experimental technique for the production and bombardment of Ar cluster ions and their application to nanofabrications are reviewed, as well as a surface analysis

1. INTRODUCTION

Ion-beam technologies are widely used in various fields such as semiconductor processing, micromachining, and surface analyzing. The rapid development of electronic and optical devices requires an accuracy within tenths of nanometers in the etching depth. In order

* E-mail address: mochiji@eng.u-hyogo.ac.jp.

to fill these needs, the ion beam energy must be significantly reduced while maintaining the ion current, which is difficult to accomplish.

Secondary ion mass spectrometry (SIMS) is a key technology to analyze the chemical species at the material surface with high sensitivity. Various efforts have been made to apply SIMS to organic substances [1-3]. Recently, it has been reported that the use of polyatomic ions and cluster ions, as the primary ions in place of the atomic ions, increases secondary ion intensity and is effective in suppressing damages to sample molecules [4-9]. When an accelerated cluster-ion collides with a solid surface it can induce multiple collisions by numerous constituent atoms, thereby making it possible to improve sputter yield and increase secondary ion intensity. The average level of kinetic energy per atom in the cluster ion is the value of the energy obtained by accelerating the cluster ion divided by the number of constituent atoms of the cluster ion (hereafter called the cluster size). For this reason, the higher the number of constituent atoms, the smaller the kinetic energy per atom will be, thus making it possible to decrease damage imparted to the sample. Until now, cluster ions of Au_n^+ (typically n=1-5), SF_5^+, and C_{60}^+ etc. have been primarily used [4-7, 9], but because of the small number of constituent atoms of these cluster ions, using applied acceleration techniques to decrease kinetic energy per atom down to ~10 eV has been difficult.

A gas cluster ion, such as the argon (Ar) cluster ion, consists of gaseous atoms aggregated by Van der Waals forces and can be made from approximately 10,000 constituent atoms [10-12]. Contamination possibility is extremely low, as the Ar cluster ion quickly evaporates after bombarding the target, and the Ar atom is chemically inert. Houzumi et al. analyzed the traces caused by Ar cluster ions bombarded on graphite [13, 14]. The kinetic energy per atom of the cluster ion used for bombardment will almost completely determine the rate at which traces are produced. A certain level, or greater amount, of kinetic energy per atom is required for causing traces, and this threshold energy is determined by the cohesive energy of the carbon atoms that constitute graphite. This result suggests that adjusting the kinetic energy per atom can suppress damage on the sample. For example, in using a polymer as a target, decreasing kinetic energy per atom to the level below the dissociation energy of the chemical bond (3~5 eV) of the polymer suppresses polymer fragmentation.

Mochiji et al. developed SIMS equipment in which the cluster size of Ar cluster ions was selected to adjust the kinetic energy per atom of the cluster ion [15]. Because the cluster size of a gas cluster ion varies substantially in the formation stage, bombarding without selecting the size of the clusters may allow the variety of different cluster sizes to affect the sample. As a result of applying the new equipment to several peptides and proteins, the intensity of fragment ions of the sample molecules was decreased by a factor of 10^2 when the kinetic energy per atom of the incident Ar cluster ions was decreased below 5 eV, and intact ions of cytochrome C (molecular weight: 12,327) and chymotripsin (molecular weight: ~25,000) were detected without using any matrix [16].

2. Experimental Set-Up

The system developed by Mochiji et al. for the generation and bombardment of Ar cluster ions is shown in Figure 1. This system was composed of four chambers: the cluster generation chamber (source chamber), the ionization chamber, the size-selection chamber, and the irradiating chamber (sample chamber). A neutral Ar cluster was generated through supersonic expansion from a nozzle with a 0.1-mm diameter aperture and a 60-mm expansion and cooling zone.

The cluster beam was collimated by a conical skimmer with an opening of 0.5 mm in diameter and then flowed into the ionizer. The neutral clusters were ionized through electron impact from a tungsten filament, and the cluster ions were then extracted at an accelerated voltage of 5~10 kV. Mono-atomic ions and small ion components were removed from the original cluster ion beam with the magnet. A cluster-size selection by the TOF (time-of-flight) technique was performed using the two pairs of ion deflectors. The continuous flow of cluster ions generated at the ionizer was chopped at the first ion deflector. The cluster ion beam pulsed by the first ion deflector contains various sizes of cluster ions, due to the size distribution of the clusters generated. The ion beam was then chopped again at the second ion deflector which was 633 mm apart from the first ion deflector. The flight time of the clusters between the two ion deflectors should be different due to the difference in cluster size. Therefore, the pulsed-ion-beam by the second ion deflector contains a specific size of cluster ions, depending on the delay time between the times to chop at the two deflectors. Figure 2 shows typical TOF spectra of the Ar cluster ion beam before and after size selection.

The stagnation pressure of the Ar gas was 1.21 MPa. The acceleration voltage of the Ar cluster ion was 5 kV. The ion beam was chopped in a width of 10 μs by the two ion deflectors. The cluster size was determined from the flight time between the first ion deflector and the ion detector (micro-channel plate, MCP), 8-cm downstream from the second ion deflector. The cluster size before size selection was widely distributed was from 500 to 7,000 atoms/cluster. The cluster size distribution after size selection was significantly narrowed, and the peak cluster-size became larger when the delay time was longer. A cluster size resolution (M/ΔM) over 10 was achieved by adjusting the pulse widths on the ion deflectors. As a result, we could control the kinetic energy per atom of incident cluster ions in the range of 1 eV and 25 eV.

The size-selected cluster ion beam entered the sample chamber through a 5-mm diameter aperture and focusing ion lens. A primary ion beam application to the samples and measurement of secondary ions were carried out under a pressure in order of 10^{-6} Pa and at room temperature.

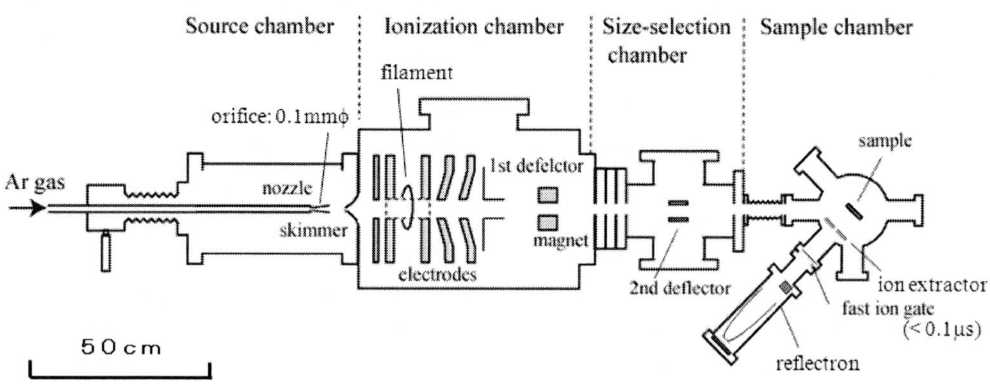

Figure 1. Schematic view of Ar cluster ion beam generation and the bombardment system.

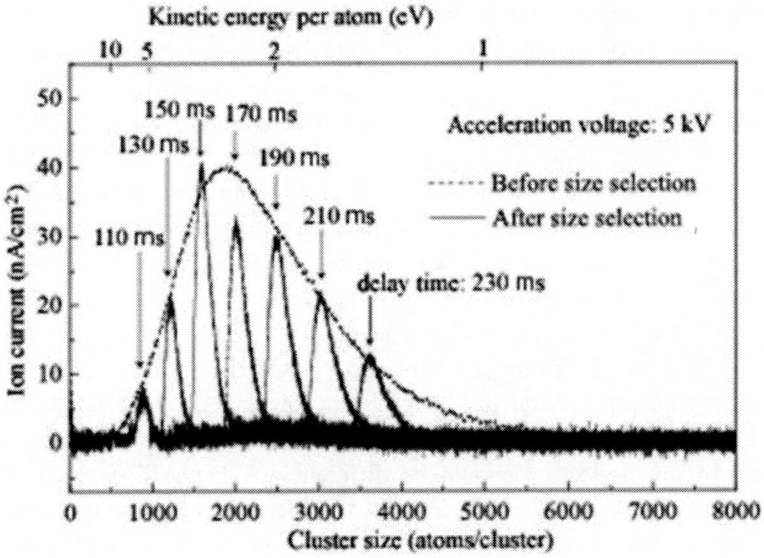

Figure 2. TOF spectra of Ar cluster ions before and after cluster-size selection.

The secondary ions are measured by a reflectron with a flight tube length of 1.8m (KNTOF-1800-SL203, Toyama, Kanagawa, Japan). For the precise measurement of ion flight time, an interleaved comb-type ion gate [17] was installed in front of the reflectron. The secondary ions passing thorough the ion gate were detected by an MCP and the ion signals were accumulated by the multi-channel scalar, MCS (P7888-1, FAST ComTec, Oberhaching, Germany). A logic-pulse signal was applied to produce high-voltage pulses to the ion gate, and this signal was used as a start signal for the TOF measurement. In these conditions, the mass resolution (m/Δm) of the reflectron was approximately 200 at m/z of 700 when the pulse width of the ion gate was 100ns.

Sample surfaces were observed before or after the ion bombardment using a scanning tunneling microscope (remodeled USM-1100, UNISOKU Ltd.) in a constant-current mode. Mechanically polished Pt-Ir tips were used. The typical tunneling current was 0.9 ~ 1.2 nA, and the bias voltage at the sample was 0.3 ~ 1.0 V. Tunneling current-bias voltage characteristics (I-V curves) were measured over both ion-bombarded and un-bombarded areas.

3. Bombarding Effect on Graphite

3.1. Defects Produced by the Bombardment

The effects of ion bombardment on graphite have been reported by other groups [18-20]. Most of these studies, however, employed mono-atomic ions such as Ar^+. Bräucle et al. analyzed defects on graphite induced by the bombardment of carbon cluster ions (dominantly C_{60}^+) [21, 22]. However, in that case, the supply of carbon atoms from the cluster itself to the defect sites may have affected the defect structure. With Ar cluster bombardment, on the other hand, the possibility of contamination is extremely low.

Tsukuda et al. investigated traces produced on graphite by Ar cluster ion bombardment and their structural changes by heating in air [23]. Figure 3 is an STM image of a graphite (highly oriented pyrolytic graphite: HOPG) surface after bombardment with Ar cluster ions of 750 atoms/cluster at 7.5 keV (E_a : 10 eV). Two types of defects were observed: crater-like defects about 10 nm in diameter and dot-like defects about 2 nm in diameter. These dot-like defects have also been reported after bombardment with other ions such as Ar^+ and Ar^{8+} [18, 19]. In contrast, the crater-like defect is uniquely produced by bombardment with gas cluster ions.

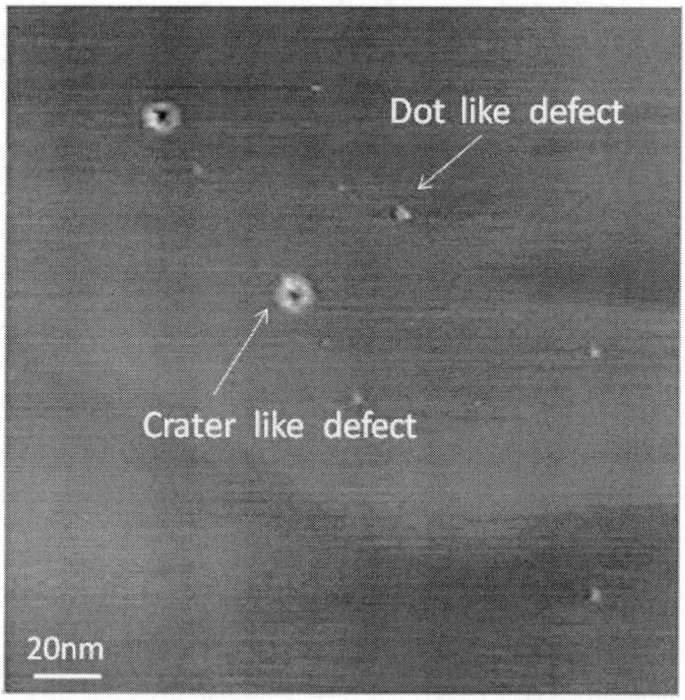

Figure 3. STM image of crater-like defects and dot-like defects on a graphite surface after bombardment with Ar_{750}^+ at 7.5 keV.

The dependence of the rate associated with the generation of these defects (the density of the defects observed divided by the ion dose) on the energy per atom of the cluster ion is shown in Figure 4. Far fewer craters than dots were produced at the energies per atom examined here. We previously reported that the crater production increases to ~80 % when the energy per atom exceeds 40 eV/atom [24].

The difference between the defect production rates can be explained as follows. When an Ar cluster ion hits the surface, the cluster is compressed at the surface and decomposed into many energetic Ar atoms.

These Ar atoms can sputter away the carbon atoms parallel to the surface (lateral sputtering), forming the crater-like defects. On the other hand, most of the dot-like defects are formed by Ar atoms trapped between the basal planes of the graphite, since most of the dots disappeared by heating at a low temperature (e.g., 560 °C, reported later). Their results indicate that Ar atoms can penetrate into the basal planes at energies below 5 eV/atom.

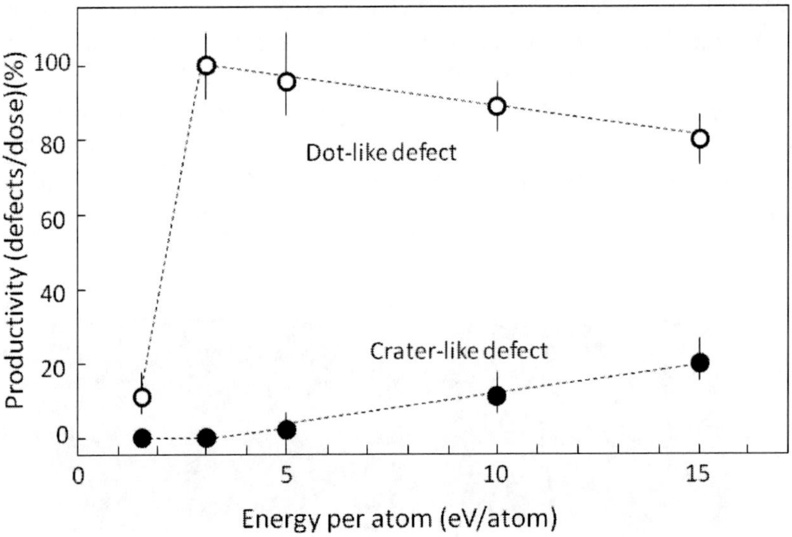

Figure 4. Dependence of defect productivity associated with the energy per atom of the cluster ion.

On the contrary, in the case of bombardment with a monoatomic ion such as Ar^+, the threshold energy for an Ar atom to penetrate into the top layer of graphite was 42 eV, according to calculations using a classical trajectory method [25]. Multiple collisions among the many Ar atoms of the cluster ion could significantly lower the energy per atom required for the penetration of the Ar atom.

3.2. Structural Changes of the Ion-Induced Defects by Heating in Air

The structure of the ion-induced defects was drastically changed by heating the sample at 560°C for 20 minutes in air. The craters completely disappeared, and the number of dot-like defects was reduced by about 90%. Instead, hexagonally shaped pits, with an average size of about 50 nm (Figure 5(a)), and irregularly shaped pits, with sizes of 5 ~ 10 nm (Figure 5(b)), appeared. A statistical study of the number of defects and pits in the STM images clarified the structural changes of the defects during heating as follows: the crater-like defects almost all became hexagonal pits. About 10% of the dot-like defects became irregularly shaped pits, about 80% disappeared, and the rest (~10 %) remained unchanged. The depth of the hexagonal pits and the irregularly shaped pits were almost uniformly 1.0 nm and 0.3 nm, respectively, which correspond to the thicknesses of 3 and 1 graphite layers, respectively. It was also confirmed that the graphite structure was fully maintained at the bottom of both pits (Figs. 5(a) and 5(b), inset). Increasing the heating time to 200 minutes did not change the pit depth, but increased the pit area by approximately 50 times.

These two types of pits are basically created by oxidative etching of dot-like defects or craters. Carbon vacancies are intensively formed by the displacement of surface carbon atoms at the craters. The etching mechanism of carbon vacancies has been explained [26] in terms of initial dissociative chemisorptions of O_2 at the dangling carbon bonds of a vacancy, followed by an activated desorption of CO or CO_2 to remove the carbon atoms. This desorption creates new dangling bonds, and the etching continues in the same carbon plane by further O_2

reactions. The etching of the dot-like defects is initiated by an oxidation of the graphite surface at a location that has been deformed by a trapped interstitial Ar atom below. The increase in reactivity at the deformed surface location results from a charge density increase in this region. The tunneling conductance near the zero bias voltage at a small number of the dot-like defects was larger than in the flat regions. Once an initial O_2 attack creates a vacancy on the basal plane, the rest of the defect can be etched in the same way as the craters. The interstitial Ar atoms seem to be randomly distributed under the basal plane, which could result in the irregularly shaped pits. On the contrary, O_2 reactions initiate more symmetrically along the crater, and successive removal of carbon atoms can form hexagonally shaped pits that reflect the graphite structure.

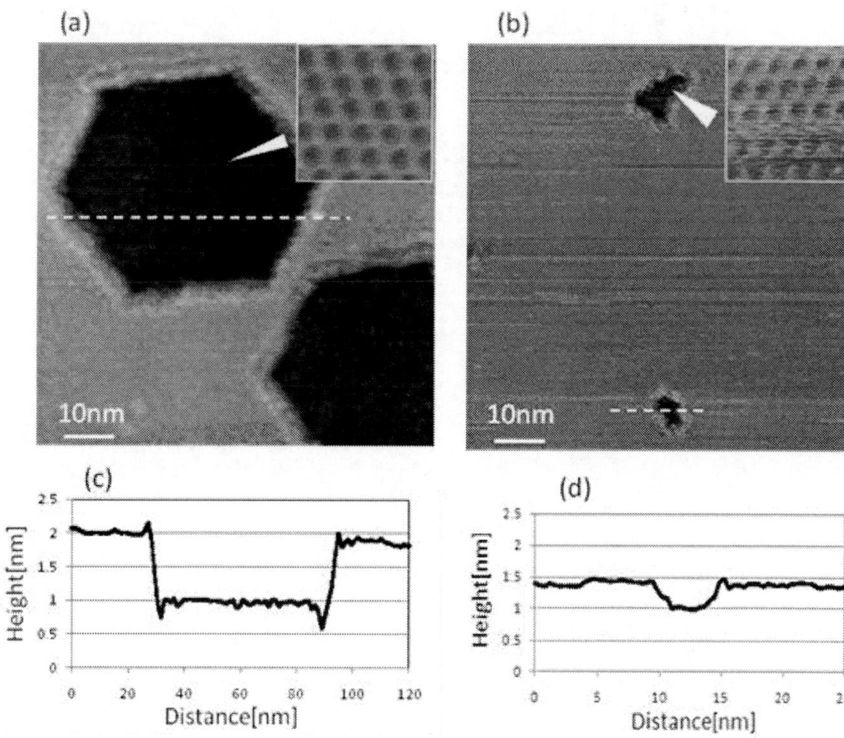

Figure 5. STM images of a graphite sample heated at 560°C for 20 minutes after bombardment with Ar_{750}^+ at 7.5 keV (a) and Ar_{2500}^+ at 7.5 keV (b), depth profiles along a dashed line over a hexagonal pit (c) and an irregulaly shaped pit (d).

4. Applications to SIMS

4.1. Detection of Intact Ions with High Mass

Mochiji et al. applied the SIMS in which Ar cluster ions were used as primary ions to some kind of organic samples including peptides and proteins. All samples were thin films formed on a silicon substrate and no matrix agent was used. The film thickness of each sample was in the range of 50~500 nm. Figure 6 shows secondary ion mass spectra produced when the same acceleration voltage of 5kV is applied to Ar cluster ions in a cluster size of 1,400 (Ar_{1400}^+) and Ar mono-atomic ions (Ar^+), and these ions bombard a 6-mer peptide of aspartic acid $(Asp)_6$ (molecular weight: 709).

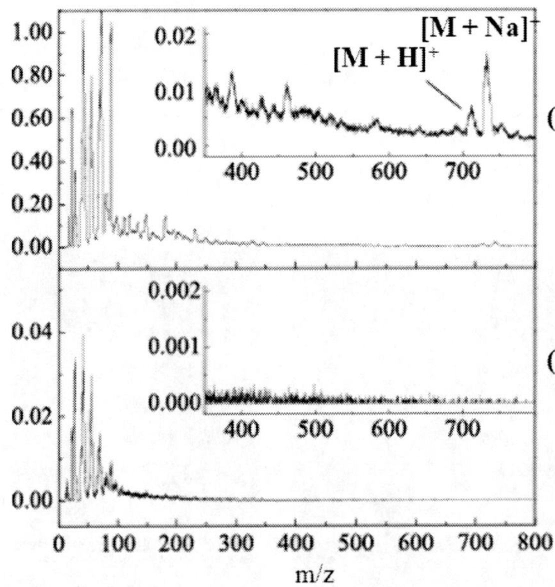

Figure 6. Secondary ion mass spectra of $(Asp)_6$ films by the bombardment of Ar_{1400}^+ (a) and Ar^+ (b) at 5 kV.

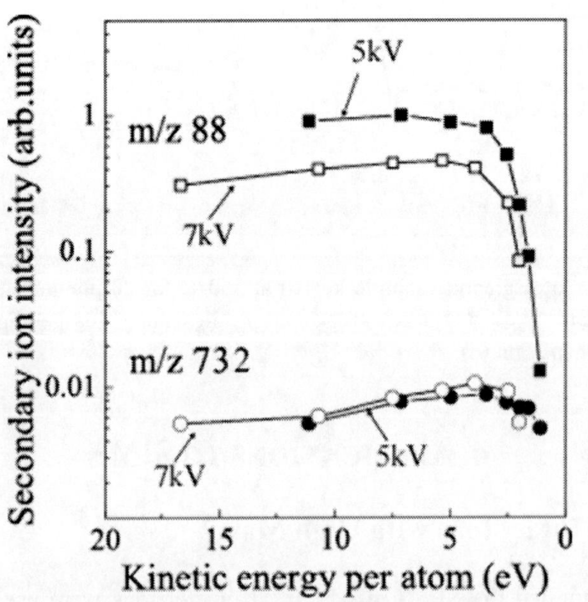

Figure 7. Secondary ion intensity of $(Asp)_6$ as a function of kinetic energy per atom.

The primary ion current applied to each sample normalizes the secondary ion intensity referred to in this chapter. With cluster ion bombardment, secondary ion intensity increased as compared with mono-atomic ion bombardment. The rate of increase was approximately 10 times for the fragment near *m/z* 100 and below, but the rate was much higher with a greater fragment. Several ion peaks appeared near *m/z* 700 in the spectrum created by cluster ion bombardment. Of these peaks, the peaks at *m/z* 710 and 732 indicate intact ions ($[M+H]^+$ and $[M+Na]^+$) with H^+ or Na^+ attached to $(Asp)_6$ molecules (Figure 6 insert). With mono-atomic ion bombardment, the signal of these intact ions was not distinguishable from the background level.

Figure 7 shows a plot for the intensity of fragment ions (*m/z*: 88) and intact ions (*m/z*: 732) when bombarded with three different acceleration voltages (5, 7.5, and 10 kV) in relation to the kinetic energy per atom. The result can be interpreted as follows. The secondary ion yield (Y_m^+) of species m, when excluding analyzer-related factors, can be expressed as a product of sputtering yield (Y_m), ionization probability to positive ions (α^+) and fractional concentration of the species m in the surface layer (θ_m) , or ($Y_m^+ = Y_m \cdot \alpha^+ \cdot \theta_m$). The ionization of biomolecules, such as peptides, often occurs by chemical interactions as represented by protonation or attachment with metal ions like Na^+, K^+ [27, 28]. The probability of these interactions hardly depends on the kinetic energy of primary ions.

Therefore, the value of α^+ may not be changed much within a small range of kinetic energy per atom. The rate of contribution to Y_m per atom decreases with the decrease in kinetic energy per atom. The decrease in kinetic energy per atom, however, means the bombardment with a larger cluster size when the acceleration voltage is constant. This means that more constituent atoms can interact with the sample surface and work together for the emission of the sample molecule from the surface. This multiple atom effect works toward increasing Y_m. Because molecule dissociation is less likely to occur with decreased kinetic energy per atom, the θ_m of small fragments becomes lower. We think that the 10^2-factor ~~or below~~ decrease in intensity of fragment ions of *m/z* 88 at 5 eV or below of kinetic energy per atom, as shown in Figure 7, resulted from substantial suppression of dissociation of the peptide. We think that the decrease in intact ion intensity below 3 eV was due to the overriding influence of the decrease in Y_m.

Figure 8 shows the secondary ion mass spectra measured with thin films of human insulin (molecular weight: 5,808), cytochrome C (molecular weight: 12,327) and chymotripsin (molecular weight: ~25,000). The cluster size of the Ar cluster ion was 1,500 atoms and the acceleration voltage was 5 kV, making the kinetic energy per atom 3.3 eV. The relative intensity of fragment ions was low in the mass spectra of all samples and clearly showed the peaks considered to be molecular ions (M^+) , doubly charged molecular ions (M^{2+}), and dimer ions ($2M^+$). It is interesting to consider how molecules can be emitted without suffering dissociation. Oshima et al. examined the effect of the direction of the incident cluster-ion beam [29]. As shown in Figure 9, the incident angle of the primary ion beam with respect to the sample surface is different, but the direction of extraction of the secondary ions is symmetrically equivalent. When the incident angle was changed from 60 degrees to 30 degrees, the intensity of the intact ions increased 10-fold, whereas the intensity of low mass fragments hardly changed.

The plot of the intensity ratio of the intact ions to the total low-mass fragment ions (m/z 1-500) as a function of the incident angle is shown in Figure 10. The intensity ratio decreased drastically between 30 degrees and 45 degrees.

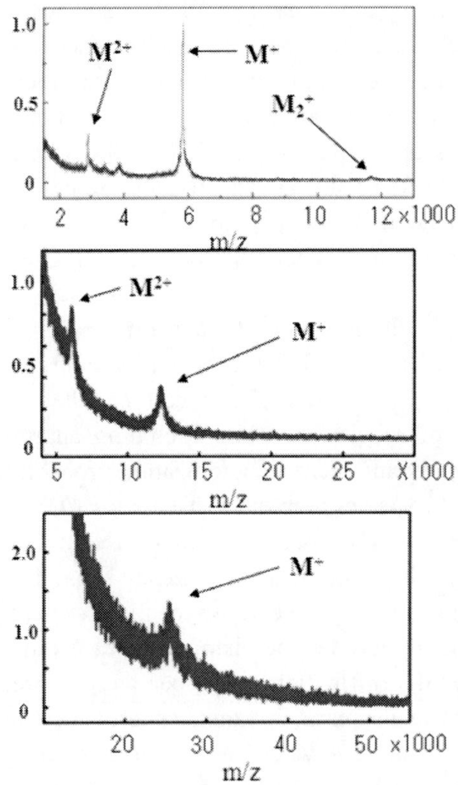

Figure 8. Secondary ion mass spectra of insulin (a), cytochrome C (b) and chymotripsin (c) by the bombardment of $Ar_{1500}{}^{+}$at 5 kV.

These results indicate that the intensity of the intact ions depends more strongly on the direction of incidence of the cluster ion beam than that of the low mass fragments. Such behavior could be related to the interaction process between the cluster ions and the insulin film. The insulin film can elastically deform at an early time during impact with the Ar cluster ions, and multiple collisions will occur between the insulin molecules and argon atoms. During such elastic vibrations of the film, there is a chance for the insulin molecules to be released without dissociation if the energy transferred to the molecule lies between the inter-molecular binding energy and the intra-molecular binding (covalent bond) energy. The inter-molecular binding energy is, in general, significantly smaller than the intra-molecular binding energy. When the cluster ions collide with the film more sharply (at a smaller angle of incidence), more molecules are freely released from the film and these molecules then move towards a vacuum. At more oblique incidences of the cluster ions, in contrast, the molecules tend to be emitted towards a direction close to the film surface and collide more often with other molecules composing the film, which may cause dissociation of the molecules. Inui et al. reported that a molecule adsorbed on a graphene sheet can be emitted without dissociation by impact with an Ar cluster [30]. In this study, the actuation of nano-drums consisting of a single graphene sheet was examined using molecular dynamics simulation. Theoretical results predicted that a quick deformation of the sheet due to impact with the cluster could break the weak bond between the adsorbed molecule and the graphene sheet and release the molecule from the surface, if the velocity of the cluster was appropriately adjusted.

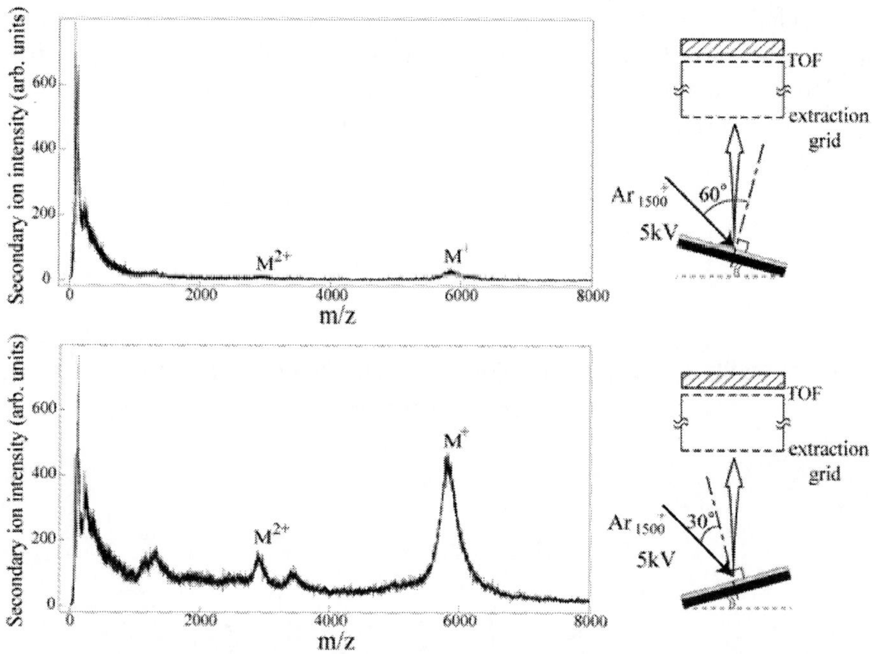

Figure 9. Secondary ion spectra of insulin thin film at different angles of incidence of the Ar cluster ions.

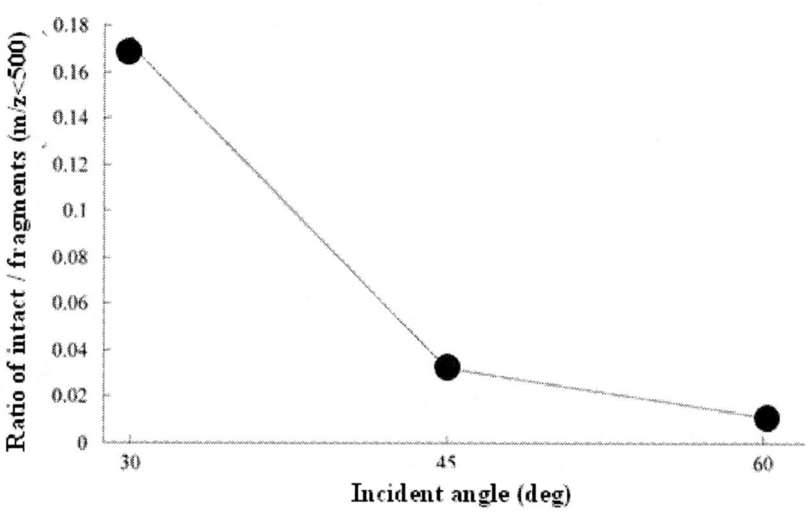

Figure 10. Intensity ratio of intact ions to the total of low-mass fragment ions (m/z:1-500) of insulin as a function of the incident angle of Ar cluster ions.

4.2. Ultra Thin Layer Analysis

Cluster projectiles cannot penetrate into the bulk as easily as atoms. Hence, the damage caused by cluster projectiles is confined to a shallower volume. Furthermore, the sputtering yields can be enhanced when an atomic projectile is replaced by a cluster ion with the same incident energy, because the cluster would be shattered upon impact and many individual atoms bombard the sample simultaneously, each creating its own cascade of moving atoms.

The simultaneous collision of many low-energy atoms caused by a cluster projectile enables effective sputter removal of surface contaminants, suggesting an improved surface sensitivity of SIMS analysis. Table 1 compares the intensities of secondary ions of a 1,4-didodecylbenzene (DDB) ($C_{30}H_{54}$, molecular weight:415) thin film bombarded by Ar_{1550}^+ and Ar^+ ion beams [31]. The secondary ion intensities were normalized by the primary ion current.

Table 1. Intensity of major fragment peaks of DDB for Ar_{1550}^+ and Ar^+ bombardment

m/z	Assignment	Secondary ion yield (normalized counts)		Ratio ($Y^+_{(Ar1550^+)}/Y^+_{(Ar^+)}$)
		$Y^+_{(Ar1550^+)}$	$Y^+_{(Ar^+)}$	
1	H^+	—	40	—
29	$C_2H_5^+$	229	22	10
43	$C_3H_7^+$, CH_3Si^+	574	32	18
45	CH_5Si^+	800	3	267
57	$C_4H_9^+$	300	23	13
59	$C_2H_7Si^+$	513	2	257
73	$C_3H_9Si^+$	6268	12	522
91	$C_7H_7^+$	121	9	13
105	$C_8H_9^+$	530	30	18
147	$C_5H_{15}OSi_2^+$	942	7	134
207	$C_6H_{19}Si_3O_2^+$	228	—	—
415	$C_{30}H_{54}^+$	910	5	182

Although the characteristic fragment ions of hydrocarbons, such as $C_2H_5^+$ (*m/z*=29), $C_3H_7^+$ (*m/z*=43), $C_4H_9^+$ (*m/z* =57), $C_7H_7^+$ (*m/z*=91), and $C_8H_9^+$ (*m/z*=105), were detected in the both spectra, the fragmentation patterns obtained by Ar_{1550}^+ and Ar^+ projectiles show remarkable differences. Intact DDB ions were detected much more for the Ar_{1550}^+ bombardment than for the Ar^+ bombardment, attributed to the low-energy irradiation effect of the gas cluster ion beam, as reported in the previous section. The most significant difference in these spectra was that the secondary ions at *m/z* =45, 59, 73, 147, 207, which were difficult to assign as hydrocarbon fragments from the DDB sample, were detected with a very high intensity in the spectrum obtained by the Ar_{1550}^+ projectile.

These fragment peaks are reasonably assigned to poly-dimethyl-siloxane (PDMS) contaminant. These contaminant species are presumably from the air because they increased after exposure to the ambient laboratory air. The peak intensity of the hydrocarbon fragment species from DDB in Ar_{1550}^+ SIMS was intensified 10 to 20 times relative to those in the Ar^+ SIMS. On the other hand, the peak intensities of the fragment species of PDMS were enhanced more than 100-500 times in the Ar_{1550}^{+} SIMS. This result indicates that the PDMS contaminants should be confined near the surface, and their fragment ion intensities are more enhanced by the sputtering with Ar cluster ions. Thickness of the contaminant layer was roughly estimated to be ~2 nm by analyzing detailed angle-dependent XPS data. Thus, a large gas cluster projectile can sputter a much shallower volume of organic material than small projectiles, resulting in an extremely surface-sensitive analysis of organic thin films.

CONCLUSION

Argon cluster ions could be a strong candidate as an ion-beam source for material processes or surface analyses. The easiness to control the energy per atom of the cluster at below 10 eV, by adjusting the acceleration voltage and the cluster size, lowers the damage on the sample induced by the ion bombardment and realizes nanometer scale fabrication. The etching depth of graphite can be controlled by adjusting the kinetic energy per atom of the bombarding cluster. Graphene, a single layer graphite, has recently attracted much attention scientifically and technologically. The graphite etching behavior obtained here may be useful in fabricating graphene with a uniformly controlled thickness.

When the kinetic energy per atom of the Ar cluster ions is controlled below 5 eV as primary ions for SIMS, intact ions of proteins with a molecular weight of 20,000 or more were successfully detected. Currently, SIMS can detect the molecular weight of intact ions up to approximately the 1,000 level. Certainly other studies have shown that larger molecular ions are detectable by making samples in the matrix, etc. [3, 29]. But the most current, pressing need is for measuring solid or thin film samples as they are, particularly with the development of bio-molecular nano-devices. Furthermore, the extremely effective sputtering of shallow volume by Ar cluster irradiation enhances the surface contaminants species in the mass spectra, which should significantly improve the surface sensitivity of SIMS.

ACKNOWLEDGMENT

The author would like to thank colleagues who offered supported for this review, especially Kousuke Moritani and Norio Inui who provided helpful discussions and input.

REFERENCES

[1] Vickerman CJ, Briggs D (2001) ToF-SIMS; Surface analysis by mass spectrometry: *IM Publications,* UK.
[2] Benninghoven A, Sichtermann, KW (1978) *Annal. Chem.* 50: 1180-1184.
[3] Barber M, Bordoil SR, Sedgwick DR, Tyler NA (1981) *Nature* 293: 270-275.
[4] Gillen G, Fahey A (2003) *Appl. Surf. Sci.* 203/204: 209-213.
[5] Wong CCS, et al. (2003) *Appl. Surf. Sci.* 203/204: 219-222.
[6] Davies N, et al. (2003) *Appl. Surf. Sci.* 203/204: 223-227.
[7] Brunelle A, et al. (2001) *Phys. Rev. A* 63: 022902.

[8] Toyoda N, et al. (2003) *Appl. Surf. Sci.* 203/204: 214-218.
[9] Wucher A (2006) *Appl. Surf. Sci.* 252, 6482-6489.
[10] Yamada I, et al. (2000) *Nucl. Instr. and Phys. Res. B* 164/165: 944-959.
[11] Toyoda N, Matsuo J, Yamada I (2004) *Nucl. Instr. and Phys. Res. B* 216: 379-389.
[12] Toyoda N, Houzumi S, Yamada, I (2006) *Nucl. Instr. and Phys. Res. B* 242: 466-468.
[13] Moritani K, et al. (2008) *J. Phys. Chem.* C 112: 11357-11362.
[14] Houzumi S, Mochiji K, Toyoda N, Yamada I (2005) *Jpn. J. Appl. Phys.*44: 6252-6254.
[15] Moritani K, et al. (2008) *Appl. Surf. Sci.* 255: 948-950.
[16] Mochiji K, Hahinokuchi M, Moritani K, Toyoda N (2009) *Rapid Commun. Mass Spectrom.* 23: 648-652.
[17] Vlasak RP, Beussman JD, Davenport RM, Enke GC (1996) *Rev. Sci. Instr.* 67: 68-72.
[18] Mochiji K, Yamamoto S, Shimizu H, Ohtani S, Seguchi T, Kobayashi N (1997) *J. Appl. Phys.* 82: 6037-6040.
[19] Kang H, et al, (1992) *Nucl. Instrum. Meth. Phys. Res. B*, 67: 312-315.
[20] Pregler SK, Hayakawa T, Yasumoto H, Kondow T, Sinnott SB (2007) *Nucl. Instrum. Meth. Phys. Res. B*, 262: 240-248.
[21] Bräuchle G, et al. (1996) *Nucl. Instrum. Meth. Phys. Res. B*, 112: 105-108.
[22] Bräuchle G, et al. (1995) *Appl. Phys. Lett*, 67: 52- 54.
[23] Tsukuda M, Yokodera Y, Moritani K, Inui N, Mochiji K (2012) *e-J Surf. Sci. Nanotech.* 10: 88-91.
[24] Moritani K, Hozumi S, Takeshima K, Toyoda N, Mochiji K (2008) *J. Phys. Chem. C*, 112: 11357-11362.
[25] Hahn JR, Kang H (2000) *Surf. Sci.* 446: L77-L82.
[26] Lee SM, Lee YH, Hwang YG, Hahn JR, Kang H (1999) *Phys. Rev. Lett.* 82: 217-220.
[27] Lange W, Jirikowsky M, Benninghoven A (1984) *Surf. Sci.* 136: 419-436.
[28] Delcorte A, Bertrand P (1998) *Surf. Sci.* 412/413: 97-124.
[29] Oshima S, Kashihara I, Moritani K, Inui N, Mochiji K (2011) *Rapid Commun. Mass Spectrom.* 25: 1070-1074.
[30] Inui N, Mochiji K, Moritani K (2008) *Nanotechnology* 19: 505501
[31] Tanaka M, et al. (2010) *Rapid Commun. Mass Spectrom.* 24: 1405-1410.

In: Argon: Production, Characteristics and Applications
Editors: B.N. Sismanoglu, H.S. Maciel et al.
ISBN: 978-1-62618-204-2

Chapter 7

ASPECTS OF MULTI CHARGED ARGON BEAM TRANSPORT

P. Beličev*[1,*]*, R. Balvanović*[1]*, V. Alexandrov*[2]*,
***N. Kazarinov*[2]* and B. Radjenović*[3]

[1]Vinca Institute of Nuclear Sciences, University of Belgrade, Belgrade, Serbia
[2]Joint Institute for Nuclear Research, Dubna, Russia
[3]Institute of Physics, University of Belgrade, Belgrade, Serbia

ABSTRACT

Some aspects of the multi charged argon beam transport are considered. The basics of the ion beam physics and paraxial ion beam optic is presented. Short survey of magnetic ion-optical elements and their transfer matrices is given. The basic properties of the analyzing section consisting of a magnetic quadrupole lens and a bidirectional analyzing magnet are described. Results of beam transport calculations along the beam transport line from the ion source up to the experimental chamber are considered. The beam transport along the analyzing section, where all ion species with different charge states are present, is subject to pronounced space charge effect. In this region the beam transport calculations are performed by using both PIC and the distribution function moments methods, due to their ability to cope with this effect. The results are presented in terms of individual ion trajectories, beam envelopes and transmission coefficients. The results of the charge state analysis by a magnetic analyzer of the argon ions, produced by ECR type ion source, are presented in terms of beam current distribution over the different ion species. After selecting the particular charge state of the argon ion at the position of the analyzing slits, the beam propagates along the transporting section of the beam lines, consisting of solenoid and quadrupole lenses, as well as bending magnets. For reason of simplicity and fast execution, the beam transport along this section was performed using the first order transfer matrix method. As an illustration of the matching function of the beam line, an example of argon ion beam axial injection into a cyclotron is presented.

---

* E-mail address: belicev@vin.bg.ac.rs.

1. Introduction

The techniques used in argon ion beams transport are very much alike those used for other charged particle beams. The guiding of the ions is achieved by subjecting them to external electromagnetic forces, being realized via ion-optical elements, similar to the optical lenses used in the transport of light. Unlike the electrostatic force, the magnetic force exerted on a moving charged particle depends on its velocity so it is reasonable to use magnetic ion-optical elements for highly energetic ions. Since argon can be ionized to a very high degree, even with moderate extraction (acceleration) voltages it can achieve energies of the order of a few hundreds of keV. In the following emphasis will be put on magnetic elements.

Usually the beam is transported from the ion source, where it is extracted from the source plasma chamber, up to the position where one can use it in different experimental set ups or inject it into an accelerator. In the former case the requirements imposed on the beam transport usually concern the minimization of the beam losses along the transport line and of the dimensions of the beam spot at the specific points, according to the needs of the experiment. In the case of injecting the beam into an accelerator, matching of the beam emittance to the accelerator acceptance is needed, which are requirements imposed on the beam properties in the phase space. The latter requirements are more difficult to be met.

In order to fulfill the requirements imposed to the beam, a typical ion beam transport line contains a combination of the following functions:

- analyzing
- transporting
- matching

All these functions are implemented via corresponding sections of the beam transport line. The basic properties of the analyzing and transporting functions will be illustrated on the example shown in Figure 1 representing a part of the FAMA facility within the TESLA Accelerator Installation [1, 2], with two ion sources M1 and M2, and two interaction chambers IC1 and IC2. The matching function of the transport line will be illustrated via the example of an injection line into a cyclotron, as shown in Figure 2.

The analyzing function of the transport line is performed after the beam extraction from the ion source. The extracted argon beam from an ion source of ECR type e.g. contains variety of argon ion species ${}^{A}\mathrm{Ar}^{k+}$, where A is the atomic mass number and k is the degree of ionization. In its natural state, argon has three stabile isotopes with $A = 36$, 38 and 40, the last one being with highest abundance of 99.6%. The value of k goes from 1 to 18 (fully stripped nucleus). Furthermore, the beam may contain ion species of other elements, oxygen e.g., since very often it is used as support gas to increase the current of the ions with particular degree of ionization [3, 4]. The purpose of the analyzing section of the line is to select specific ion species from the ion population in the beam and prepare them for further transport. This analysis can be achieved due to the dispersive properties of the magnetic/electric fields when they are applied perpendicular to the ion trajectories.

The most important ion-optical element in this section is the analyzing magnet, representing a sector dipole magnet with homogeneous or non homogeneous field, directed perpendicularly to the propagation of the ion beam. In addition to this a pair of vertical slits at the object and image points of the system are needed. By choosing appropriate intensity of the magnetic field in the analyzing magnet, and appropriate openings of the slits, one can separate the desired ion species for further usage. Strictly speaking, by an analyzing magnet the ions can be separated with respect to their mechanical momentum per unit charge (p/q) – also known as magnetic rigidity, while the use of electrostatic separator (a cylindrical condenser e.g.) allows separation with respect to ion's double kinetic energy per unit charge ($2E_k/q$) - also known as electrostatic rigidity [5-7]. A specific version of the analyzing section, realized with a bidirectional constant field analyzing magnet AM1, and a magnetic quadrupole lens QM, is shown in Figure 1.

As mentioned before, the main role of the section performing the transporting function is to provide transport of the analyzed beam between two positions with as high as possible transmission, and control of the beam spot dimensions at particular points along the beam transport line – usually at the target and diagnostic boxes.

In order to obtain high beam transmission one should reduce the interactions of the ions with the vacuum tubes, chambers and apertures along the beam transport lines to a degree as low as possible. This is achieved by restricting the beam envelope (transversal beam dimensions) to about 80% of the beam transport line clearance. Speaking about providing high beam transmission, one should also consider the effect of ion interactions with the residual gas within the beam line. This is particularly important in the case of highly charged argon ions, where the interaction cross section for ion charge exchange is very high. This aspect suggests that the transport of highly charged argon ions should be performed via relatively short lines, otherwise vacuum levels of 10^{-8} mbar or better are necessary, which considerably increase the cost of the vacuum equipment.

A variety of ion-optical elements capable of focusing the beam are used for this purpose. Only two of these elements, very often used in transport lines, will be considered here: the quadrupole lens and the solenoid lens.

The quadrupole lens is an element consisting of two pairs magnetic dipoles, rotated for 90^0 with respect to each other. A transverse cross section of a possible realization is given in Fig. 3 b). The lens exerts force on the ion due to the presence of field gradient in the plane transverse to the beam propagation. This is known as the strong focusing principle, being invented with the advent of AVF cyclotrons [8]. Theoretically, the gradient of the magnetic field is constant around longitudinal axis of the lens. It is worth mentioning that the quadrupole lens shows inherently both focusing and defocusing action: while the ions are focused in the horizontal plane e.g., in the vertical plane they are simultaneously defocused. Due to this property this lens is seldom used as a single one. In order to obtain focusing effect in both planes they are gathered in groups of 2 and 3 – called doublets and triplets, respectively [9].

The solenoid lens focuses the ions by creating a field whose direction is nearly coaxial with the solenoid axis. Depending on the specific application, these lenses can be long or short, with or without iron yoke. An illustration of possible construction is depicted in Figure 3 a). Strictly speaking, focusing is not an appropriate word to be attributed to this lens. As a matter of fact, due to the relatively small angle between the ion velocity and the vector of the magnetic field the ions have spiral-like trajectories around the lines of the magnetic field. Only at the entrance/exit of the solenoid the lens exerts stronger focusing action due to diverging of the magnetic field lines at these positions. According to this, the solenoid lens focuses the beam only at its edges, while its action between them can be described as confining. As a consequence, short solenoids are used for beam focusing, and the long ones for guiding the beam along longer distances without considerable change in its transverse dimensions.

Comparing quadrupole to solenoid lenses one can say that the advantages of the former is lower consumption of electric power since they work at lower magnitude of the magnetic field. On the other hand, the solenoid lens preserves the axial symmetry of the beam which is quite cumbersome task for the quadrupole lenses, especially in their doublet configuration. In the case when the transport line should change its direction, it is done by use of sector dipole magnets, when they perform only bending but not analyzing action [10].

Figure 1 shows one possible realization of a transporting section consisting of solenoid lenses, quadrupole lenses, analyzing and bending magnets. The calculations of the argon beam transport through this section will be discussed later.

The matching function of the beam transport line concerns the properties of the beam represented in the phase space [11-13]. Depending on the type of the accelerator and its particular construction, the injected external beam should satisfy a set of requirements in order to be successfully accelerated. Usually the ions of the beam should occupy defined part of the phase space, possessing particular distribution. In general this requirements concern six dimensional (6D) hyperspace while in reality they are reduced to 4D or even to a set of 2D phase space planes. Common matching ion-optical elements used for this purpose are quadrupole and solenoid lenses. In theory, the number of the used elements is proportional to the number of the requirements imposed to the beam parameters, while in practice (due to the limited space and budget) the number is often minimized. An illustration of the matching beam transport line is given in Figure 2. The line consists of an ion source IS producing multiply charged heavy ions, which are analyzed by AM analyzing magnet and injected into a cyclotron. The matching elements are solenoids S1, S2 and S3, and quadrupole lenses Q1 and Q3. The calculations of the argon beam transport through this section will be discussed later.

2. Basics of Ion Beam Transport

Not very strict definition of ion beam will be that it is a collection of ions all moving in nearly same general direction with nearly same momentum, and having small transverse separation. The physical processes within the beam are due to mutual interactions between the ions as well as with the external electromagnetic fields. The simplest model of an ion beam is the one with mutually non interacting particles moving in external electromagnetic fields with non relativistic velocities, propagating in vacuum. More complex models are those describing high current relativistic beams propagating in plasma or gas [14, 15]. In the following the basics of the simplest model will be reviewed.

The beam at any moment can be represented by a set of points in 6D phase space. The coordinates in this space are three positions: x, y and z, and three momenta p_x, p_y and p_z. Roughly in the center of the occupied 6D region a so called reference point is chosen, representing the beam reference (or central) particle with its reference position vector R_0 and reference momentum vector p_0. The trajectory $R_0(t)$ of the reference particle is referred to as reference one.

It is common practice to describe the beam by a moving coordinate system usually referred to as optical coordinate system. It is bound to the reference trajectory of the beam with x, y and z axes in the directions of the normal, tangent and bi-normal unit vectors of that trajectory. For the sake of convenience, whenever it is possible, x and y is chosen to be horizontal and vertical, respectively. In the field of beam transport it is accepted that particles are defined by six coordinates: x, x', y, y', l and δ where $x'=atan(p_x/p_z)\approx p_x/p_z$, and $y'=atan(p_y/p_z)\approx p_y/p_z$, with p_x, p_y and p_z being the particle's momentum components. The l coordinate is the path length difference between an arbitrary and the reference particle in the beam, and $\delta=\Delta p/p_0$ is the fractional momentum deviation of an arbitrary particle from the reference one [5]. Although the angular variables x' and y' are not canonical in Hamiltonian sense, they are widely adopted due to their simplicity and possibility to be experimentally determined [13, 15].

The equations that govern beam evolution in the phase space can be easily linearized leading to the well-known paraxial ion optics formalism [5, 7]. This formalism allows the beam in the phase space to be represented by a 6D hyper-ellipsoid. According to the Liouville's theorem [12], the hyper-volume of the 6D beam hyper-ellipsoid remains constant. Moreover, the areas of the 2D projections (ellipses) of the 6D hyper-ellipsoid on the x-x', y-y' and l-δ coordinate planes are also conserved for conservative systems. The ellipses' areas divided by π are referred to as beam 2D emittances, and are considered to be the indicators of the beam quality.

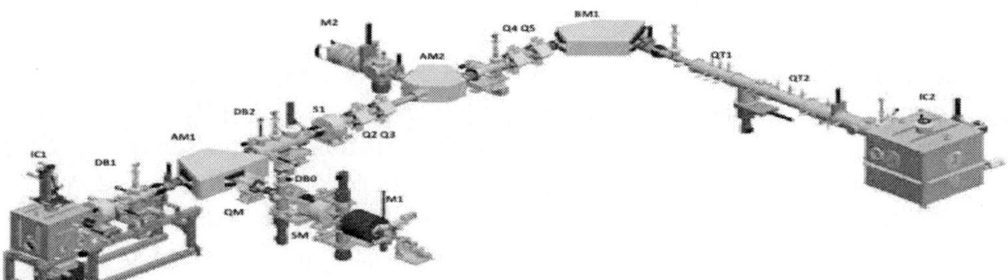

Figure 1. Illustration of an ion beam transport line with analyzing and transporting sections. The analyzing section between diagnostic boxes DB0 and DB1/DB2, consists of a bidirectional analyzing magnet AM1 and quadrupole lens QM. The object defining slits are situated into DB0 box, while the analyzing slit is in DB1/DB2 box. The transporting section starts at DB2 box and goes up to the target point inside IC2 experimental chamber, containing quadrupole and solenoid lenses and bending magnets.

The 6D beam ellipsoid is defined by the matrix equation $\{q_i\}\sigma^{-1}\{q_i\}^T=1$, where $\{q_i\}=[x,x',y,y',l,\delta]^T$ is phase space position vector of the particle within the beam (superscript ^{T} stands for matrix transposition), and σ is a 6×6 real, positive definite, symmetric matrix, often called beam sigma matrix. The hyper-volume of the beam ellipsoid is given by $(3\pi/4!)det[\sigma]$, imposing $det[\sigma]>0$. The diagonal elements σ_{ii} define the extensions of the hyper-ellipsoid in the direction of the coordinate axes as $q^2_{i\_max}=\sigma_{ii}$, while the correlation coefficients $r_{ij}=\sigma_{ij}/(\sigma_{ii}\sigma_{jj})^{1/2}$ define its orientation. As for the 2D projections of the beam ellipsoid, the ellipses are given by $\sigma_{ii}q_j^2-2\sigma_{ij}q_iq_j+\sigma_{jj}q_i^2=1$, with the areas $A_{ij}=\pi\varepsilon_{ij}=\pi(\sigma_{ii}\sigma_{jj}-\sigma_{ij}^2)^{1/2}$, where ε_{ij} represents corresponding 2D beam emittance. Very often for the x-x', y-y' phase planes, the following alternative presentation of the phase space ellipses is used:

$\gamma_q q^2+2\alpha_q qq'+\beta_q q'^2=\varepsilon_q$, where q stands for either x or y. Coefficients α_q, β_q and γ_q are called Courant-Snyder or Twiss parameters, and they are related with the elements of the beam sigma matrix [5, 7].

2.1. Transfer Matrix Formalism

Within the framework of paraxial ion optics formalism a 6×6 transfer matrix has been associated to each ion-optical element possessing mid-plane field symmetry, of the following general form [5, 7]:

$$\mathbf{R} = \begin{bmatrix} R_{11} & R_{12} & 0 & 0 & 0 & R_{16} \\ R_{21} & R_{22} & 0 & 0 & 0 & R_{26} \\ 0 & 0 & R_{33} & R_{34} & 0 & 0 \\ 0 & 0 & R_{43} & R_{44} & 0 & 0 \\ R_{51} & R_{52} & 0 & 0 & 1 & R_{56} \\ 0 & 0 & 0 & 0 & 0 & 1 \end{bmatrix} \tag{1}$$

According to this matrix formalism, the phase space coordinates of the ion $\{q_i\}_1$ after passing through the ion-optical element are simply calculated as $\{q_i\}_1=R\{q_i\}_0$, where $\{q_i\}_0$ are the ion's coordinate at the entrance of the element. This relation is easily extended to a line of N ion-optical elements with corresponding transfer matrices R_k to the following form $\{q_i\}_1= R_N R_{N\text{-}1}\ldots R_1\{q_i\}_0$. In the cases when the particle trajectories in the horizontal (x-z) and vertical (y-z) plane of an ion-optical non-dispersive element are decoupled, the R matrix reduces to two 2×2 matrices, and the particle coordinates at the exit of the element are given by:

$$\begin{bmatrix} x_1 \\ x_1' \end{bmatrix} = \begin{bmatrix} R_{11} & R_{12} \\ R_{21} & R_{22} \end{bmatrix} \begin{bmatrix} x_0 \\ x_0' \end{bmatrix}; \quad \begin{bmatrix} y_1 \\ y_1' \end{bmatrix} = \begin{bmatrix} R_{33} & R_{34} \\ R_{43} & R_{44} \end{bmatrix} \begin{bmatrix} y_0 \\ y_0' \end{bmatrix} \tag{2}$$

Important property of the transfer matrix is that its determinant is equal to one. It is important to note that this matrix formalism can be applied not only to describe the motion of an individual ion but also of the beam as a whole. Knowing the beam σ matrix at the element entrance, its transformation at the exit is given by $\sigma_1=R\sigma_0R^T$. This equation allows calculation of the beam envelopes along the beam line.

In the following the transfer matrices for some basic ion-optical elements are given.

Drift Space

This represents a limited space in vacuum with length L, free of electromagnetic fields. The ion trajectories are straight lines, and the transfer matrix, being identical in both planes, is given by:

$$\mathbf{R}_{x,y} = \begin{bmatrix} 1 & L \\ 0 & 1 \end{bmatrix} \quad (3)$$

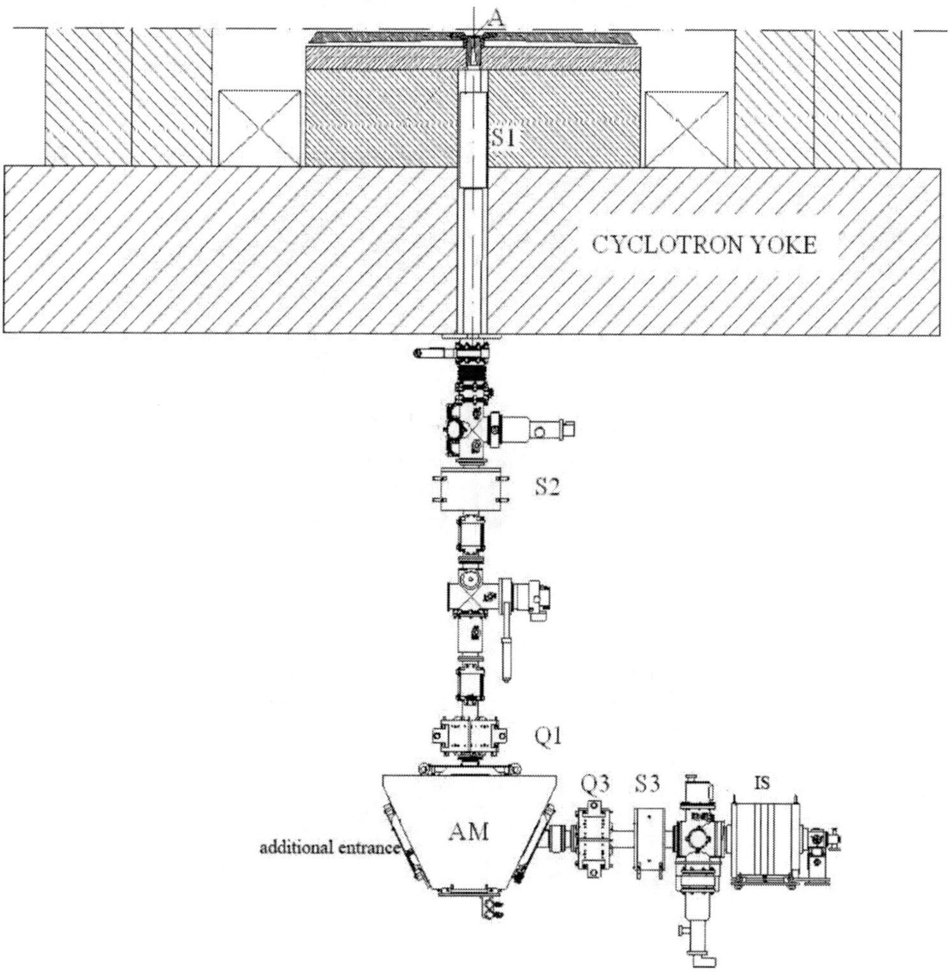

Figure 2. Layout of the injection section of a transport line: A – inflector entrance; IS – ECR type ion source; S1, S2, S3 – solenoid lenses, Q1, Q3 – quadrupole lenses; AM – bidirectional analyzing magnet.

Quadrupole Lens

$$\mathbf{R}_{qF} = \begin{bmatrix} \cos(k_q L) & \frac{1}{k_q}\sin(k_q L) \\ -k_q \sin(k_q L) & \cos(k_q L) \end{bmatrix}; \quad \mathbf{R}_{qD} = \begin{bmatrix} \cosh(k_q L) & \frac{1}{k_q}\sinh(k_q L) \\ k_q \sinh(k_q L) & \cosh(k_q L) \end{bmatrix} \quad (4)$$

where R_{qF} and R_{qD} stand for the focusing and defocusing plane of the quadrupole lens, respectively. Parameter k_q is defined as $k_q^2 = G/(B\rho)$ where, G is the magnetic induction gradient in the lens and $(B\rho)$ is the magnetic rigidity of the ion. The length of the reference trajectory through the lens is equal to L [5, 7].

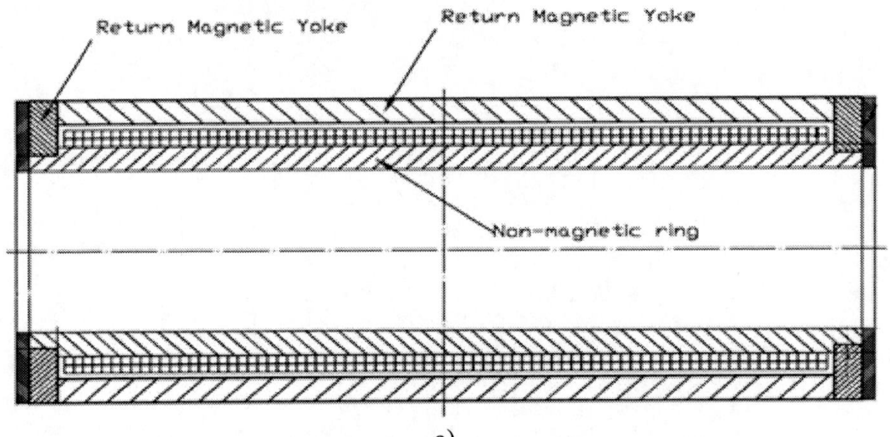

a)

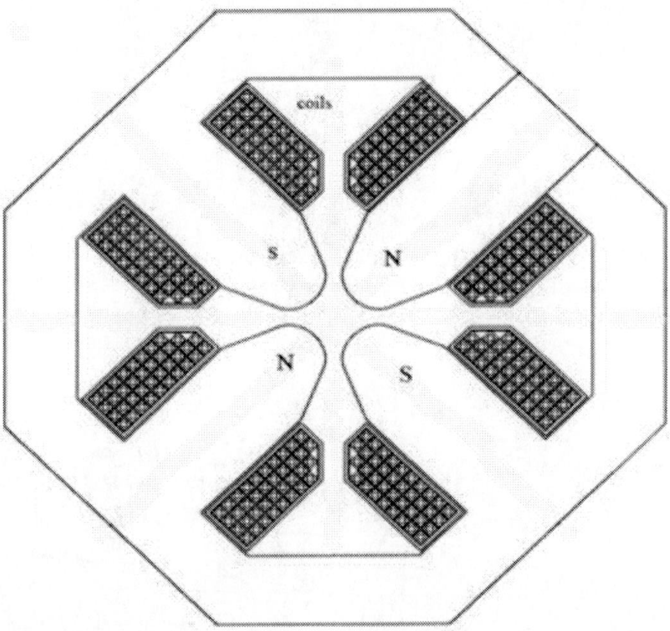

b)

Figure 3. a) Axial cross section of a solenoid lens, b) transverse cross section of a magnetic quadrupole lens.

Solenoid Lens

$$\mathbf{R}_{sol} = \begin{bmatrix} \cos^2(k_s L) & \frac{1}{k_s}\sin(k_s L)\cos(k_s L) & \sin(k_s L)\cos(k_s L) & \frac{1}{k_s}\sin^2(k_s L) \\ -k_s \sin(k_s L)\cos(k_s L) & \cos^2(k_s L) & -k_s \sin^2(k_s L) & \sin(k_s L)\cos(k_s L) \\ -\sin(k_s L)\cos(k_s L) & -\frac{1}{k_s}\sin^2(k_s L) & \cos^2(k_s L) & \frac{1}{k_s}\sin(k_s L)\cos(k_s L) \\ k_s \sin^2(k_s L) & -\sin(k_s L)\cos(k_s L) & -k_s \sin(k_s L)\cos(k_s L) & \cos^2(k_s L) \end{bmatrix} \quad (5)$$

The dimension of the solenoid transfer matrix R_{sol} is 4×4 due to the coupled motion of the ions in the transverse phase space. Parameter k_s is defined as $k_s = B_0/(2\ B\rho)$, where B_0 is the magnetic induction on the solenoid axis, and $(B\rho)$ is the magnetic rigidity of the ion. The length of the reference trajectory through the lens is equal to L [5, 7].

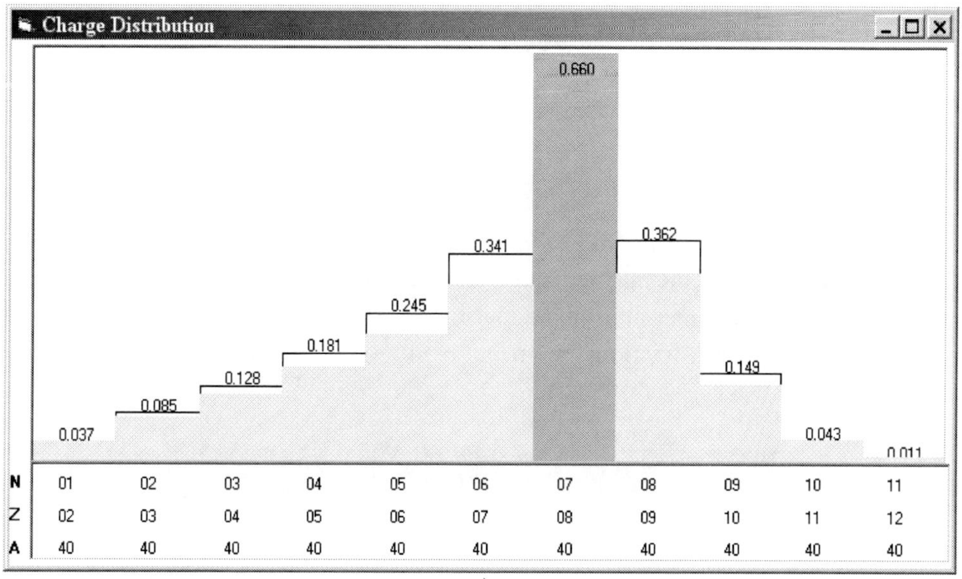

a)

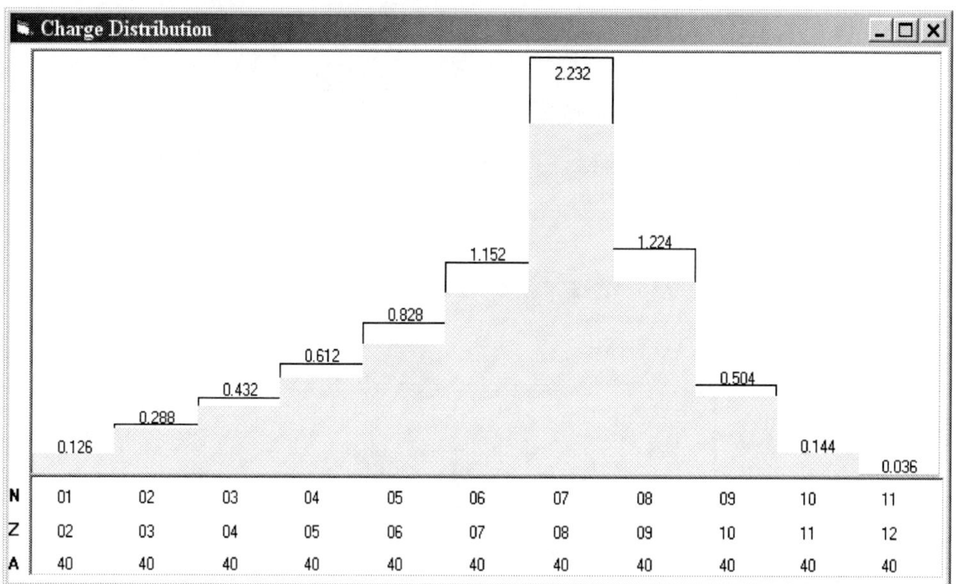

b)

Figure 4. Adopted ion current distributions for extracted argon ion species from the ion source optimized for delivering a) 200 keV $^{40}Ar^{8+}$, and b) 300 keV $^{40}Ar^{12+}$ as reference ions. The figures at the top of the bins indicate the currents for different ion species, in mA. Z and A stand for the ion's degree of ionization and mass, respectively.

Sector Dipole Magnet with Homogenous Field

This element is important due to its analyzing and bending properties.

$$\mathbf{R}_b = \begin{bmatrix} R_{11} & R_{12} & R_{16} \\ R_{21} & R_{22} & R_{26} \\ R_{61} & R_{62} & R_{66} \end{bmatrix} = \begin{bmatrix} \cos\theta & R\sin\theta & R(1-\cos\theta) \\ -\dfrac{1}{R}\sin\theta & \cos\theta & \sin\theta \\ 0 & 0 & 1 \end{bmatrix}$$
$$\mathbf{R}_{nb} = \begin{bmatrix} R_{33} & R_{34} & R_{36} \\ R_{43} & R_{44} & R_{46} \\ R_{61} & R_{62} & R_{66} \end{bmatrix} = \begin{bmatrix} 1 & R\theta & 0 \\ 0 & 1 & 0 \\ 0 & 0 & 1 \end{bmatrix} \tag{6}$$

where indices b and nb stand for the bending and non-bending plane, and R and θ are the bending radius and angle of the magnet. Due to its dispersive properties, expressed by R_{16} and R_{26} matrix elements, the dimension of the matrix is 3×3. It is easy to see that this element separates ions with different fractional momenta $\delta = \Delta p/p_0$ only in the bending plane, while in the non-bending one it acts as a drift space.

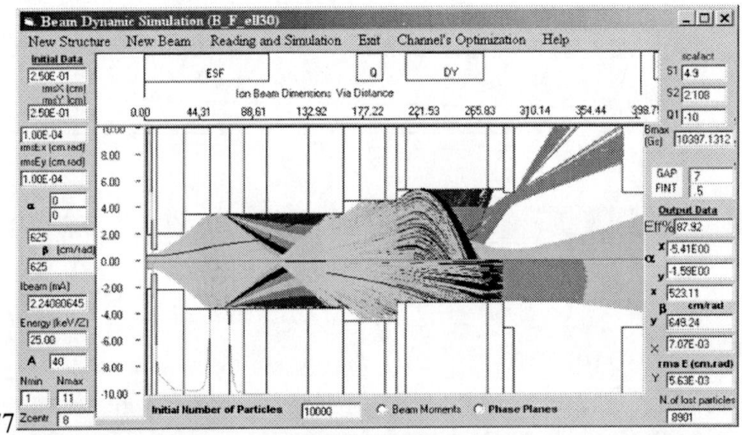

77

a)

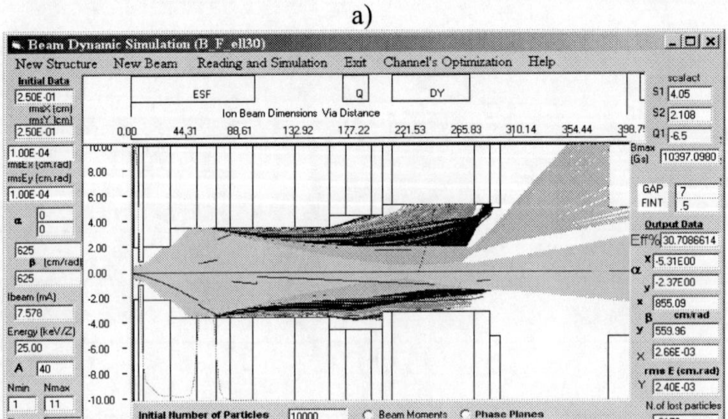

b)

Figure 5. Ion trajectories of all argon ion species along the analyzing section with parameters set for a) 200 keV $^{40}Ar^{8+}$, and b) 300 keV $^{40}Ar^{12+}$ as reference ions. Results obtained using PIC method.

The above matrices are valid for sector magnet with no pole face rotation angles, meaning that the reference trajectory of the beam enters/exits perpendicularly to the magnet poles. In the case when the pole faces are rotated, the edges of the magnet simultaneously focuse/defocuse the beam in the bending/nonbending plane, depending on the sign of the pole face rotation angle. This effect is modeled as if a thin lens was positioned at the magnet edge, with the following transfer matrices [5, 7]:

$$\mathbf{R}_b = \begin{bmatrix} 1 & 0 & 0 \\ \dfrac{\tan u}{R} & 1 & 0 \\ 0 & 0 & 1 \end{bmatrix} \quad \mathbf{R}_{nb} = \begin{bmatrix} 1 & 0 & 0 \\ -\dfrac{\tan u}{R} & 1 & 0 \\ 0 & 0 & 1 \end{bmatrix} \tag{7}$$

where R is the bending radius of the magnet, and u the pole face rotation angle that can take positive or negative values. In order to obtain the full transfer matrix, corresponding matrix multiplication for the exit edge, sector magnet and entrance edge should be performed.

The shortly reviewed method of paraxial ion optics presents efficient tool for fast and efficient analysis of the ion beam transport in wide range of applications. However, one should have in mind that this formalism deals with the rather simple beam model which is not adequate for treating intense beams with pronounced space charge effect [9, 11]. In this case more complex methods, such as particle-to-particle, particle-in-cell (PIC) and the distribution function moments (in the further text will be referred shortly as moments) methods e.g. should be used [14, 16-18]. Unlike the matrix formalism method, which in some simpler cases allows computation "by hand", implementation of these methods calls for serious computational power. In the following, beside the matrix formalism, results obtained by using PIC and moments methods will be presented.

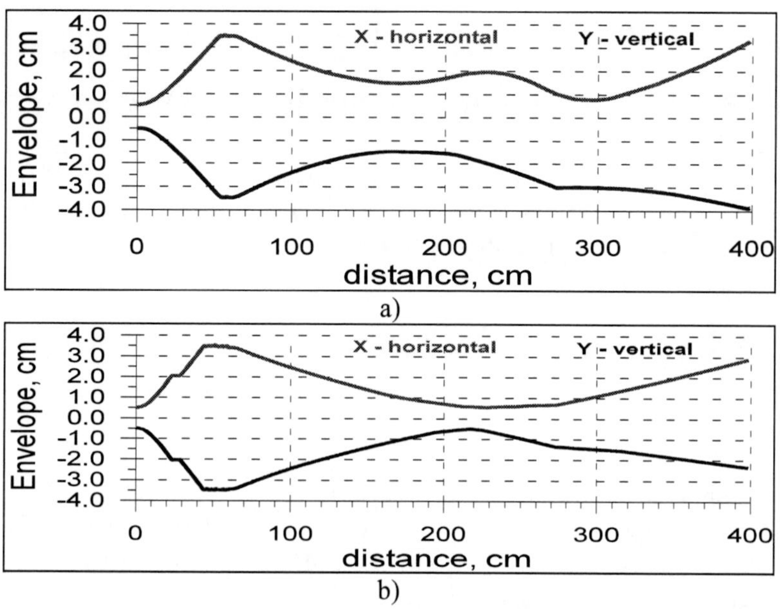

Figure 6. RMS envelopes along the analyzing section for a) 200 keV $^{40}Ar^{8+}$, and b) 300 keV $^{40}Ar^{12+}$ beam. Results obtained using moments method.

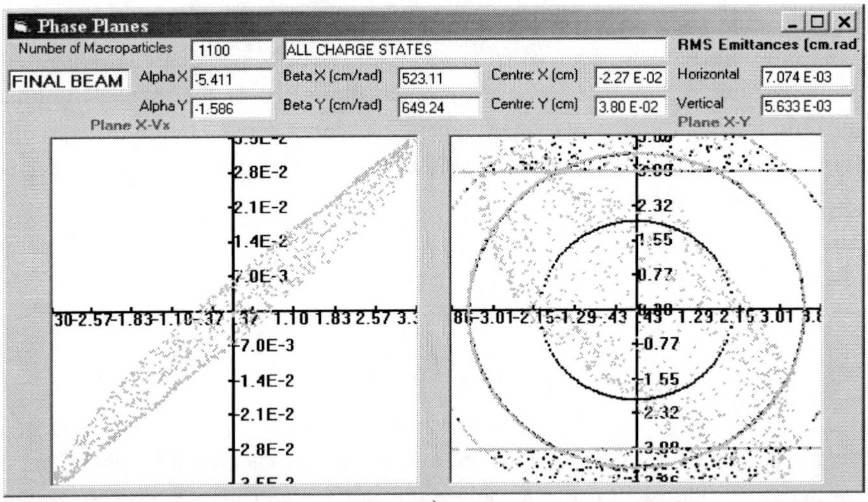

a)

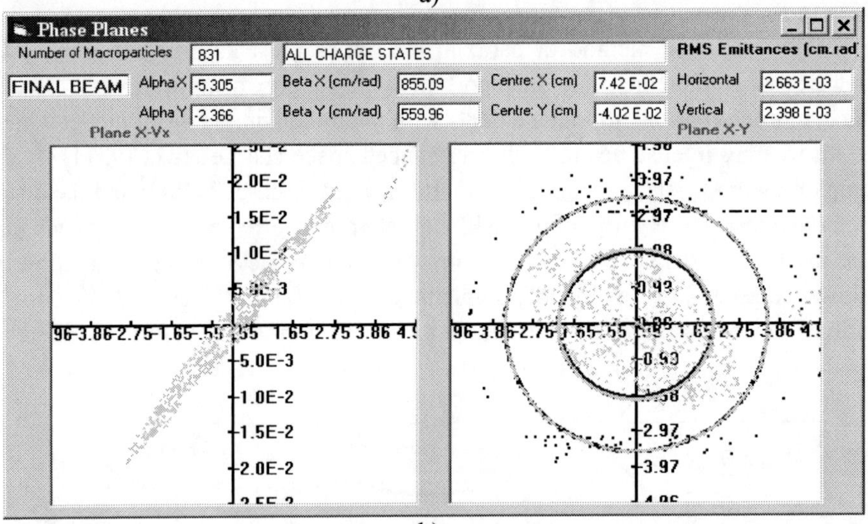

b)

Figure 7. 2D phase space portraits at the end of the analyzing section for a) 200 keV $^{40}Ar^{8+}$, and b) 300 keV $^{40}Ar^{12+}$ beam. The circular shapes in the x-y plane diagram represent the ions lost in interaction with the walls of the transport line tubes. Results obtained using PIC method.

3. Examples of Argon Ion Beam Transport

3.1. Ion Beam Analysis

Analysis of the argon ion beam and separation of ion species according to the degree of ionization is the first step to be done after its extraction form the ion source. The analyzing section starts immediately after the solenoid lens SM and goes up to the diagnostic boxes DB1 or DB2 equipped with vertical slits, as shown in Figure 1. The section consists of horizontally defocusing quadrupole lens QM, bi-directional 90^0 analyzing magnet AM1 and number of drift spaces between the ion-optical elements. Within the frame of the paraxial ion optics, this section is defined by its transfer matrix obtained by the corresponding matrices of the containing elements. The analyzing mechanism can be explained starting from the explicit expression for the ion coordinate in the bending plane: $x_1=R_{11}x_0+R_{12}x'_0+R_{16}\delta$, where indices 0 and 1 stand for the entrance and exit point of the analyzing section. In order to separate ions with different relative momenta δ, the section should provide the so called object to image mapping in the bending plane. Mathematical expression of this statement, for ions entering the section at its central point $(x_0=0)$, is $x_1=R_{16}\delta$ with restriction of $R_{12}=0$, which a well known requirement for the point-to-point imaging in the bending plane [5, 7]. The matrix #element R_{16} is known as dispersion coefficient D, defining the resolution of the system and is proportional to the bending radius of the analyzing magnet. Higher dispersion coefficient values provide larger separation of the ions with different relative momenta. The condition $R_{12}=0$ allows variation of the free parameters in the line, which are the magnetic field in the quadrupole lens QM and the lengths of the drift spaces. The magnetic field in the analyzing magnet AM1 depends on the argon ion magnetic rigidity $(B\rho)$ is equal to $B=(B\rho)/R$, with R being the bending radius of the magnet.

Although the paraxial ion optics gives sound explanation of the analyzing mechanism, it is not adequate enough for practical transport calculations. This is due to the fact that the beam propagating along this section contains different ion species and has relatively high current (a few mA e.g.) which leads to pronounced space charge effect [19].

In the following, the results of the argon beam transport along the analyzing section, using PIC and moments methods are given.

The adopted initial parameters of the argon beam extracted from the M1 ion source (ECR type) are as follows: emittance $\varepsilon_x=\varepsilon_y=250(\pi\times\text{mm}\times\text{mrad})$, maximal dimensions $x_m=y_m=2.5\text{mm}$, ellipse orientation in phase space $\alpha_x=\alpha_y=0$ (double waist). Usually Gaussian or Kapchinskij-Vladimirskij distribution of the particles is assumed and adequately sampled [20, 21]. In the simulation 10^4 macro-particle histories were used.

The case of two reference argon ion species, 200 keV $^{40}Ar^{8+}$ and 300 keV $^{40}Ar^{12+}$, is considered. This means that the free parameters of the analyzing section elements, as well as those of the ion source, were chosen to be optimized for these ions. The distributions of the argon ion species with respect to their ionization degree in the extracted beam are given in Figures 4 a) and b). These distributions can be obtained ether by simulation of the beam extraction from the ECR type ion source, or by measurement of the ion beam currents after its analysis. The first approach presents a problem by itself [22, 23], while the second is not very reliable since the distribution after the beam analysis is not necessarily identical to the one immediately at the source extraction electrodes. The difference between the two distributions is due to the fact that the ion source parameters are optimized for maximizing the current of the desired reference ion (compare 11 μA to the optimized 36 μA for $^{40}Ar^{12+}$). Using these starting distributions, the trajectories of all ion species are calculated, using PIC method. The results are shown in Figures 5 a) and b). It is visible that all ion species but the referent ones are lost by hitting the walls of the transport line. Even for the reference ions some losses are present due to insufficient focusing power of the solenoid lens SM. For the sake of comparison, the calculations have been performed by using the moments method too. As a result the root mean square (RMS) envelope of the beam is obtained. The results shown in Figures 6 a) and b) are in good agreement with those obtained by PIC method. In Figures 7 a) and b) the phase portrait in the horizontal plane as well as the profile of the beam at the end of the analyzing section are given, obtained also by PIC method. These results can be used as a starting point for further beam transport calculations using some other, simpler methods, such as paraxial optics e.g. Although the elements of the analyzing section are set to provide optimal conditions for the referent ions, full transmission of the beam seldom can be achieved. In Figures 8 a) and b) the transmissions of all ion species along the section are shown. It can be seen that all non-reference ions are lost, while the transmission of the reference ones varies from about 30% to 90%. The reason for this is the fact that the system shown in Figure 1 is designed for production and transport of very wide range of ions with respect to their masses, charges and energies. In that situation it is almost impossible to achieve maximal performance in the whole range.

In Figure 9 the values of the currents of all ion species, separated by the analyzing section, are shown. The currents are measured by a Faraday cup situated in the DB1 diagnostic box (Figure1), immediately after the analyzing slits. One should note that this distribution can be optimized for particular ion species by changing the ion source parameter such as the magnetic fieled in its plasma confining area, the injected microwave power and the flow of the supporting gas (usually oxygen).

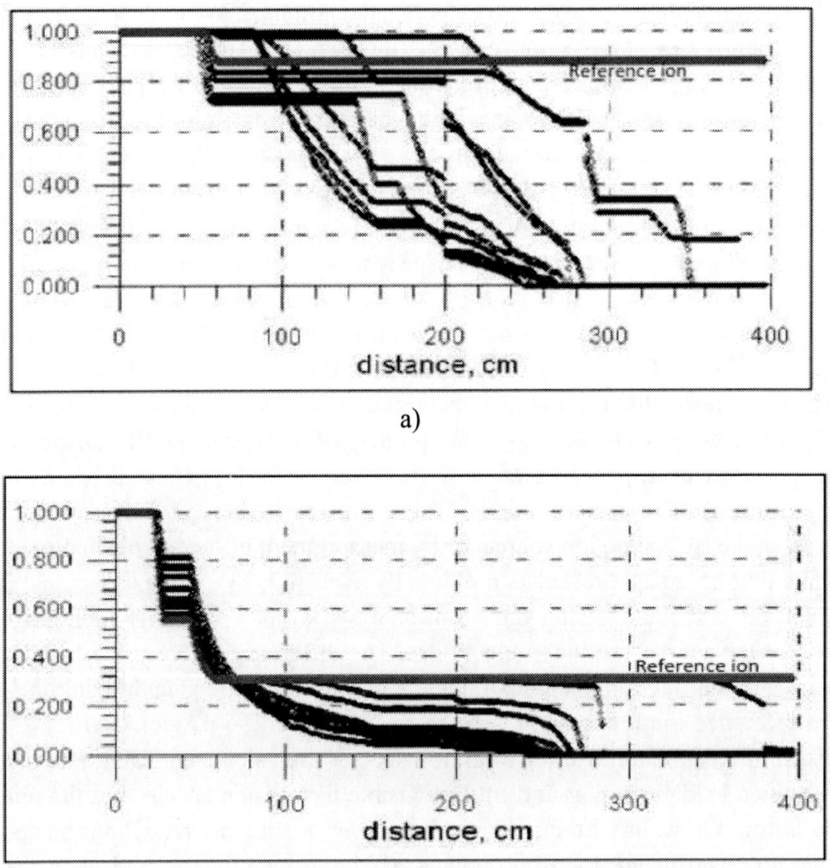

a)

b)

Figure 8. Transmission of all argon ion species along the analyzing section with parameters set for a) 200 keV $^{40}Ar8^{+}$, and b) 300 keV $^{40}Ar^{12+}$ as reference ions. Results obtained using PIC method.

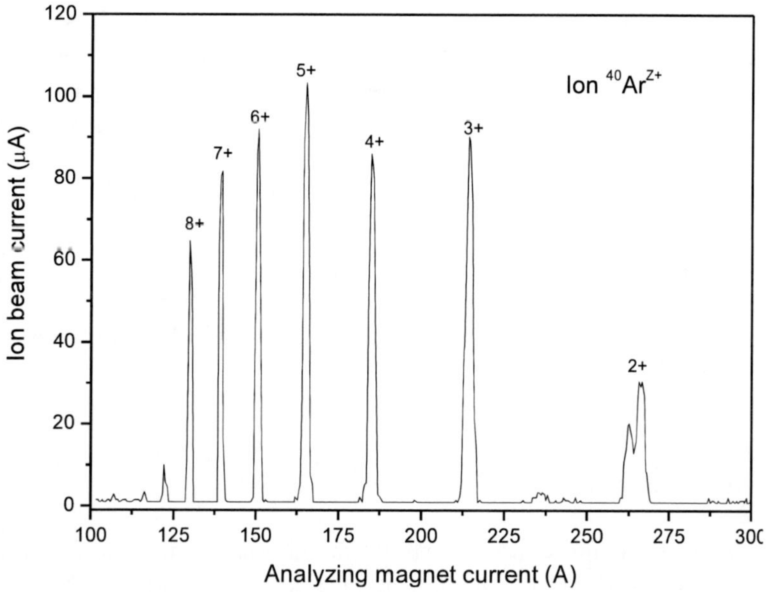

Figure 9. Beam current distribution for separated argon ion species. The currents are measured by a Faraday cup installed in the diagnostic box DB2 in Figure 1, after the beam analysis.

3.2. Transport of the Analyzed Beam

After the separation of the ion species by the analyzing section, further transport of the beam becomes simpler due to the fact that beam current decreased and only one ion species exits with well defined beam parameters. As it has been already mentioned, the function of the transporting section is mainly to restrict the dimensions of the beam envelops within the clearance of the vacuum tubes and chambers along the transport line, and to provide particular beam spot dimensions at particular positions. This problem, for relatively long lines with numerous ion-optical elements, involves some optimization aspects beside the calculation of ion trajectories and beam envelope. Namely, with increase of the number of the elements, the number of the variable parameters (field intensity, gradients, lengths, etc.) becomes so high that implementation of some optimization algorithms is almost unavoidable. Unlike PIC and moments methods, the formalism of transfer matrix, reviewed in the previous section, turns out to be very convenient for implementation of optimization algorithms. Examples for such implementation are the well known TRANSPORT [24] and Beamline Simulator [25] codes being widely used in designing beam transport lines. This reason, together to the fact that this type of calculation takes less computational time and resources, makes the matrix formalism first choice in analyzing of the transport sections. The relative simplicity and fast response of this approach is very convenient for acceptance calculations [26], beam line design calculations [27] as well as in the control of the ion beam transport lines when the operator, changing routinely particular lens parameters, wants immediately to observe the effects of his action on the beam spot or envelope [28, 29].

The staring beam parameters for this type of calculation are usually estimated by the results at the final point of the analyzing section, similar to those shown in Figures 7 a) and b), obtained by more precise methods. This is illustrated in Figures 10 a) and b), where the starting phase space ellipses for 300 keV $^{40}Ar^{12+}$ beam are shown. The results of beam envelope calculation through the transporting section of the line presented in Figure 1, performed by the Beamline Simulator code, are shown in Figures 11 a) and b). The section starts at the analyzing slit in the diagnostic box DB2 and ends at the target in the experimental chamber IC2. In order to reduce the electric power consumption, in the calculation the solenoid lens S1 is replaced by a quadrupole lens. One should note that the analyzing magnet AM2 (Figure 1) in this calculation is not taken as an active element, since it is a part of the analyzing section of the M2, which is a source of light ions.

Results similar to these are used for quick adjusting the beam line parameters and as preliminary calculations for the design, providing good initial data for using more precise methods.

3.3. Matching Transport Calculations

The matching calculations are typical when an ion beam should be injected into accelerator. Different types of accelerators require specific conditions to be met by the injected beam. With certain degree of simplification, the problem reduces to the matching of the beam emittance to the accelerator acceptance. Depending on how demanding the requirements are, the matching process involves 2D, 4D or even 6D emittances.

When multi charged argon beams are considered, the matching ability of the appropriate transport line section should be quite high. Namely, the appropriate accelerators in this case are of the heavy ion type, designed to accelerate ions in relatively large range of masses and charges. In the case of research cyclotrons e.g., even protons could be included when the matching section may become extremely complex.

In the example shown in Figure 2, a matching section consisting of two quadrupole lenses Q1 and Q3 and three solenoid lenses S1, S2 and S3. Since this section partly passes through the axial bore of the cyclotron yoke, where considerable intensity of the cyclotron stray magnetic field exists, the ions are subjected to the influence of additional focusing action. The intensity of the axial component of this magnetic field, for three different excitations of the cyclotron main coils, is given in Figure 12. In the case of axial injection into cyclotrons, due to relatively small diameter of the axial bore, the matching section should also restrict the beam envelope to the imposed limits.

The flexibility of the transfer matrix formalism with respect to its implementation into appropriate optimization algorithms was already mentioned when transporting section was discussed. Since the matching type of calculations deals with considerable number of free matching parameters, the use of this formalism is a good choice for initial calculations.

A matching calculation has been performed for 87 keV $^{40}Ar^{6+}$ beam along the line shown in Figure 2. The initial beam is defined by its emittance of 150 (π×mm×mrad), double waist (α_x=α_y=0) and maximal dimensions of x_m=y_m=7mm. The imposed matching requirements to be fulfilled are the following: 1) the beam profile in the region where the longitudinal magnetic field exists (within the axial bore and below the yoke) should be kept as axially symmetric as possible, 2) at the spiral inflector entrance, the relation β_x [mm/rad] = β_y [mm/rad] = $2R_m$ [mm] should hold, and the beam should have double waist (α_x= α_y=0), where α and β are Twiss parameters, and R_m is the magnetic radius of the inflector, 3) The beam envelope should not exceed ~80% of the dimensions of the apertures defined by the vacuum chambers within which the beam transport occurs. It should be mentioned that these requirements are imposed by the results of the ion beam dynamics in the cyclotron central region and its spiral inflector [30-33].

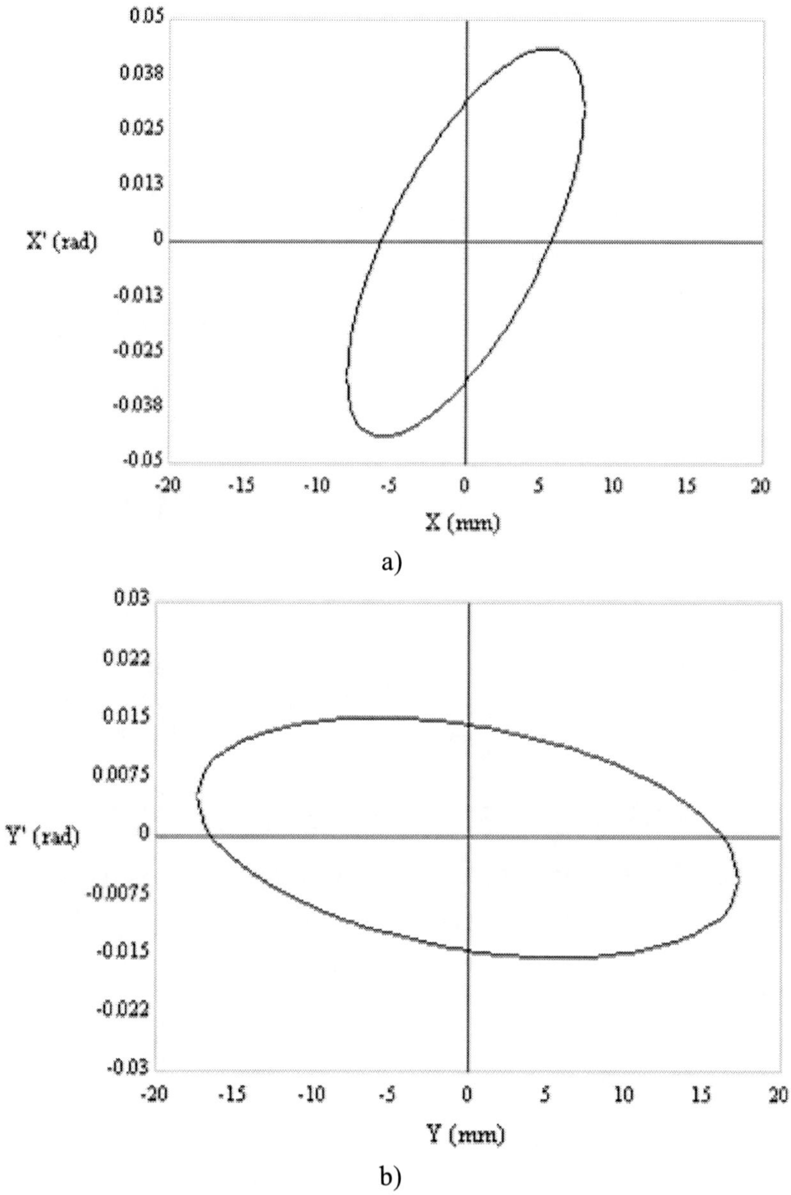

Figure 10. Adopted initial horizontal a) and vertical b) phase space ellipses used as input data for beam transport calculation along the transporting section shown in Figure 1.

The calculations are performed using the TANSPORT code, and the stray magnetic field along the yoke bore was represented by a series of 20 mm long virtual solenoid lenses with appropriate magnetic field. The obtained beam envelope is shown in Figure 13, indicates that the first and third matching conditions are fully met. The second condition, with the given configuration of the section, can be met with discrepancy of about 20%.

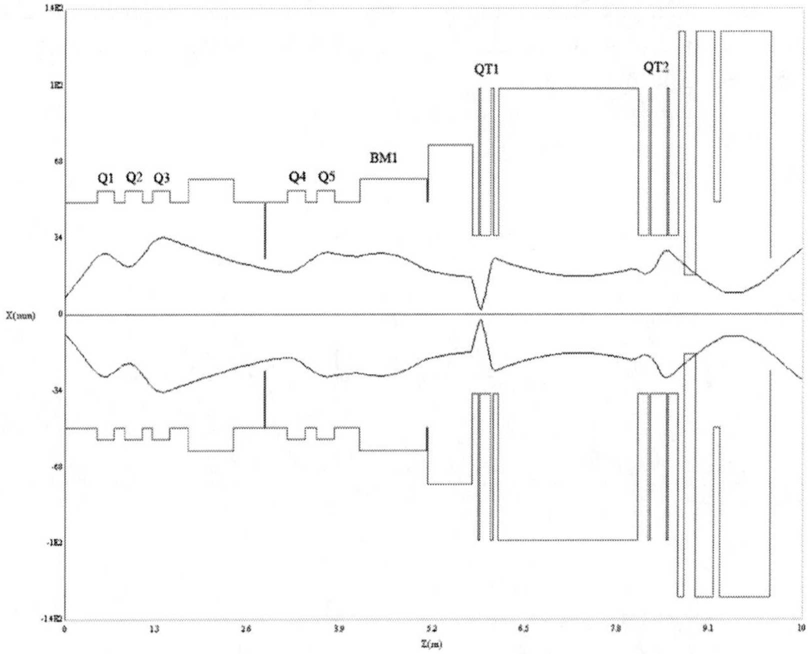

a)

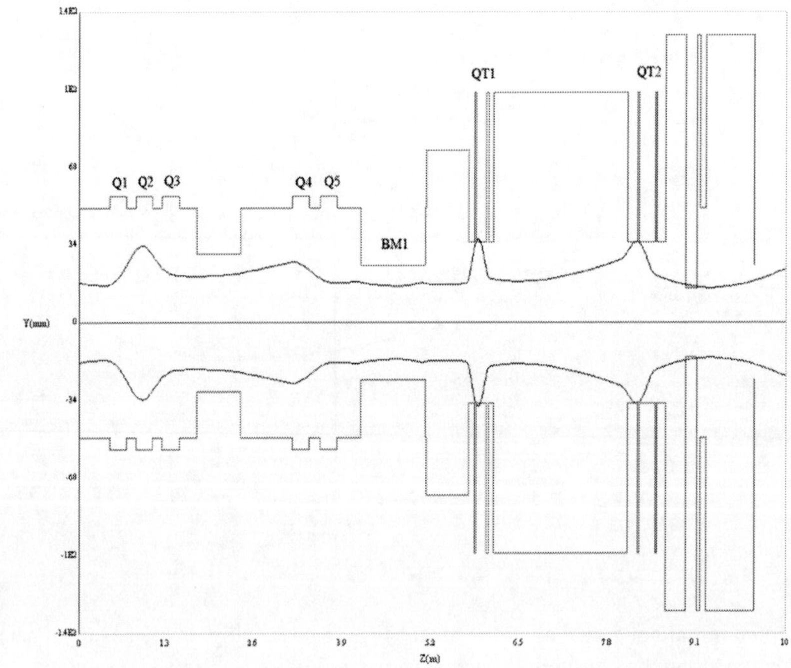

b)

Figure 11. Horizontal a) and vertical b) 300 keV $^{40}Ar^{12+}$ beam envelopes along the transporting section of the line shown in Figure 1.

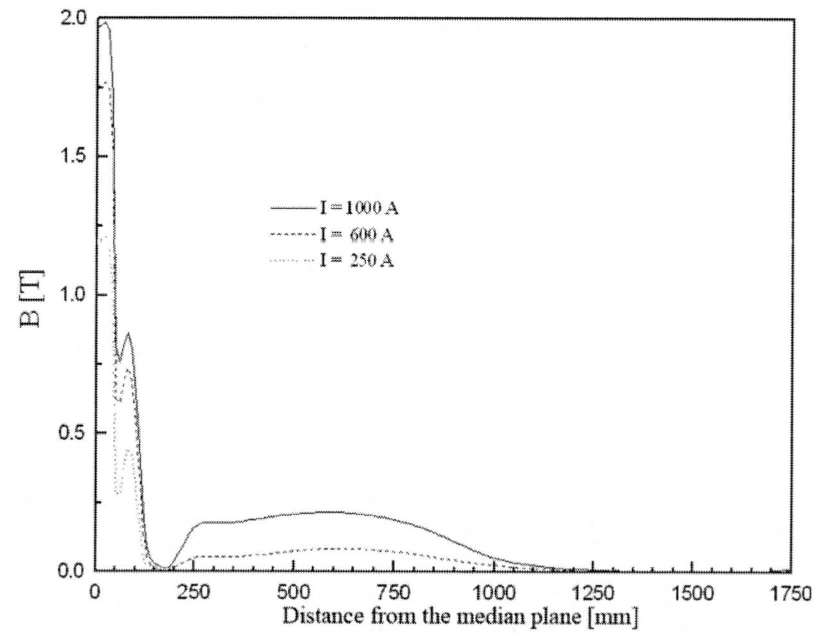

Figure 12. Intensity of the axial stray magnetic field along the yoke bore, for three excitation currents of the cyclotron main coils.

In the case of injecting high current beams, the results obtained by this method should be improved by using more adequate models.

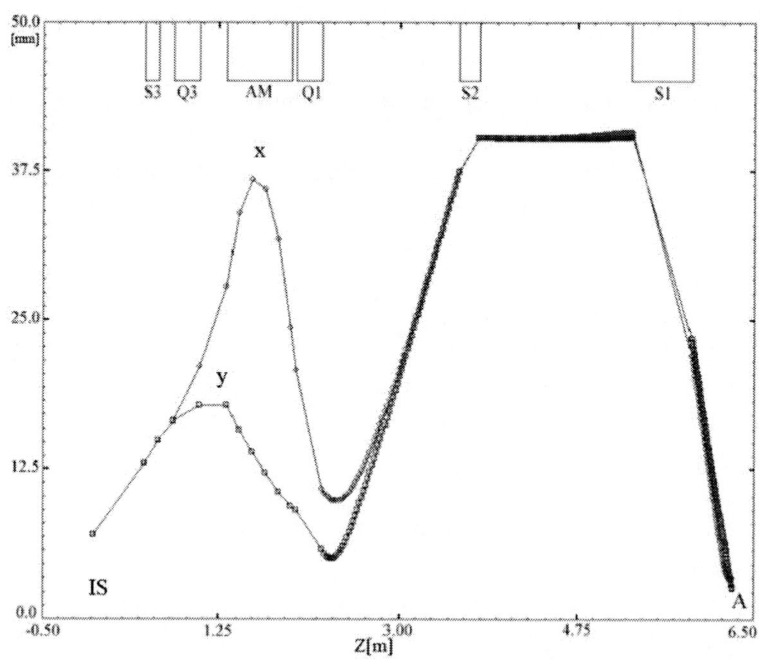

Figure 13. Envelopes of the 87 keV 40Ar6+ ion beam along the matching section of a cyclotron axial injection transport line shown in Figure 2.

CONCLUSION

The field of ion beam transport has considerable importance in many scientific and technological applications. According to the functions performed by the transport lines with respect of the effect they exert on the beam, three main types of transport sections have been identified: analyzing, transporting and matching one. Before getting acquainted with their properties in more details, the basics of ion beam physics and ion beam optics have been reviewed. The basic terms of the beam representation in the phase space, such as optical coordinate system, beam emittance, phase space ellipse etc. have been explained, as well as the equations governing the beam parameter transformations under the influence of the ion-optical elements.

Due to its relative simplicity and wide applicability, special attention has been paid to the paraxial ion optics theory and its implementation into the transfer matrix formalism. The properties and functions of the most commonly used ion optical elements, quadruple and solenoid lenses as well as dipole sector magnets, have been explained and explicit expressions of their transfer matrices have been given. The reason for restricting only to magnetic elements is the fact that they are more efficient with highly energetic and highly charged ions, such as multi-charged argon ions e.g. The limitations of the paraxial ion optics, such as its inability of treating the space charge effect e.g., has been underlined. In those cases other ion beam models and numerical algorithms should be used, such as PIC and moments methods.

The analyzing section of the beam transport line has been treated in considerable details, both for its inevitability in the ion production devices and for the complexity of the calculation. Usually the ion beams passing this section consists of variety of ion species and have relatively high currents (several mA) leading to pronounced influence of the space charge forces. This is the reason for rare usage of the paraxial optics in analysis of this type of sections. As an illustration of this, beam transport calculation through a concrete analyzing section has been performed by using PIC and moments methods.

The functions of the transporting and matching sections have been discussed on the examples of a typical transport line for performing ion implantation experiments and the transport line for axial injection of the beam into a cyclotron. In these cases the transfer matrix formalism is very adequate since it can be implemented into appropriate optimizing algorithm, which is very useful property for this kind of calculations.

The material in this chapter provides the reader with basic knowledge of the ion beam transport theory in general, and multi charged argon beam transport in particular. It can serve as good starting point in this field and gives directions for solving potential practical problems in experiments with ion beams.

ACKNOWLEDGMENT

The work has been supported by the Ministry of Education and Science of Serbia through the projects No. III45006 and O171037.

REFERENCES

[1] Dobrosavljević, P. Beličev, V. Jocić, N. Nešković, I. Trajić, V. Vujović, Lj. Vukosavljević, Facility for Modification and Analysis of Materials with Ion Beams (FAMA), *Proceedings of the 19th International Conference on Cyclotrons and their Applications,* 2010, Lanzhou, MOPCP028, p. 107.

[2] Nešković N., Ristić-Đurović J., Vorojtsov S. B., Beličev P., Ivanenko, S. Ćirković, A. S. Vorzhtsov, B. Bojović, A. Dobrosavljević, V. Vujović I. A., Čomor J. J., Pajović S. B., *Nukleonika*, 48, S135, (2003).

[3] Brown I. G., *The Physics and Technology of Ion Sources* (John Wiley & Sons, New York, Chichester, Brisbane, Singapore, 1989).

[4] Wolf Bernhard, *Handbook of Ion Sources* (CRC Press, Boca Raton, New York, London, Tokyo, 1995).

[5] Carey D. C., *The optics of charged particle beams* (Harwood Academic Publisher, Chur, London, Paris, New York, 1987).

[6] Liebl H., *Applied Charged Particle Optics* (Springer, Berlin, Heidelberg, New York, 2008).

[7] Wollnik H., *Optics of Charged Particles* (Academic Press, Orland, San Diego, New York, Austin, Boston, London, Sydney, Tokyo, Toronto, 1987).

[8] Livingood J. J., *Principles of Cyclic Particle Acceleratos* (van Nostrand, Princeton, New Jersey, 1961).

[9] Septier A., *Focusing of Charged Particles* (Academic Press, New York and London, 1967).

[10] Livingood J. J., *The Optics of Dipole Magnets* (Academic Press, New York and London, 1969).

[11] Reiser M., *Theory and Design of Charged Particle Beams* (Wiley-VCH, 2008).

[12] Septier A., *Applied Charged Particle Optics* (Academic Press, New York, London, Toronto, Sydney, San Francisco, 1980).

[13] Lichtenberg J. A., *Phase-Space Dynamics of Particles*, (John Wiley & Sons, Inc., New York, London, Sydney, Toronto, 1972).

[14] Hockney R. W., Eastwood J. W., *Computer Simulation Using Particles* (Adam Higler, Bristol and Philadelphia).

[15] Goldstein H., *Classical Mechanics* (Adison-Wesley, Reading, Menlo Park, London, Amsterdam, Don Mills, Sydney, 1980).

[16] Radjenović B., Beličev P., Preliminary Calculations of Space Charge Effects in the Spiral Inflector of the VINCY Cyclotrone, *23th Summer School and International Symposium on the Physics of Ionized Gases - SPIG*, 2006, Kopaonik, Serbia, p. 467.

[17] Radjenović B., Radmilović-Radjenović M., Beličev P., *WSEAS Transactions on Information Science & Applications*, 3, 869, (2006).

[18] Aleksandrov V., Kazarinov N., and Shevtsov V., Multi-Component Ion Beam Code MCIB04, *Proceedings of the 19th Russian Particle Accelerator Conference (RuPAC 2004)*, 2004, Dubna, Russia, p. 201.

[19] Alexandrov V., Bogomolov S., Kazarinov N., Shevtsov V., Belichev P., Neshkovich N., Dobrosavlevich A., *Physics of Particles and Nuclei Letters*, 9, 437, (2012).

[20] Lund S. M., *Physical Review Special Topics – Accelerators and beams*, 12, 114801, (2009).

[21] Beličev P., Altiparmakov D. V., Bojović B., Sampling of beam particles normally distributed over four-dimensional phase space, *20th Summer School and International Symposium on the Physics of Ionized Gases - SPIG,* 2000, Zlatibor, Yugoslavia,p. 191.

[22] Madin J., *Etude Experimentale et Simulation des Conditions d'Extraction d'un Faisceau d'Ions Multicharges d'une Source a Resonance Cyclotronique Electronique*, PhD Thesis, Universite de Caen, 1996.

[23] Šiljegović M., Dobrosavljević A., Beličev P., *Review of Scientific Instruments*, 75, 1506, (2004).

[24] Brown K. L., Rothacker F., Carey D. C., Ch. Iselin, TRANSPORT – A Computer Program for Designing Charged Particle Beam Transport Systems, *SLAC Report No. 91* (1977).

[25] Dehnel K., Dehnel M., *Using Beamline Simulator* V1.3, http://www. dehnel.com.

[26] Beličev P., Šiljegović M., Dobrosavljević A., Nešković N., *Review of Scientific Instruments*, 75, 1678, (2004).

[27] A. Dobrosavljević, N. Nešković, P. Beličev, J. J. Čomor, V. Vujović, R. Balvanović, Ristić-Djurović J., *Nuclear Instruments and Methods* A, 597, 136, (2008).

[28] Balvanović R., Radjenović B., Beličev P., Nešković N., *Nuclear Instruments and Methods* A, 607, 322, (2009).

[29] Balvanović R., Beličev P., Radjenović B., *Nuclear Instruments and Methods* A, 690 17, (2012).

[30] Vorozhtsov S. B., Perepelkin E. E., Vorozhtsov A. S., Beličev P., Nešković N., Rajčević M., Ion Beam Dynamics Simulations for the Vincy Cyclotron, *The XX Russian Accelerator Conference (RuPAC2006),* September 10 - 14, 2006, Novosibirsk, Russia, p. 316.

[31] Tomic S., Nešković N., Beličev P., Samsonov E. V., Ion beam dynamics in the acceleration region of the VINCY Cyclotron, *Proceedings of the 15th International Conference on Cyclotrons and their Applications*, 1998, Caen, p. 540.

[32] Beličev P., Jocić V., Nešković N., Rađenović B., Rajčević M., Perepelkin E. E., Vorozhtsov A. S., Vorozhtsov S. B., Spiral Inflectors and Electrodes in the Central Region of the VINCY Cyclotron, *Proceedings of the 18th International Conference on Cyclotrons and their Applications*, 2007, Giardini Naxos, p. 400.

[33] Beličev P., Altiparmakov D. V., *Nuclear Instruments and Methods* A, 456, 177, (2001).

In: Argon: Production, Characteristics and Applications
Editors: B.N. Sismanoglu, H.S. Maciel et al.
ISBN: 978-1-62618-204-2

Chapter 8

MODELING OF THE BREAKDOWN VOLTAGES IN ARGON DISCHARGES WITH IMPROVED SECONDARY ELECTRON EMISSION MODELS

M. Radmilović-Radjenović*[*,1], *B. N. Sismanoglu*[2], *B. Radjenović*[1] *and K. G. Grigorov*[1,3]

[1]Institute of Physics, University of Belgrade, Belgrade, Serbia
[2]TechnologicalInstituteofAeronautics (ITA-DCTA), São José dos Campos/SP, Brazil
[3]Space Research and Technology Institute, Sofia, Bulgaria

ABSTRACT

This chapter relates on results obtained by both theoretical and simulation studies based on the breakdown dc voltage, as well as on the microwave discharges in argon atmosphere with and without magnetic field presence.

Emphasizes were given on the role of the secondary emission processes on the breakdown voltage. Calculations were performed using a Particle-in-cell/Monte Carlo collisions (PIC/MCC) code with the secondary emission model adjusted to accountthe energy and angular dependence of the electron yieldat large separations, as well as the secondary electron emission enhanced by field emission micrographs.

For high frequency discharges, the breakdown voltage - frequency dependence wasanalyzed and a simple scaling law was suggested. In crossed or parallel electric and magnetic fields, the breakdown voltagewas determined taking into account the electron yield variationswith magnetic field. The presented results indicate that PIC/MCC code with improved secondary emission model provides a good physical description of the plasma processes for various gap sizesin presence of electric fieldunder the simultaneous action of both electric and magnetic fields.The simulation data obtained are ingood agreement with the experimentalresults. Apart from their theoretical significance, the obtained results could be useful for determining the minimum ignition voltages in

[*] E-mail address: marija@ipb.ac.rs.

microplasma sources, as well as the maximum safe operating voltage and critical dimensions in other micro-devices.

1. Introduction

Being one of the oldest problems in plasma science, gas breakdown has been a subject of numerous investigations due to its relevance for both industrial applications [1]-[7] and a necessity of gaining a better understanding of the complex mechanisms of gas discharge phenomena [8]-[14]. Renewed interest in breakdown phenomena, especially breakdown in small gaps, emerged from the possibility of lower facility and process costs for a variety of plasma processing and micro-manufacturing techniques currently performed at low pressures [15]-[20]. At the same time, direct current (DC), pulsed DC and radio-frequency (RF) discharges are widely used in the microelectronics industry, in plasma display panels, for depositing thin films, for semiconductor processing, surface modification, waste treatment, etc. [20]-[28]. In order to optimize the plasma technological processes, it is often necessary to know the gas breakdown conditions in the discharge device. Therefore, it is of considerable interest to simulate and/orto measure the breakdown voltage curves.

For large scale systems, the experimentally observed Paschen law has been satisfactorily explained by the Townsend theory [29]. Paschen curve dictates the dependence of the breakdown voltage for a particular gas on the product of the gap distance and the pressure. This breakdown voltage curve represents the voltage needed to achieve a balance between the number of electrons lost by diffusion and drift in the inter-electrode gap and the number of secondary electrons generated at the cathode or in the gas phase [30].Typically, the Townsend mechanism is valid for*pd* products less than $1000\ Torr \cdot cm$ or gaps around a centimeter atthe pressure of one atmosphere. For the gaps of the order of a few micrometers, however, deviations from the well-known Paschen lawhave been reported [31]-[35]. The rapid fall of thebreakdown voltage with decreasing the gap is attributed to the high electric fields generated in micro-gaps which enhance the secondary emission leading to a lowering of the breakdown voltage and a departure from the Paschen curve [36]-[38].

Argon is very important as a benchmark gas in discharge studies. Numerous experimental and theoretical papers have been published on the breakdown voltage characteristics in argon discharges. However, as far as argon breakdown is concerned, agreement between the measured and calculated datahas not been achieved until recently. According tothe revised model, different sources of secondary electron emission, nonhydrodynamic processes close to absorbing and reflecting metal boundaries,space charge (SC) effects on Townsend discharges and also both linearand nonlinear dependence of secondary electron yields strongly influences the basic discharge parameters [39].

In particular, a constant secondary electron yield (for8% argon taken from beam surface experiments) should not be used in modeling of the DC breakdown and collisional plasmas. Instead, more appropriate models for the secondary electron production are needed.

Even at relatively simple case of pure argon discharge, the collisional processes taking place among the three major species (electrons, neutrals and ions) are quite complicated.

Hence, rather than trying to describe the plasma through analytical means, it is more fruitful to use computer simulation techniques to gain information on some of the plasma properties.In the past few decades, computer modeling and simulation have extensively developed as effective tools that complement laboratory experiments and analytic models. Furthermore, the difficulties in achieving well-defined experimental conditions and the limited diagnostic techniques, especially for small scale discharges, favor the investigation of nano-scale systems with simulation tools [40]-[45]. In this chapter, Particle-in-cell/Monte Carlo collision code (PIC/MCC) code codes have been extensively used in order to explore the breakdown mechanism in direct current (DC), radio frequency (RF) and microwave discharges in large and in small scale systemsunder external magnetic field influence. Simulation conditions were mainly based on the available experimental circumstances. The simulation results for the breakdown voltage curves,presented here, clearly demonstrate that it is possible to obtain satisfactory qualitative agreements with the experimental data by using improved models for the secondary electron emission processes.

2. Electrical Breakdown

Transforming the initially non conducting gas into a conducting medium is called breakdown and comprises set of transient processes,poorly understood even in the present days. In general, breakdown will occur if two conditions are fulfilled: there must be suitably placed initiatory electrons and a mechanism of ionization must exist to produce amplification of ions or electrons which compensate the loss by diffusion. The Townsend mechanism leads in successive ionizations of gas molecules inducing the gas breakdownprocess at large separations. Avalanches, however, cannot be built up in the same way at micrometer separations when the gas breakdown is initiated by the secondary emission process rather than processes in the gas. Paschen's law reflects the Townsend breakdown mechanism in gases, that is, a cascading of secondary electrons emitted by collisions in the gap. The significant parameter is *pd* (the gap distance *d* times the pressure *p*). The Paschen curve is typically roughly u-shaped with a minimum breakdown voltage at a specific *pd* and increasing voltages at both increasing and decreasing values of *pd*. This breakdown voltage curve represents a balance between the number of electrons lost by diffusion and drift in the inter-electrode gap and the number of secondary electrons generated at the cathode [46]. In general, over a wide range of pressures and electrode separations, the probability of ionization per electron-neutral collision and the probability of the secondary electron production at the cathode by ion impact are proportional to the reduced electric field *E/N* (electric field to the gas number density ratio) and lead to the well-established *pd* similarity law [47]. When applied to breakdown at small gaps corresponding to atmospheric pressure, Townsend mechanism however, shows certain limitations [48]. The high electric fields obtained in small gaps combined with the lowering of the potential barrier seen by the electrons atthe cathode lead to the onset of ion-enhanced field emissions and lowering of the breakdown voltages [48].

2.1. DC Discharges

The DC discharges operate under conditions when secondary electron production at the cathode surface is the key mechanism in discharge maintenance [49]. To be a self-sustainable discharge, the generation of electron-ion pairs must be sufficient to compensate the loss of charged particles by collision processes such as attachment,drift and diffusion to the walls. At low pressures the ionization process becomes ineffective since the probability for electron-neutral collisions is negligible, whereas at higher pressures elastic collisions prevent electrons from gaining sufficient energies for ionization and theion-neutral collisions are also responsible for increasing the ion losses to the walls. DC breakdown criteria lead tothe expression for the breakdown voltage [49]:

$$V_{DC} = \frac{B_k^k pd}{\left(\ln A_k / \Gamma + \ln pd\right)^k}, \tag{1}$$

with $\Gamma = \ln(1 + 1/\gamma)$ where γ is the second Townsend coefficient, p is the gas pressure, d is the gap spacing and $A_k$and $B_k$are numerical constants [50]. The index k has been empirically found to be 1for the molecular gases and 2 for the atomic gases [51]. Regardless of the sort of gases, k=1 is conventionally used for the purpose of mathematical simplicity, which causes disagreement between the observed and predicted breakdown criteria for argon discharges [51]. From eq.(1) it is trivial to find the minimum and inflection point of the Paschen curve. It can be noticed that ratio $(pd)_{inflection}/(pd)_{minimum}=e$ is independent on$k$, while $V_{inflection}/V_{minimum} = e[k/(k+1)]^k$, where e is the base of natural logarithm. The minimum on the Paschen curve occurs whenthe electron mean free path is just barely sufficient to allow electrons to gain the ionization energy.

2.2. Microwave Discharges

As long as the electrons and ions have sufficient time to transit the electrode gap within an AC cycle, the gas breakdown mechanism in low frequency alternating electric fields is essentially the same as for DC fields, i.e. it is controlled by secondary electron emission due to ion impact. At sufficiently high frequencies, however, ions are not capable to respond to the ac field and electrons are "trapped" in an oscillatory motion within the inter-electrode gap. When this occurs, the electron loss is dominated by diffusion and a significant reduction of the breakdown voltage is observed as compared to the DC case [51]-[54]. As in the case of DC discharges, however, departures from the large scale similarity laws are expected with the onset of field emissions in small gaps.

According to the fluid approach in the diffusion-controlled regime, the microwave breakdown criterion is defined by:

$$\nu_i = D / \Lambda^2, \tag{2}$$

where ν_i is the ionization rate (frequency) and Λ is the characteristic diffusion length [2-5, 51]. The electron diffusion coefficient D is taken to be the transversal one D_τ in DC field, since the transverse and longitudinal diffusion at very high frequencies (higher than *200 MHz*) become almost isotropic with values nearly equal to D_τ when $\omega \to 0$ [51,55]. Using expression $\nu_i = \alpha\mu E_e$, where μ is the electron mobility, the criteria can be rewritten in the form:

$$\frac{\varepsilon}{\Lambda^2} = \frac{\alpha E_r}{\left(1+\omega^2/\nu^2\right)^{1/2}}, \tag{3}$$

where $\varepsilon = D_\tau / \mu$ is the characteristic energy and in this study is assumed to be independent of the electric field strength.

When an AC electric field is presented in a weakly ionized gas where the dominant collision process is electron-neutral collisions with collision rate ν, in order to accountthe frequency effect in total, it is convenient to use the effective field strength defined by:

$$E_e = \frac{\nu}{\sqrt{\nu^2+\omega^2}} E_r. \tag{4}$$

As can be noticed ineq. (4), E_e becomes E_{DC} at $\omega = 0$ and $\nu E_r/\omega$ at $\omega \to \infty$. Substituting expression for the first Townsend's coefficient $\alpha = A_k p e^{-B_k\sqrt{p/E_{DC}}}$ in which E_{DC} is replaced by E_e given by eq. (4), we can obtain expression for E_r. Then, differentiating it with respect to p we obtain parameters at the minimum pressure as listed in Table 2 in which $\sigma = 1+(\omega/\nu_m)^2$ and $\tau = 1+2(\omega/\nu_m)^2$, where ν_m is the collision frequency at $p=p_m$. Calculations for the inflection point using the condition $\frac{d^2E_r}{dp^2}=0$ are complicated, so we assume that $(E_r/p)_i = B_2^2/10.5$, for k=2 and $\nu >> \omega$.

2.3. RF Discharges

The RF discharges can be described by the so called "oscillation amplitude limit" which occurs when the oscillation amplitude x_0 of a charged particle, given by the expression (see, for example, [56]):

$$x_{0e,i} = \frac{e_{e,i}E_0}{m_{e,i}\omega\left(\nu_{ce,i}^2+\omega^2\right)^{1/2}}, \tag{5}$$

becomes comparable to a relevant chamber dimension *d*, i.e. $x_0 \approx d/2$, with subscripts *e* and *i*related to electrons and ions, respectively. In RF discharges electrons can attain oscillatory energy from the AC field. When an electron undergoes a collision, its oscillatory motion is disturbed and its momentum is randomized. In the kinetic theory, the RF gas breakdown is described by the Kihara equation [57]:

$$e^{\frac{B_0 p}{2E}} = A_1 pL\left(1 - \frac{E / B_0 p}{C_2 L / \Lambda}\right), \tag{6}$$

where $E = E_{RF} / \sqrt{2}$ is the effective RF field, p is the gas pressure, L is the electrode spacing, Λ is the vacuum wavelength of the RF field and finally, A_1, B_0 and $C_2$are the molecular constants.

In order to determine the position of the inflection point, turning point as well as the minimum on the breakdown curve, we find the first dU_{RF} / dp and second derivatives d^2V_{RF} / dp^2. Then, by using the following conditions $\frac{dV_{RF}}{dp} = 0$, $\frac{dV_{RF}}{dp} = \infty$ and $\frac{d^2V_{RF}}{dp^2} = 0$ for the minimum, for turning and inflection point on the breakdown curve, respectively we obtain the following expressions [56]:

$$\left(\frac{V_{RF}}{pL}\right)_{minimum} = \frac{B_0 / \sqrt{2}}{1 + (\lambda / 2C_2 L)}, \tag{7}$$

$$\left(\frac{V_{RF}}{pL}\right)_{inflection} = \frac{B_0 / \sqrt{2}}{1 + (\lambda / 2C_2 L)}, \tag{8}$$

$$\left(\frac{V_{RF}}{pL}\right)_{turning} = \frac{B_0}{2\sqrt{2}}\left(\sqrt{1 + \frac{8LC_2}{\lambda}} - 1\right). \tag{9}$$

Lowering the pressure, the RF voltage first decreases and passes through an inflection point and a minimum on the breakdown curve and then it grows and approaches the turning point on the breakdown curve.

2.4. Discharges in Crossed and Parallelelectric and Magnetic Fields

It was shown that the magnetic field strongly affects the characteristics of the left-hand side of the breakdown curve [58,59]. In the low-pressure regime, the breakdown voltage decreases by increasing the magnetic flux density and theposition of the breakdown minimum is shifted due to the magnetic field. Since the mass of the positive ion is many times greater than the mass of the electron, the magnetic field exert more appreciable influence on the electron motion than on the motion of the positive ions. Therefore, the expression for the first Townsend's coefficient in the presence of both electric and magnetic fields changes, while the expression for the second Townsend's coefficient $\gamma_i$remains constant [58,59]:

$$\left(\frac{\alpha}{p}\right)_{E,B} = A_k\sqrt{1+C\left(\frac{B}{p}\right)^2} \times \exp\left[-B_k\left(\frac{p\sqrt{1+C\left(B^2/p^2\right)}}{E}\right)^{1/k}\right], \quad (10)$$

where B represents the magnetic flux density , $C = \left[(e/m)(\lambda_e/v)\right]^2$, λ_e is the electron mean free path in the gas at the pressure of *1 Torr* and v is the electron velocity.Recently, more accurate description of the breakdown of a gas under the influenceof the both electric and magnetic fields has been provided taking into account not only the influenceof themagnetic field on the first but also on the second Townsend's coefficient.Based on the previous studies [60]-[62], the effect of the magnetic field on the breakdown mechanism can be considered as an increase of the pressure from $p$to the so called "equivalent" pressure p_e in accordance to:

$$p_e = p\sqrt{1+C\frac{B^2}{p^2}}. \quad (11)$$

When simulating the pressure dependence on the breakdown voltage accounting both the magnetic and the electric field influence, the expression takes the form [62]:

$$V_{E,B} = \frac{B_k^k pd\sqrt{1+C\frac{B^2}{p^2}}}{\left[\ln A_k/\Gamma_k + \ln\left(pd\sqrt{1+C\frac{B^2}{p^2}}\right)\right]^k}. \quad (12)$$

The derived equations above lay on the assumption that only ions are responsible for the secondary electron emission when striking the cathodesurface initiating so the breakdown process. Recent studies show that it is possible to find correlation between the binary beam data for secondary electron emissionand the gas breakdown phenomenaif assuming all the relevant processes such as photo emission, metastable concentration, fast neutrals as well as ions. Assuming that $dV_{E,B}/dp = 0$, the gas pressure and the breakdown voltage at the minimum on the breakdown curve can be found:

$$p_{\mathrm{minimum}} = \frac{\sqrt{e^{2k}\Gamma_k^2 - cd^2B_k^2A_k^2}}{A_kd}, \quad (13)$$

$$\left(V_{E,B}\right)_{\mathrm{minimum}} = \left(\frac{B_k}{k}\right)^k\frac{e^k}{A^k}. \quad (14)$$

If compare the two sets of equations, i.e. (11)-(12) and (12)-(13) one can see that even if the position of the minimum breakdown voltage (MBV) at crossed magnetic field is shifted toward the lower pressure, the value of the MBV remains the same.

3. Simulation Technique

The simulation codes used in this work are based on one-dimensional 2d3v electrostatic particle-in-cell (PIC) codes [63] with Monte-Carlo (MC) treatment of collisional processes [64]. These codes have been well documented in numerous publications [63]-[68], so only a brief description of the code is given here.

The particle-in-cell (PIC) method permits the statistical representation of general distribution functions in phase space. The use of fundamental equations retains the full nonlinear effects, and space charge and other collective effects can be included self consistently by coupling charged particles to the field equations via the source terms. The particle treatment enables the incorporation of relativistic effects. In general, the PIC method employs the fundamental equations without much approximation, allowing it to retain most of the physics.

In the PIC method, a group of electrons (or ions) are represented by a single simulation particle (super-particle or pseudo particle). Each superparticle is composed of a large number of real particles-electrons and ions and the motion of each particle are assumed to follow Newton's law. It is assumed that such super-particles have no inter-particle interaction except for very short-range interactions corresponding to collisions: Forces exerted on super-particles originate from self-consistent electromagnetic fields. In this way, only collective motions (except for collisions) are modeled by a PIC simulation.

If the model is collisional, Monte Carlo collision (MCC) scheme is applied.Using the knowledge of collision cross sections for all possible collisions and generating a sequence of random numbers, the MCC method can be used to determine whether a specific collision takes place during a given time interval; if it occurs, the particle velocity is revised accordingly. There is an efficient scheme to perform this task, designated as the null collision algorithm (detailed descriptions of the scheme can be found, for example, in [64]). As we use about a million super-particles for each target species, computing the collision probabilities for all the particles becomes very expensive.

A more efficient method for null collision probability have been used assuming that all the traced particles could be modeled by one independent energy, hence the collision probability too. To manage this, a null-collision is considered which probability is a constant a value decreased only by the true collisional probability of each particle. Hence, such a null-collision does not introduce any change to the involved particles lot and the code should evaluate only fractions of them for each time step.

All simulations have been carried out with argon as a filling gas. Neutral atoms were assumed to be uniformly distributed throughout the simulation domain with a Maxwellian velocity distribution at room temperature (*0.026 eV*).

The collision processes considered in the simulation include momentum-transfer, excitation and ionization collisions for electrons and momentum-transfer and charge exchange for ions. Calculations were performed by using well established cross sections for argon [69]. Depending on the gas pressure, in our calculations, the time step d*t* was varied between $10^{-18}\,s$ and $10^{-12}\,s$ which consumed a lot of time for running each case as well as the initial number of 10^{6} computer particles.

In order to determine the breakdown voltage, we use the fact that the breakdown is not an instantaneous phenomenon, it is occurs over a finite period of time which is determined by the balance between creation of charged species by ionization and their losses via collisional processes and diffusion to the walls.

For each value of the gas pressure, the time servitude of the electron density was observed and depending on its increasing or decreasing nature, through trial and error process, the breakdown voltage could be found with an accuracy of *0.5V*.

Thus, for each value of the gas pressure a number of calculations were needed. Our modifications and improvement to this code include the implementation of the energy and angle dependence of the electron yield per ion [56], the ion-enhance field emission mechanism, as well as the effect of the magnetic field on the secondary electron production as will be described in the following section.

4. Secondary Emission Model

4.1. The Yield Per Ion

It is well known that the onset of the breakdown is determined by the so called γ-effectsat the cathode. The secondary electron emission from a surface under the action of an ion is described by the electron yield per ion γ_i-quantifying the number of secondary electrons produced at the cathode per ion. In majority studies in the past, it was often assumed that γ_i has a constant value (i.e. independent of the ion energy or the local field in front of the cathode)causing deficiencies in determining plasma characteristics, especially breakdown voltage [70]-[72]. Actually, the coefficient γ_i depends on the ion energy, reduced electric field as well as the cathode material. We have included the energy dependence of the coefficient accordingthe expression [39,73]:

$$\gamma_i(\varepsilon_i) = \frac{0.006 \cdot \varepsilon_i}{1+(\varepsilon_i/10)} + 1.05 \cdot 10^{-4}\,\frac{(\varepsilon_i - 80)^{1.2}}{1+(\varepsilon_i/8000)^{1.5}}, \tag{15}$$

where ε_i represents the incident ion energy.

Figure 1 shows the experimental beam data (solid curve) and the fitted one with γ_i= 0.08 for argon.

In expression (15), for ion energies below *80 eV*, there is no contribution of the second term. For the ion energies higher than *80 eV*, the yield increases significantly and exceeds the assumed average of *0.08*[39,73].

Following the angular dependence of the secondary emission coefficient suggested by Thierberger et al. [74] and Thomas [75] we suggest the following angular dependence of the electron yield per ion γ_i [56]:

$$\gamma_i(\varepsilon_i,\theta) = \gamma_i(\varepsilon_i)\cos^{-1}\theta, \quad \text{for}\ \theta < 60^0 \tag{16}$$

In order to avoid extremely high values of the yield at higher angles of incidence and singularity that may happened at the incident angle of 90^0:

$$\gamma_i(\varepsilon_i,\theta) = \gamma_i(\varepsilon_i)\cos^{-1}(60^0), \quad \text{for}\ \theta \geq 60^0 \tag{17}$$

where θ is the angle of incidence measured with respect to the surface normal.

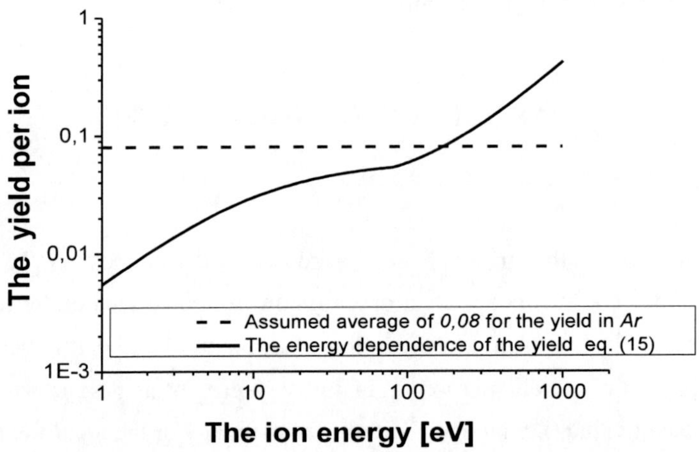

Figure 1. The electron yield per Ar^+ incident on various dirty metal surfaces versus particle energy [39].

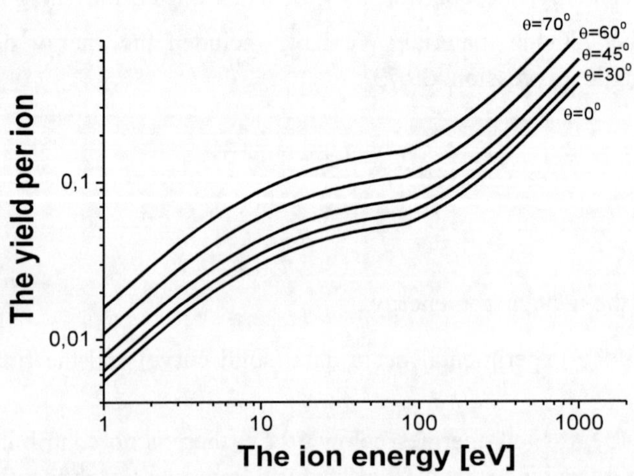

Figure 2. The yield per ion as a function of the ion energy for several angles of incidence on a metal electrode.

Variation of the yield with respect to the angle of incidence can be seen from Figure 2. Increasing the angle of incidence significantly increases the electron yield affecting the value of the breakdown voltages. Analyzing the data shown in Figure 2, one may conclude that for the more accurate calculations, instead of the constant yield, the energy and angular dependence of the yield should be taken into account.

4.2. Field Emission

Current models of secondary electron emissions in low fields require ions to be neutralized as a first step in the emission process. Therefore, single-charged ions are capable of ejectingno more than one electron. The mechanism, however, is quite different in the presence of high electric fields [34]-[38]. When the electric field near the cathode is sufficiently large, electron tunneling from the metal to the gas phase must be taken into account and the previously described mechanism of the ion induced secondary electron emission, however, is not applicable in the presence of high electric fields. Furthermore, as an ion approaches the cathode, it lowers the potential barrier seen by the electrons in the metal as shown in Figure 3, resulting in an ion-enhanced electron field emission [76].

Another improvement to the code includes the implementationof the ion-enhanced field emission mechanism that becomes very important at small gaps between the electrodes.

The effective yield of electrons per ion defined as the ratio of the field-emission electron current density to the incoming positive-ion current density, at the breakdown, is given by the expression [76]:

$$\gamma = Ke^{-D/E}, \tag{18}$$

where K and D are material and gas dependent constants and E is the electric field near the cathode.

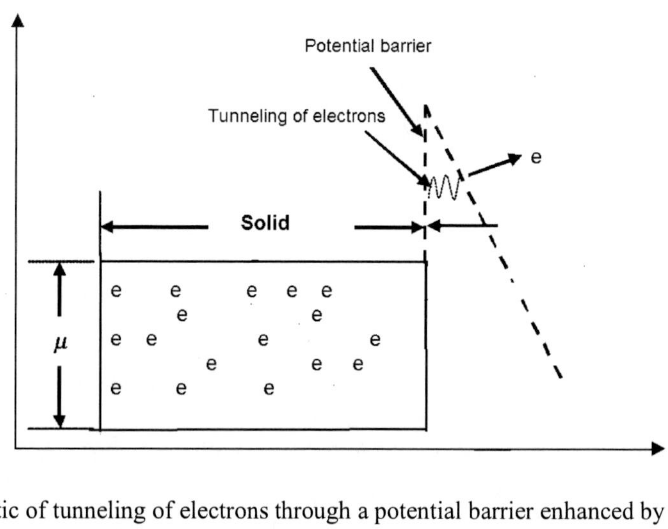

Figure 3. Schematic of tunneling of electrons through a potential barrier enhanced by field emission.

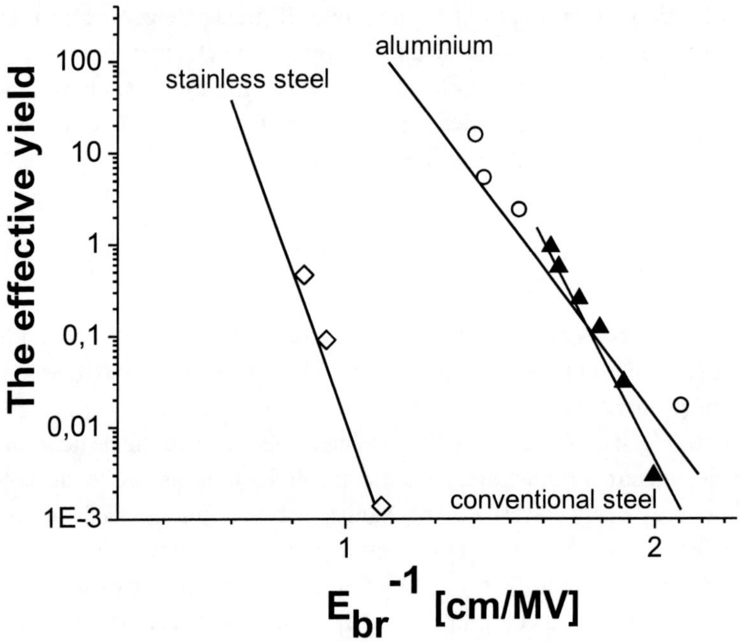

Figure 4.The effective yield against the inverted electric field. Symbols represent experimental data from [76] compared with the theoretical prediction (18).

When the electric field values in the near cathode region becomes higher than the threshold value given by *D*, the electron yield per ion increases rapidlyas depicted in Figure 4.

From the slopes of the straight lines the constant *D*was found to be equal to *26, 11* and *9.3 MV/cm* for stainless steel, conventional steel and aluminum, respectively [76].

4.3. Secondary Emission Model in Crossed and Parallel Magnetic and Electric Fields

Since the magnetic field leads to an increase of the pressure from p to the "equivalent" pressure p_e, the coefficient γwill bealso influenced by the presence of the magnetic field. The expression that describes the secondary electron yield in crossed electric and magnetic fieldsis [58,59]:

$$\gamma_{E,B} = \gamma - \frac{A'E}{p}\left(1 - \frac{1}{\sqrt{1 + C\left(B^2 / p^2\right)}}\right), \tag{19}$$

where γ is the effective yield of secondary electrons in the absence of a magnetic field and suitable constants A' and C.

Figure 5 shows the dependence of the effective yield on:a) *E/p* ratio and b) magnetic field.

The experimentaldata taken from [58] (symbols) satisfactorily agree with the theoretical predictions obtained by expression (19) (curves). As can be seen from Figure 5a, for the crossed electric and magnetic fields, the effective yield decreases with increasing the magnetic field.

a)

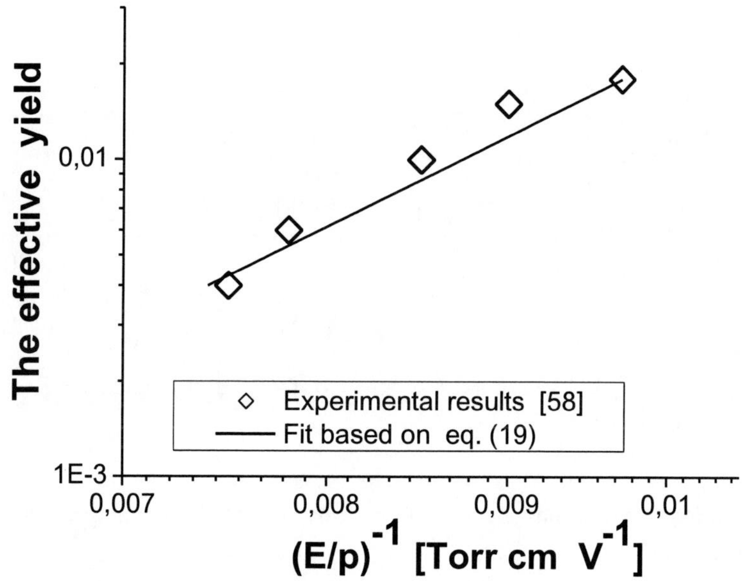

b)

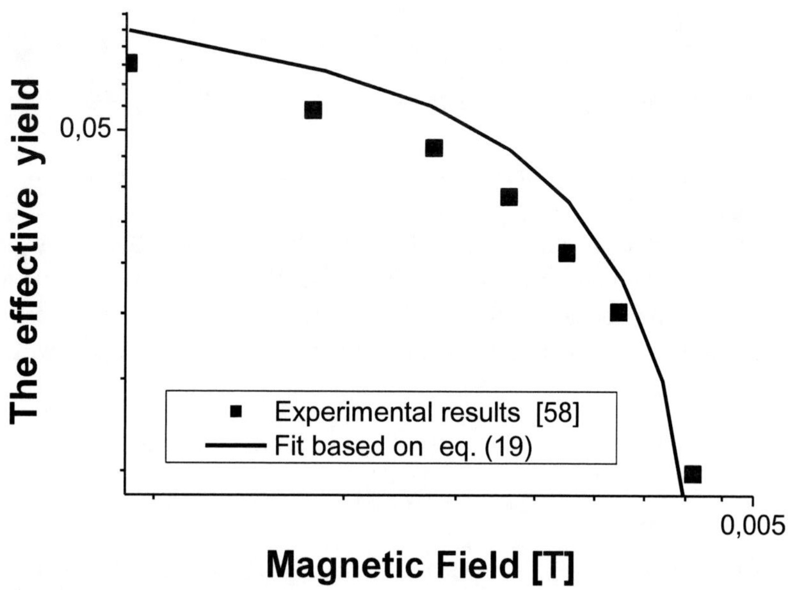

Figure 5. The effective yield as a function of: a) $(E/p)^{-1}$ and b) the magnetic field in a crossed electric and magnetic fields.

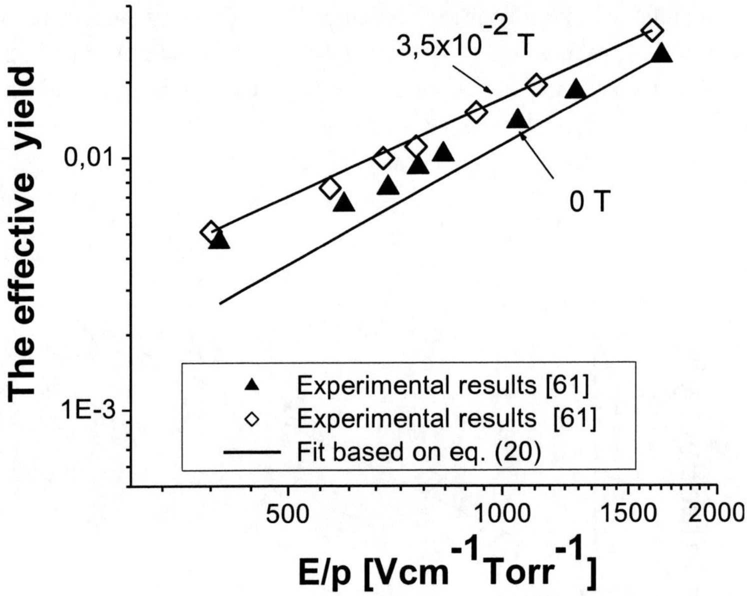

Figure 6. The effective yield versus the *E/p* ration for the magnetic field of: *0T* (solid symbols) and *3,5x10*$^{-2}$*T* (open symbols).

When the magnetic field is parallel to the electric, an empirical expression for the secondary electron yield has been obtained by fitting the experimental data published Petraconi *et al* [61] :

$$\gamma_{E,B} = \gamma - \frac{C'E}{p}\left(1 - \frac{1}{\sqrt{1 + C''\left(B^2 / p^2\right)}}\right), \tag{20}$$

where C' and C'' are appropriate coefficients.

Figure 6 shows the secondary emission coefficientγ on the E/p ratio in argon for two values of the magnetic field.

The experimental results based on electric field evaluations only are given by open squares, while solid symbols correspond to the results based on measurements in presence of the magnetic field.Theoretical predictions are presented by solid lines corresponding to the theoretical curves obtained by using equation (20) and satisfactorily agree with the available experimental data.

5. Results

This section presents our PIC/MCC simulation results for the breakdown voltage characteristic in various types of argon discharges. Since the calculations were mainly performed under experimental conditions, the simulations results will be compared not only with the available experimental data but also with the theoretical predictions (described in section 2).

5.1. DC Discharges

In DC discharges, the charge production at the electrode and the electron emission properties of the cathode are of primary importance. The emitted electrons from the cathode material under gas discharge conditions are perhaps among the less known phenomena in the gas discharge physics. At the same time the results of modeling calculations sensitively depend on the electron yield data. In this part results onthe modeling of the DC breakdown voltage characteristics will be presented and discussed.

Figure 7 contains the simulated breakdown voltage curves for argon for the distance between the electrodes of*1,1cm*. The simulation resultswere obtained by using a constant electron yield (solid squares) and expression (15) (solid triangles)represents the experimental data published in [77] (open symbols). In both cases a good agreement between the experimental and simulation results is achieved taking into account the electron yield energy dependence.If a constant yield isused a lower breakdown voltage is obtained. On the other handthe theoretical curve (dash line) is located in the higher pressuresregion than the measured and simulated one due to the fact that relation (1) has been derived to the admission that the electron yield γ is a constant. Similar trends between simulationand experimental resultsare presentedin the Figure 8. The lower breakdown voltage obtained in the simulations as comparedto the experimental data is attributed to the differences between the simulation and experimental conditions.

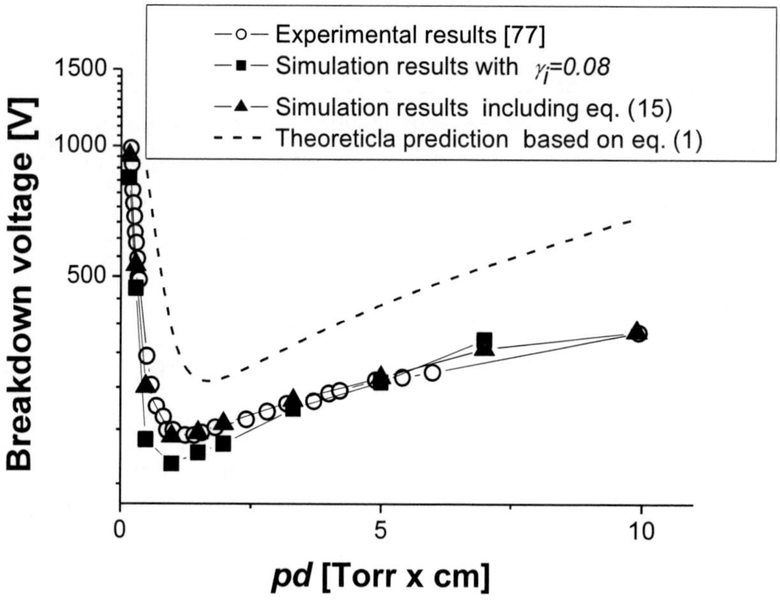

Figure 7. Breakdown voltage as a function of *pd* production in DC argon discharges for the gap size of *1.1 cm*.

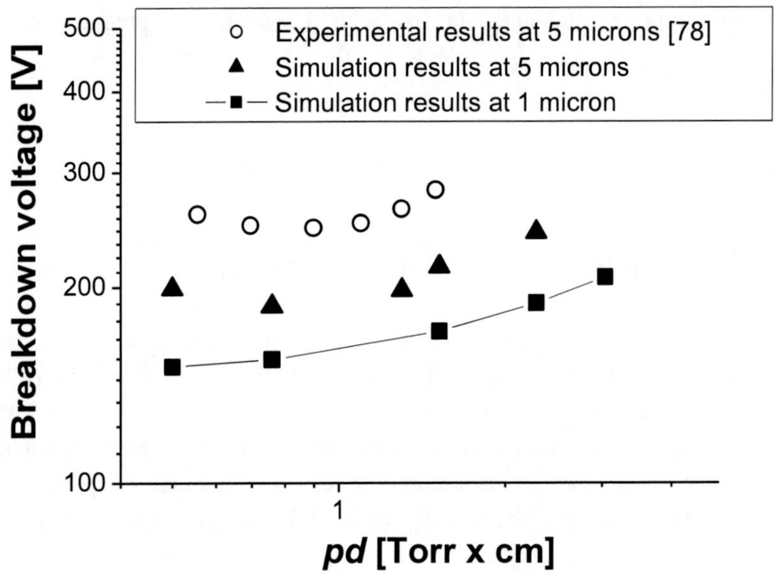

Figure 8. Breakdown voltage versus the pressure–gap spacing product for argon in *5μm* and *1μm* gaps.

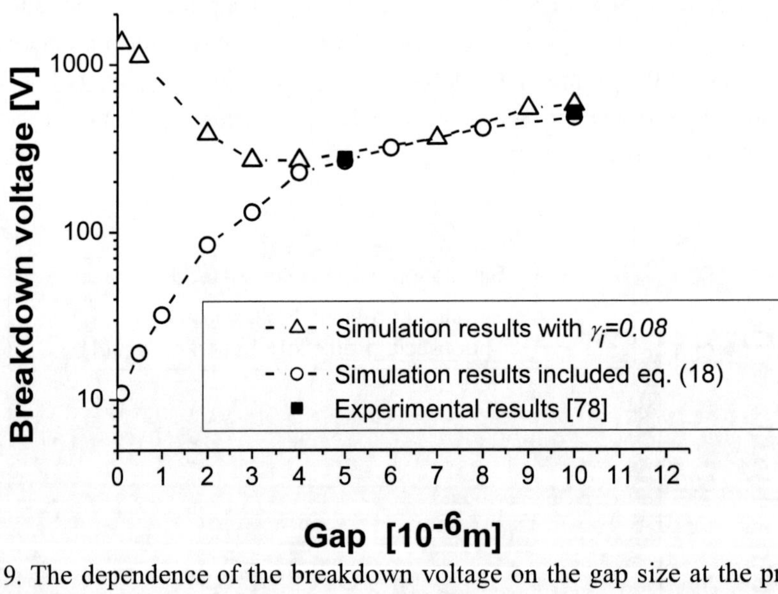

Figure 9. The dependence of the breakdown voltage on the gap size at the pressure of *1 atm.*

Despitethe same gap size used, the gas temperature and the electrode materialare unknown.For the *5μm*, the minimum breakdown voltageoccurs at *pd* value of *8 Torr cm* . Since field emission effect is expected for the gaps less than *5μm,* simulation results obtained by eq. (18) for the *1μm* (solid squares) confirm that the breakdown voltage decreases with decreasing the gap.

In Figure 9 two curves are plotted. The first one(circles) is obtained by using theenergy dependence of the yield per ion (15) that does not account for the ion-enhanced fieldemissions. The second one (squares) includes field emissions effect (18). For relatively large gaps, a good agreement is found between both simulations. For gaps smaller than

5μm,however, significant differences can be observed since the breakdown is no longer controlled by the processes within the gas. At pressure of *1 atm*, the electron mean path isof the order of a few micrometers, so at small inter-electrodespacing breakdown is initiated by the secondary emission processes instead of a gas avalanche process. The rapid fall of the breakdown voltage with decreasing gap size is associated withthe presence of high electrical fields.

5.2. Microwave Discharges

Studies of the microwave discharges are relatively easily than DC or low-frequency discharges mainly for two reasons: the firstone is that the electrode does not play any role in high-frequencies discharge and the second one isthat an electron does not undergo a net drift motion to receive energy from the high-frequency field. For breakdown consideration, the crucial loss mechanisms are attachment to molecules and diffusion out of the field region. For argon as non-attaching gas, the breakdown criterion is simple: rate of electrons production by ionization must exceed rate of loss of electrons by diffusion.

PIC/MCC simulation results (solid symbols) and the theoretical prediction based on eq. (4) (dash curve) for the pressure dependence of the breakdown voltage in argon at *2. 45 GHz* and a *600 μm* gap are compared in Figure 10 [79]. From a good agreement between them confirm that the expression (4) for theelectric field derived from the diffusion-controlled model satisfactorily described microwave discharges even at the pressures of a few hundred *Torr*. PIC/MCC simulation results (solid symbols) are compared with and a satisfactorily agreement have been found between them.

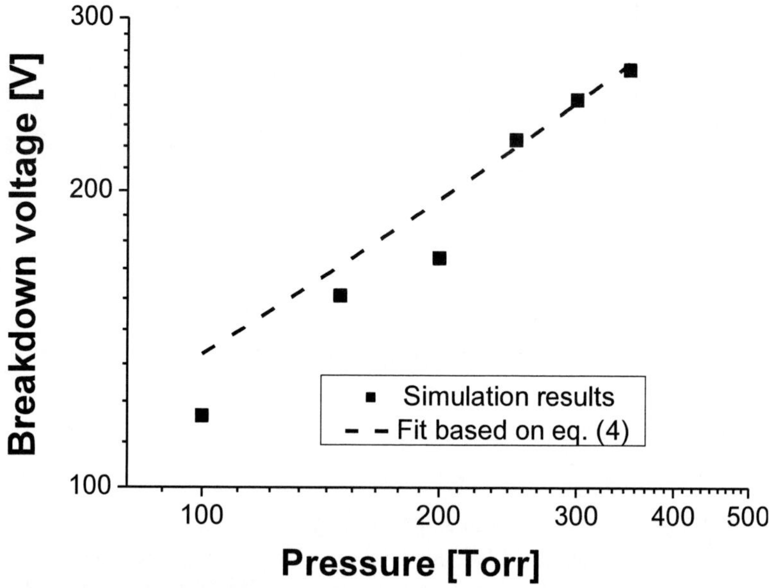

Figure 10. The breakdown voltage as a function of the pressure in argon microwave discharges at *2.45 GHz* and a *600μm* gap size.

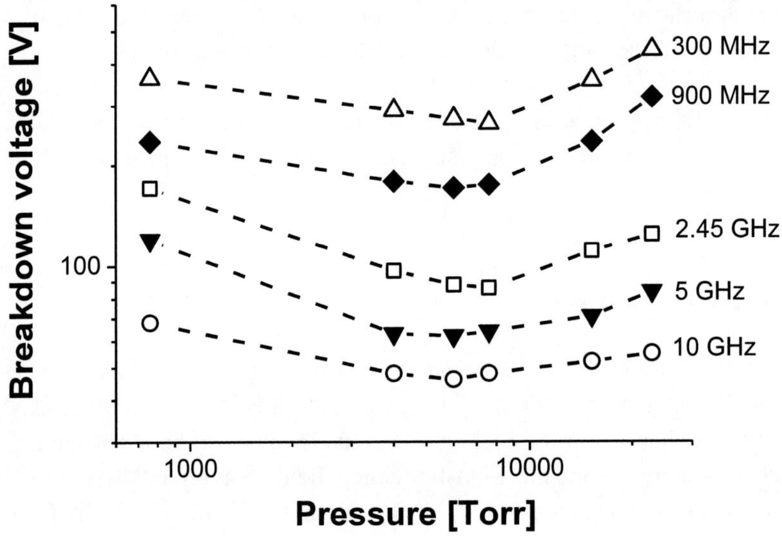

Figure 11. The breakdown voltage versus the gas pressure in argon microwave discharges at different frequencies.

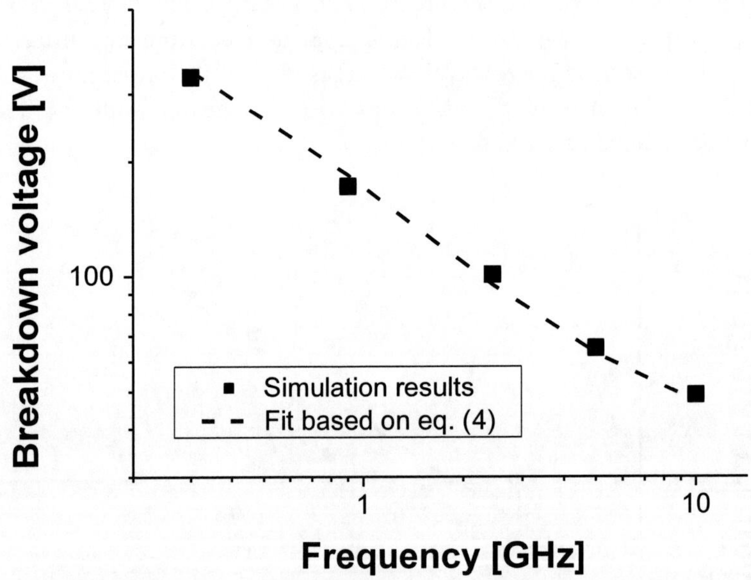

Figure 12. Breakdown field strength as a function of microwave frequency in argon at the pressure of *10 atm* and the gap spacing of *1μm*.

The effect of ion-enhanced field emissions in high frequency discharges is illustrated by the simulations for a *1μm* gap size [36]. Frequency was varied from*300MHz* up to *10 GHz* and the results are presented in Figure 11. As expected, the breakdown voltage decreases with increasing frequencyand shows a minimum at a pressure around *8 atm* for the simulated conditions.

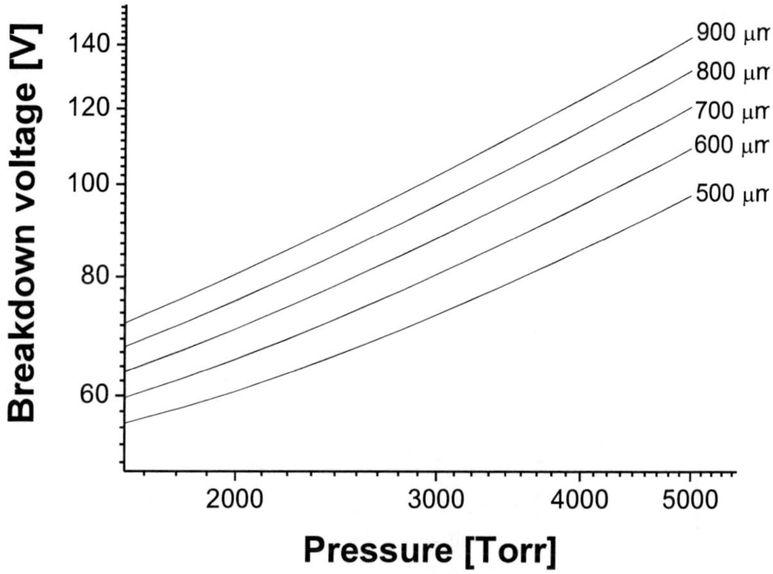

Figure 13. The pressure dependence of the breakdown voltage in argon microwave discharges at *2.45 GHz*.

The frequency effect on the microwave breakdown voltage can be resolved from Figure 12. The simulation results were obtained at the pressure of *10 atm* and gap size of$1 \mu m$ by varying the frequency from *300 MHz* to *10 GHz*[37]. The breakdown voltage represents a strong function of the frequency and decreases with increasing the frequency. Theoretical prediction (4) is shown by dash line and fits well the simulation results (solid symbols).

The breakdown in high frequency field has features that require specifications since microwave discharge properties essentially change with respect to pressure as demonstrated in Figure 13.Calculations were performed for the gaps of the order of a few hundred micrometers by varying thegas pressurefrom 1600*Torr* to *5000 Torr* [80]. As can be noticed from Figures 11 and 13, the breakdown voltage is slightly increases at higher pressures.

5.3. RF Discharges

The secondary emission processes strongly affect the slope of the RF breakdown voltage.In this chapter standard description of RF breakdown is given by Kihara's theory (see subsection 2.3. eq. (6)) [57] and we shall use it here in order to compare to these results withour PIC simulations. For this purpose, we primarily performed calculations focusing on the conditions to the left of the breakdown Paschen minimum where secondary emission processes dominate.

In Figure 14 we compare experimental results[81] (solid symbols) with our simulation (open symbols) for breakdown voltage as a function of the argon gas pressure RF discharges at *13.56 MHz* at the electrode separation of *1.4 cm*.

A good agreement between the experimental data (solid symbols) and the simulation results (open circles) were obtained when assuming both the energy and the angular dependence of the yield per ion (see eqs. (15)-(17)). The latter agreement is especially true for the left side curve's minimum where the secondary electron emission determinates the breakdown.

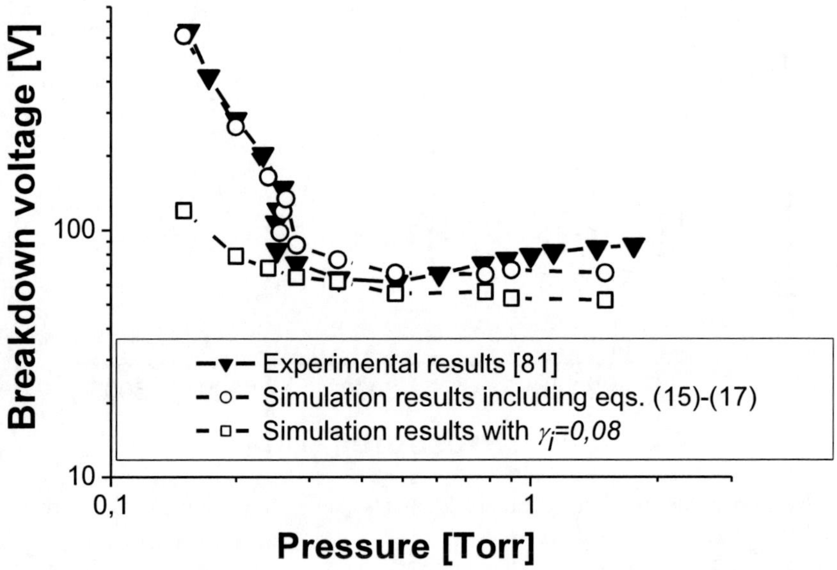

Figure 14. The pressure dependence of the breakdown voltage in argon RF discharges at *13.56* MHz and the electrode separation of *1.4 cm*.

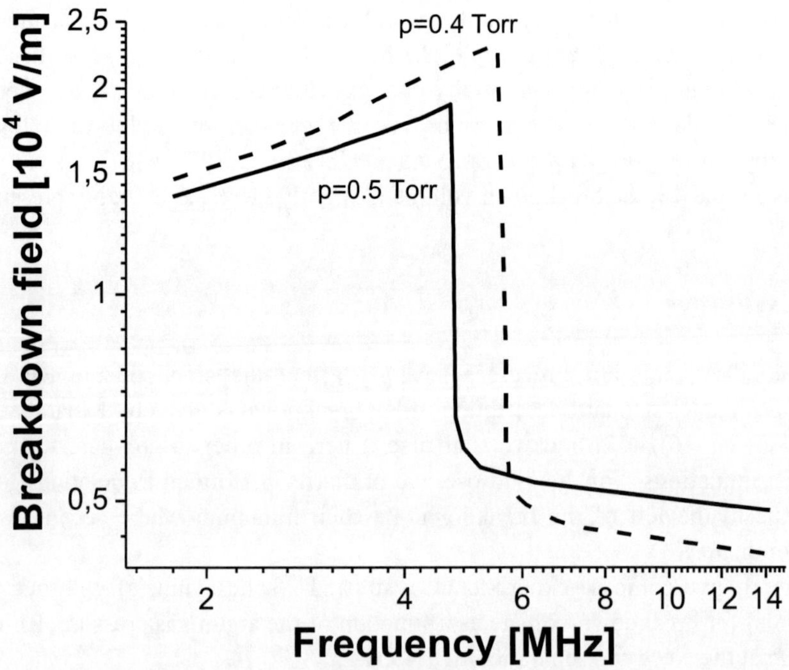

Figure 15. The frequency dependence of the breakdown field strength at gas pressure of: *0.5 Torr* and *0.4 Torr*.

For the region of the right-hand branch of the breakdown curve, one can assume that the secondary electron emission processes do not participate in the RF gas breakdown, so distinctions between the experimental and simulation results are not due to the proposed model. On the other hand, the simulations performed with the constant yield of *0.08* (open squares) seriously differ from the experimental data, especially at lower pressures.

The expressions (7)-(9) can be usedfor scaling the pressure and the breakdown voltage corresponding to the minimumignition voltage at the inflection point of the RF breakdown curve.

At the electrode separation of *1.4 cm*, values of the breakdown voltage corresponding to the turning point of *148 V* (from the experimental results) and *138V* (from the simulation results), for the same pressure of *0.26 Torr* are in disagreement with the very high voltage of *348 V* obtained by using molecular constants [57]. Similar, for the minimum of the breakdown curve, the experimental results provide a value of *61 V* while simulation results give the value of *66 V* and both differ from the very high breakdown voltage of around *339 V*[57]. Disagreements between our simulation results and those obtained by theoretical expressions (7)-(9) with molecular constants [57] can be explained by the fact that secondary emission process is not included in the phenomenological approach (6) which also mentions that the values of molecular constants published in [57] have to be modified.

The frequency effect on the breakdown field (defined as the ratio of the breakdown voltage and the gap size) is illustrated in Figure 15. At very low frequencies, both electrons and ions respond to the ac fields quickly and the oscillation amplitude for both species is well beyond the oscillation amplitude limit (*x*oe,i>>*L*/2) [82]. Both species therefore reach the electrodes at the ends. Thus, as in the DC discharge, the plasma is also sustained by secondary electrons produced by ion bombardment of the cathode. The only difference is that the cathode and anode alternate in each half cycle of the wave, so that the electron–ion bombardment also switches round each half cycle. As the frequency increases, the ion response to the ac field becomes more sluggish, while the electrons follow the ac fields without any difficulty. Thus, the ion current to the cathode decreases so that both ion bombardment and secondary electron production go down. To compensate this reduced secondary electron production, the breakdown field rises. As the frequency is increased further, the maximum deviationof the electrons also goes below oscillation amplitude limit (*x*oe<*L*/2). At thatpoint, the heavy loss of electrons ceases and the breakdown field strength drops abruptly by a very large amount, as can be seen from Figure 15. In order to determine the scaling law for the dependence of the breakdown voltage on the frequency, calculations were performed keeping the RF current constant and varying the frequency from *13.56* to *54.4 MHz* at gas pressure of *250 mTorr*. Simulation results (open symbols) and their comparison with the experimental data [83] (solid symbols) are shown in Figure 16. Although similar trends are observed, simulation results are systematically lower than the experimental data due to the fact that experimental conditions are unknown and differ from the simulation [83].

As expected, the breakdown voltage decreases with increasing the frequency and both experimental and simulation results pointed out that the breakdown voltage is a strong function of the frequency for all the current levels. The experimental data are shown in Figure 16, where for the three highest current values obey $V_{RF} \propto f^{-1.4}$ dependence, while for the lowest current value give $V_{RF} \propto f^{-1.2}$ dependence.

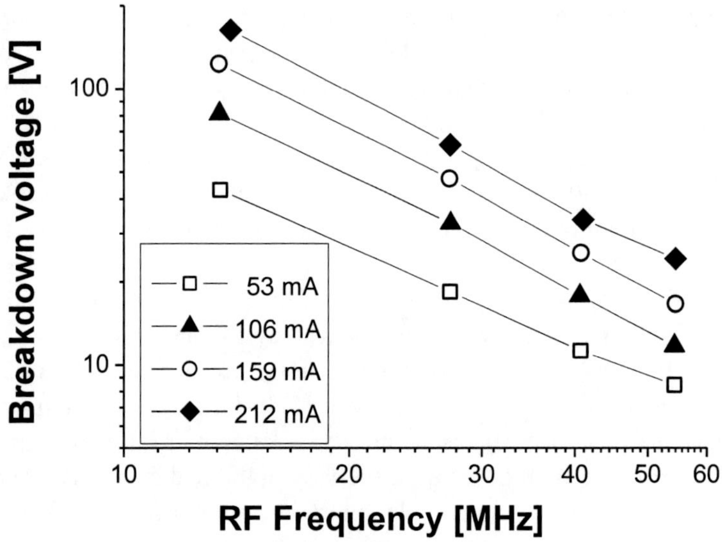

Figure 16. RF breakdown voltage versus the frequency at gas pressure of *250 mTorr* and electrode separation of *2 cm* for four different values of the current.

For the three highest current values the simulation results show the same frequency dependence of the breakdown voltage as obtained from the measurements and a slightly weaker dependence described by $V_{RF} \propto f^{-1.1}$ was found for the RF current of *0.53 mA*.

5.4. Discharges in Crossed and Parallelelectric and Magnetic Fields

PIC/MC code with improved secondary emission model (see section 2.4) has been also used for modelingthe breakdown voltage under the simultaneous action of both the electric and magnetic fields. The influences of the magnetic field on the electrical breakdown characteristics in both crossed and parallel electric and magnetic fields were considered separately.

In a crossed electric and magnetic field, when magnetic field is1.21×10^{-2} *T*, the pressure dependence of the breakdown voltage is shown in Figure 17. Results of measurements [60] (open squares) better agree with the simulation results achieved with improved secondary emission model eq. (19) (solid circles) than the calculations obtained with constant yield (solid diamonds) and theoretical predictions based on eq. (12) (solid line).

Values of the breakdown voltage corresponding ofthe minimum on the breakdown curve at the pressure of *0.41 Torr*are: *372V* (from the experimental results taken from [60]), *368V* (from the simulation results with improved secondary emission model) and the lower value of *255V* obtained from expression (14).

Variations of the breakdown voltage by changing the value of the magnetic field are given in Figure 18. The simulation results obtained by the old secondary emission model (solid diamonds) show different tendencies than the experimental data [60] (open squares) and the simulation results achieved by the new secondary emission model (solid circles). The results of the simple phenomenological model represented by equation (12) provide better agreement with the experimental data at higher pressures.

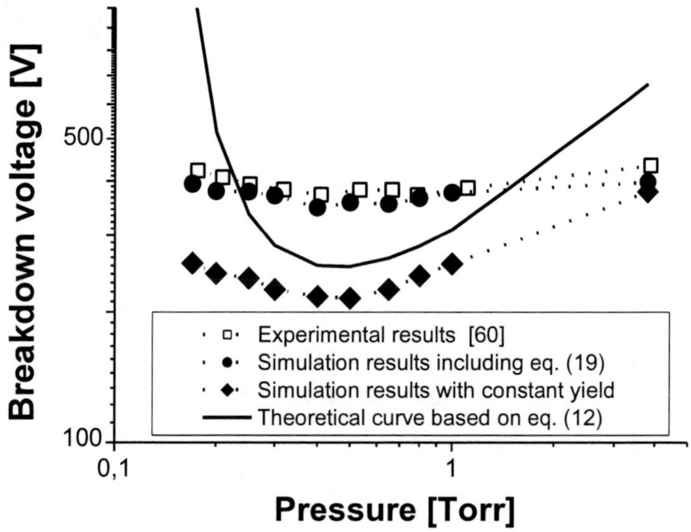

Figure 17. The breakdown voltage curve in a crossed electric field and magnetic field of *1.21x 10* $^{-2}$ *T.*

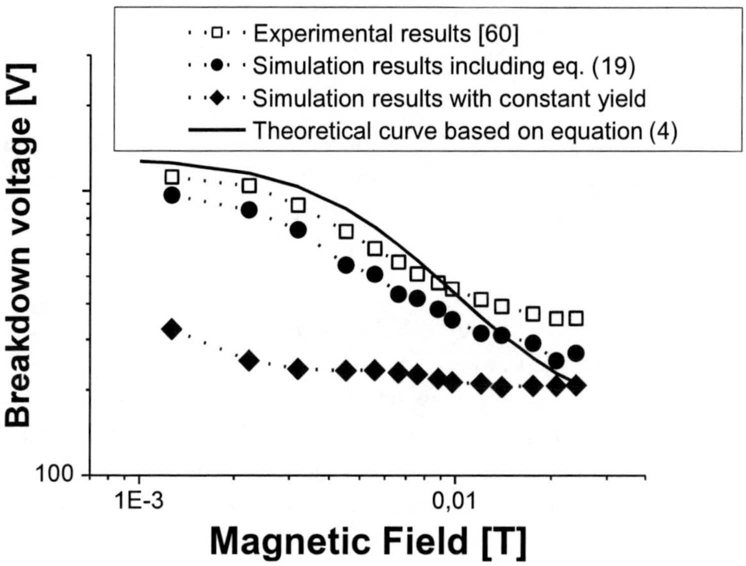

Figure 18. The breakdown voltage against the magnetic field in acrossed electric and magnetic field at the pressure of *0.18 Torr*.

Thiscan be explained by the fact that at higher pressures much more collisions take place,approaching conditions similar to the assumptions grounded on the phenomenological model.

In the case of parallel electric and magnetic fields, the breakdown voltage curve is displayed in Figure 19. Experimental data taken from [61] (solid symbols), show better agreement with the simulation results obtained with the new secondary emission model (open circles) then those obtained using constant yield (open circles). Systematic departures from the experimental data are the results of the different conditions employed in simulations and measurements.

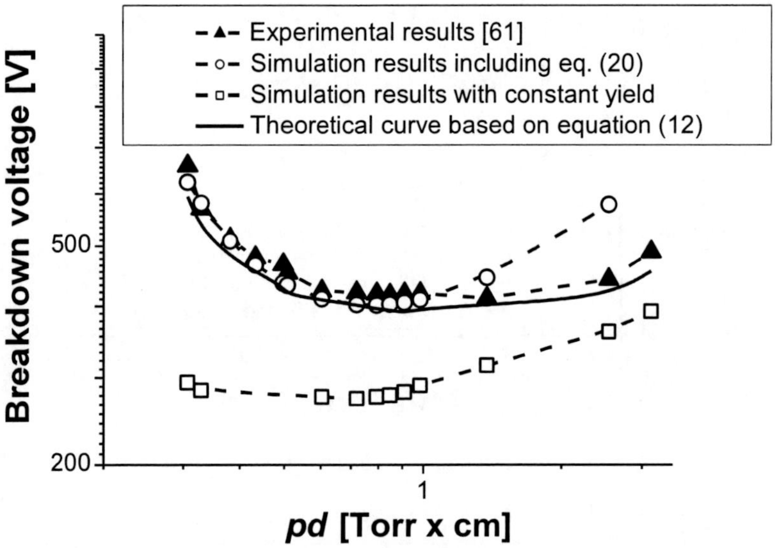

Figure 19. The breakdown voltage versus the pd product when electric fieldis parallel to the magnetic field of $3.5x\ 10^{-2}$ T.

As it can be noticed from Figures 19, the presence of magnetic field extends the validity of equation (12) to the higher pressure. A possible explanation lies in thefact that the effect of the magnetic field is equivalent to anincrease in pressure.

Conclusion

This chapter contains a broad overview of the breakdown mechanism in argon discharges focusing on the role of the secondary emission process in determining the breakdown voltage. In particular, it was demonstrated that application of appropriate expressions for the secondary electron yield in the plasma modeling may lead to a good agreement with the experimental data.

According to the well-known Townsend theory, an electricspark can occur only if free electrons are accelerated by anelectric field and gain enough energy between successive collisionswith neutral atoms. Ionization process leads to releasingof additional electrons which are also accelerated and collide with theneutrals atoms causing more ionization. The resulting avalanche leads to a spark. In DC and RF discharges, in large scale systems, PIC/MCC calculations were performed using both a constant energy and angular dependent yield per ion. Simulation results obtained with the yield per ion of *0.08* are in better agreement with the experimental results inDC discharges than in RF discharges. Onepossible explanation may be the fact that the assumed average of *0.08* for the yield per ion in argon better corresponds to the Volt-Ampere characteristics in DC discharges, while the kinetic in RF discharges is better described by the energy dependence of the yield.

If the electric field atthe surface of a conductor is sufficiently large some of the conduction electrons in the metal lattice are pulled out into the gap resulting in the field emission.The obtained simulation results clearly show that in small gapswhen the electric field near the cathode is sufficiently large,electron tunneling from the metal to the gas phase needs to betaken into account through the implementation of the expression (18) in the PIC/MCCcode instead of the old secondary emission model.

In high frequency discharges, the breakdown voltage is a strong function of frequency and obeys simple scaling law decreasing with increasing the frequency.

The breakdown voltage also depends on the pressure and has increasing tendency at higher pressures.

The influences of the magnetic field on the electrical breakdown characteristicsin both crossed and parallel electric and magnetic fields have been studied separately. In particular, the previously developed theoretical approach for the description of the breakdown features in crossed electric and magnetic fieldshas been modified to be applicable in studies of breakdown phenomena in a parallel electric and magnetic field. The secondary emission model in the case of parallel electric and magnetic fields is suggested for the first time in [62]. Appropriate coefficients that appear in the expressions forthe ionization coefficient, the secondary electron yield andthe breakdown voltage in the presence of both electric andmagnetic fields have been determined based on the available experimental data.The obtained simulation results confirm that the replacement of the old secondary emission model (i.e. constant electron yield) in the PIC/MC code with the new one based on the simple phenomenological model representedby relations (19) and (20) in our paper provides better agreement with the available experimental data.

A good agreement between simulation results presented here, and the available experimental data indicate that we are capable to model DC, RF and microwave discharges as well as discharges under the simultaneous action of electric and magnetic fields both at large and microcaps by using PIC/MC code with proper choice of the secondary emission model.

ACKNOWLEDGMENTS

The work has been carried out under Ministry of Education and Science Republic of Serbia O171037 and III41011 projects. The partial financial support of FAPESP (author B. N. Sismanoglu process number 2012/13064-4) is strongly acknowledged as well as the financial support of CAPES (author K.G.Grigorov process number 83581/2012).

REFERENCES

[1] F. Charbonnier, 1998 *J. Vacum Sci. Technol. B* 16, 880 (1998).
[2] A. Neumann, *J. Appl. Phys.* 90, 1 (2001).
[3] I. Tsuyohito and T. Kazuo, *Appl. Phys. Lett.* 80, 2648 (2002).
[4] J. Xiong, Z. Zhou, X. Ye, X. Wang and Y. Feng, *Sensors Actuators A*108, 134 (2003).
[5] S. Yang, J. K. Lee, S. W. Ko, H. C.Kim and J. W. Shon, *Contrib. Plasma Phys.* 44, 536 (2004).

[6] J. Chen, J.G. Zhao, D. B. Wu, W. Liu and T. Zhang, *Vacuum*86, 189 (2011).
[7] L. Schwaederlé, M. K. Kulsreshath, L. J. Overzet, P .Lefaucheux, T. Tillocher and R. Dussart, *J. Phys. D: Appl. Phys.* 45, 065201 (2012).
[8] M. J. Druyvesteyn and F. M. Penning, *Rev. Mod. Phys.* 12, 87 (1940).
[9] R. G. Meyerand and A. F. Haught, *Phys. Rev. Lett.* 11, 401 (1963).
[10] Yu. D. Korolev and G. A. Mesyats, *Physics of pulsed breakdown in gases,* URO-PRESS, Yekaterinburg (1998).
[11] X. Zheng, G. H. Chen, Z. Li, S. Deng and N. Xu, *Phys. Rev. Lett.* 92, 106803 (2004).
[12] E. Wagenaars, M. D. Bowden and G. M. W. Kroesen, *Phys. Rev. Lett.* 98, 075002 (2007).
[13] M. Klas, Š. Matejčík, B. Radjenović and M. Radmilović-Radjenović, *Europhys. Lett.*95, 35002 (2011).
[14] F. Liu, W. Wang, X. Chang, Z. Wu, L.He, Z. Li, Z.He and R. Liang, *Europhys.Lett.*97, 6500 (2012).
[15] J.M. Torres and R. S. Dhariwal, *Nanotechnology*10, 102 (1999).
[16] D. Ono, D.Y. Sim and M. Esahi M, *J. Micromech. Microeng.* 10, 445 (2000).
[17] C. Oliveira, J. A. Souza Corrêa, M. P. Gomes, B. N. Sismanogluand J. Amorim,*Appl. Phys. Lett.*93, 041503 (2008).
[18] Z. Lj. Petrović, N. Škoro, D. Marić, C. M. O. Mahony, P. D.Maguire, M. Radmilović - Radenovićand G Malović, *J. Phys. D: Appl. Phys.*41, 194002 (2008).
[19] B. N. Sismanoglu, C. L. A. Cunha, M. P. Gomes, R. Caetano and K. G. Grigorov, *Brazilian Journal of Physics*40, 459 (2010).
[20] M. Radmilović-Radjenović, B. Radjenović, M. Klasand Š. Matejčik, *Micro and Nano Letters*7, 232 (2012).
[21] P.K. Chu, J.Y. Chen, L.P. Wang and N. Huang, *Materials Science and Engineering R*36, 143 (2002).
[22] A. Bogaerts, E. Neyts, R. Gijbels, J. van der Mullen, *Spectrochimica Acta Part B*57, 609 (2002).
[23] T. Makabe and Z. Lj. Petrović, *Plasma Electronics: Applications in Microelectronic Device Fabrication,* Taylor and Francis, Boca Raton (2006).
[24] K. H. Becker, K. H. Schoenbach, and J. G. Eden, *J. Phys. D: Appl. Phys.* 39, R55 (2006).
[25] R. Wolf, A. C. Sparavigna, *Engineering* 2,397 (2010).
[26] Y. Wang and X. Y. Shen, Journal *of Macromolecular Science, Part B: Physics* 51,662 (2012).
[27] B. Radjenović and M. Radmilović-Radjenović, *Central European Journal of Physics* 92, 265 (2012).
[28] S. Samukawa, M. Hori, S. Rauf, K. Tachibana, P. Bruggeman, G. Kroesen, J. C. Whitehead, A. B. Murphy, A. F. Gutsoland S. Starikovskaia, *J. Phys. D: Appl. Phys.*45, 253001 (2012).
[29] J. S. E. Townsend, *Electrons in Gases,* :Hutchinson's, London (1947).
[30] F. Paschen, *Wied. Ann.* 37, 69 (1889).
[31] F. L. Jones and C. G. Morgan, *Phys. Rev.* 82, 970 (1951).
[32] L. H. Germer, *J. Appl. Phys.* 30, 46 (1959).
[33] P. Kisliuk, *J.Appl. Phys.* 30 , 51 (1959).
[34] J.M. Torres and R. S. Dhariwal , *Microsystem Technologies* 6, 6 (1999).

[35] C. H. Chen, J. A. Jeh and P. Wang, *J. Micromech. Microeng.* 16, 1366 (2006).
[36] M. Radmilović-Radjenović, J. K. Lee, F. Iza and G. Y. Park, *J. Phys. D:Appl. Phys.* 38, 950 (2005).
[37] M. Radmilović-Radjenović and BranislavRadjenovic, *Plasma Sources Sci. Technol.*16, 337 (2007).
[38] M. Klas, Š. Matejčík, B. Radjenović and M. Radmilović-Radjenović, *Physics Letters A*376 , 1048 (2012).
[39] A.V. Phelps and Z. Lj. Petrović, *Plasma Sources Sci. Technol.*8, R21 (1999).
[40] M. M. Turner, Phys. Rev. Lett.75, 1312 (1995).
[41] D. J. Economou, *Thin Solid Films*365, 348 (2000).
[42] H.C. Kim and J.K. Lee, *Phys. Rev. Lett.*93, 085003 (2004).
[43] H.C. Kim, F. Iza, S.S. Yang, M. Radmilović-Radjenović and J.K. Lee, *J. Phys. D: Appl. Phys.*38,R283 (2005).
[44] B. Radjenović, J.K. Lee, and M. Radmilović-Radjenović, *Computer Phys. Communications*174, 127 (2006).
[45] N. Yu. Babaeva, N. Ning, D. B. Graves and M. J. Kushner, *J. Phys. D: Appl. Phys.*45, 115203 (2012).
[46] J. M. Meek and J. D. Craggs, *Electrical breakdown of gases,* Oxford University Press, Oxford (1953).
[47] L. Loeb, *Fundamental Processes of Electrical Dishcarges in Gases,* J. Wiley and Sons, Inc., New York (1939).
[48] M. Radmilović-Radjenović and BranislavRadjenovic, *Europhys. Lett.*83, 25001 (2008).
[49] M. Radmilović-Radjenović, Z. Lj. Petrović, G. N. Malović, D. Marić and and B. Radjenović, *Czech. J. Phys.*56, B996 (2006).
[50] Yu. P. Raizer, *Gas Discharge Physics,* Springer, Berlin (1991).
[51] M. Sato, *Bull. Yamagata Univers.*25, 119 (1999).
[52] S. Krasik, D. Alpert and A. O. McCoubreyn, *Phys. Review* 76, 722 (1949).
[53] W.P. Allis and S.C. Brown, *Phys. Review* 87, 419 (1952).
[54] A.V. Gurevich, N.D. Borisov and G.M. Milikh, *Physics of Microwave Discharges*, Gordon and Breach, New York, (1997).
[55] Spectroscopy Letters.
[56] M. Radmilović-Radjenović J. K. Lee, *Physics of Plasmas* 12, 063501 (2005).
[57] T. Kihara, *Rev. Modern Physics* 24, 45 (1952).
[58] S. N. Sen and A. K. Ghosh, *Proc. Phys. Soc. B* 79, 293 (1962).
[59] S. N. Sen and A. K. Ghosh, *Proc. Phys. Soc.* 80,909 (1962).
[60] S. Z. Li and H. Uhm, *Phys. Plasmas* 11, 3443 (2004).
[61] G. Petraconi, H. S. Maciel, R. S. Pessoa, G. Murakami, M. Massi, C. Otani, W. M. I. Uruchi and B. N. Sismanoglu, *Braz. J. Phys.* 34, 1662 (2004).
[62] M. Radmilović-Radjenović and B.Radjenović, *J. Phys. D:Appl. Phys.* 39, 3002 (2006).
[63] C. K. Bridsall, *IEEE Trans. Plasma Sci.*19, 65 (1991).
[64] V. Vahedi and M. Surendra, *Comput. Phys. Commun.* 87, 179 (1995).
[65] J. P. Verboncoeur, M. V. Alves, V. Vahedi, and C. K. Birdsall, *J. Comput. Phys.*104, 321 (1993).
[66] J. P. Verboncoeur, A. B. Langdon and N. T. Gladd, *Comput. Phys. Commun.* 87, 199 (1995).
[67] S. Hamaguchi, *IBM J. Res. Develop. B* 43, 199 (1999).

[68] D. L. Bruhwiler, R. E. Giacone, J. R. Cary, J. P. Verboncoeur, P. Mardahl, E. Esarey, W. P. Leemans and B. A. Shadwick, *Phys. Review Spec. Topics – Accelerators and Beams*4, 101302 (2001).
[69] J.K. Lee, N. Yu. Babaeva, H.C. Kim, O.V. Manuilenko, and J.W. Shon, *IEEE Trans. On Plasma Science*32, 1 (2004).
[70] J.P. Boeuf, *Phys.Rev.A*36, 2782 (1987).
[71] T.H. Chung, H.J. Yoon, T.S. Kim and J.K. Lee, *J. Phys. D: Appl. Phys.*29, 1014 (1996).
[72] M. Soji and M. Sato, *J.Phys.D: Appl.Phys.*32, 1640 (1999).
[73] A. V. Phelps, L. C. Pitchford, C. Pedoussat and D. Donko, *Plasma.Sourc.Sci.Technol.*8, B1 (1999).
[74] P. Thieberger, A. L. Hanson, D. B. Steski, S. Y. Z. V. Zajic, and H. Ludewig, *Phys. ReviewA*61, 1 (2000).
[75] E.W. Thomas, *Nuclear Fusion, Spec. Issue* 94, 94 (1984).
[76] W.S. Boyle and P. Kisliuk, *Phys. Review* 97, 255 (1955).
[77] V. A. Lisovskiy, S. D. Yakovinand V. D. Yegorenkov, *J. Phys. D: Appl. Phys*. 33, 2722 (2000).
[78] T. Ito, Izaki and K. Terashima, *Thin Solid Films* 386, 300 (2001).
[79] M. Radmilović-Radjenović and B.Radjenović, *Inter. Review. of Physics* 1, 96 (2007).
[80] M. Radmilović-Radjenović and B.Radjenović, *Spectroscopy Letters*44, 146 (2011).
[81] N. Yu.Kropotov, Yu. A.Kachanov, A. G.Reuka, V. A.Lisovskiy, V. D.Erorenkov, and V.Farenik, *Sov. Tech. Phys. Lett.*14, 159 (1988).
[82] A.Ganguli and R. D.Tarey, *Current Science* 83, 279 (2002).
[83] M. J. Colgant, M. Meyappan, and D. E. Murnick, *Plasma Sources Sci. Technol.*3, 181 (1994).

In: Argon: Production, Characteristics and Applications ISBN: 978-1-62618-204-2
Editors: B.N. Sismanoglu, H.S. Maciel et al.

Chapter 9

CHEMISTRY STUDIES OF LOW-PRESSURE ARGON DISCHARGES: EXPERIMENTS AND SIMULATION

R. S. Pessoa[1,2]*, B. N. Sismanoglu[1], M. P. Gomes[1], H. S. Medeiros[1], J. C. Sagás[1], M. Roberto[1], H. S. Maciel[1,2] and G. Petraconi[1]

[1]Technological Institute of Aeronautics (ITA-DCTA), São José dos Campos, Brazil
[2]IP and D, University of Paraiba Valley (UniVap), São José dos Campos, Brazil

ABSTRACT

This chapter presents the chemistry studies of low-pressure electrical discharges generated with argon gas. This noble gas is widely used in gas-discharge lamps, as sputtering gas for deposition and/or etching processes, arc welding, among others. Especially, in the semiconductor industry, this gas is present in most of the processing steps of a chip or a micro device. The magnetron sputtering reactor, designed for the deposition of thin films, and the reactive ion etching reactor, designed for etching process of materials at micro/nano scale level are among the most common equipment that use argon-based plasma environment. Moreover, microplasma reactors can currently generate plasmas at pressures of the order of tens of Torr. In order to tune or better understand these plasmas, it is necessary to investigate the chemistry occurring in their generation and during the self-sustained discharge. Experimental diagnostic tools, such as Langmuir probe and optical emission spectroscopy, are used to determine plasma/gas parameters namely electron density, electron temperature, argon ion density, metastable species, gas temperature, etc. In argon discharges, several excited and metastable species are observed, and they affect the physics and chemistry of the medium due to occurrence of multistep and Penning ionization processes. These ionization modes appear commonly when discharge parameters such as pressure or power are increased, but are strongly dependent on the concentration and distribution of the metastable species as well as reactor geometry. Thus, simulation tools are essential to complement the experimental data and explain the discharge mechanisms. In this work, some primarily results of plasma simulations with collisional-radiative model are presented. Moreover, a revised set of collisional-radiative reactions are presented and discussed. Finally, for the case of

* E-mail address: rspessoa@univap.br.

magnetron sputtering plasma, a brief discussion about the influence of argon gas incorporation in the properties of the grown thin films by this technique is presented.

1. Introduction

Before starting the sections of this chapter, it is necessary to present a little history, properties, and applications of the argon gas. Argon is a member of the noble gas family, which is located in group 18 on the periodic table. It was discovered by Henry Cavendish in 1785 and was named argon, which is derived from the Greek word "argos" meaning inactive, a reference to the fact that the element undergoes almost no chemical reactions. Cavendish mixed air with oxygen and performed several electrical discharges over a mixture of potassium hydroxide. Through this process, he converted nitrogen from the air into potassium nitrate. Then, he removed the excess oxygen, but a small amount of gas still remained, which was not larger than 1/120 of the original volume of air. He guessed that another element was in the air, but he was unable to identify what it was [1, 2]. It was not until 1894 that Lord Rayleigh and Willian Ramsay repeated the experience of Cavendish, removing the nitrogen with heated magnesium. With use of spectroscopy, the residue revealed an unknown spectrum, and it was shown that this new gas was not combined with the nitrogen. The newfound element was thus named argon and was the first member of the noble gases to be discovered. The symbol for argon is now Ar, but up until 1957 it was A [1, 3, 4]. In earth's atmosphere (air), Ar contributes with 0.93% (9,300 ppm) of total composition, making it the third most abundant gas [1, 5]. The main isotopes found in the atmosphere are: ^{40}Ar (99.6%), ^{36}Ar (0.34%), and ^{38}Ar (0.06%) [1]. Its main characteristics include: (i) it is a colorless, odorless and tasteless gas; (ii) its density is 1.784 grams per liter, while the density of air, for comparison, is about 1.29 grams per liter; (iii) it changes from gas to liquid at 87.3 K, and then it changes from a liquid to a solid at 83.8 K; (iv) the ionization potential is 15.76 eV. For more characteristics, see reference [1]. The noble gases have a particularly stable electronic configuration, comprising fully filled s and p valence orbital. This makes these elements relatively non-reactive, and they exist at room temperature as monatomic gases. However, they have the capability to form some compounds. For example, the creation of argon fluorohydride (HArF), a stable compound of argon with fluorine and hydrogen, was reported by researchers at the University of Helsinki in 2000 [1, 6]. Although the neutral ground-state chemical compounds of Ar are presently limited to HArF, Ar can form clathrates with water when its atoms are trapped in a lattice of the water molecules [7]. Argon-containing ions and excited state complexes, such as ArH^+ and ArF^*, respectively, are known to exist [8, 9]. Theoretical calculations have predicted several Ar compounds that should be stable [1, 10], but for which no synthesis routes are currently known. Argon is produced industrially by the fractional distillation of liquid air in a cryogenic air separation unit, a process that separates liquid nitrogen, which boils at 77.3 K, from argon, which boils at 87.3 K, and liquid oxygen, which boils at 90.2 K [1]. Chapter 1 of this book and reference [5] presents more details about the Ar production process. Currently, about 700,000 tonnes of Ar are produced worldwide [1].

Ar gas is widely used both in industry and in scientific research. Its main application is in situations where materials need to be protected from oxygen or other gases. For example, it is used in the lighting industry for the filling of incandescent and fluorescent light bulbs.

Because it does not react with the filament, even at high temperatures, it protects it. Welding industry uses Ar as a shielding gas to protect metal from oxidation during welding. Moreover, Ar is effective in the field of thermal insulation and can be used as filler for windows with double glazing. Because of its low thermal conductivity, this gas improves insulation substantially and hence increases windows thermal efficiency [11]. Additionally, in the semiconductor industry, this gas is present in most of the processing steps of a chip or micro device. The main techniques used in these steps arc thermal annealing, atomic layer deposition, plasma assisted deposition, and etching of materials. [12-14]. In the field of scientific research about special Plasma Technology, Ar-based plasma is used as a source of energetic species that interacts with either solid material or precursor gas, creating diverse species, which can be used in thin film deposition, etching of materials, or activation of the surface of different materials as a pre-treatment to improve adhesion [14-16]. Furthermore, Ar can be used to "spectroscopically probe" the plasma, providing useful diagnostics such as information on the electron temperature and neutral species density (for the last case when Ar is mixed with molecular gas) [17]. In Ar-based plasmas, another important parameter is the number density of Ar metastable levels [18, 19]. Metastable levels play an important role in the energy kinetics, as they are in practice energy reserves [20]. The goal of this chapter is to review and present some recent studies about the chemistry of low-pressure discharges generated with Ar gas. For this, section 2 presents the main reactions of low-pressure Ar plasma chemistry as well as a revised set of collisional-radiative reactions. The next two sections are dedicated to showing the main experimental and simulation results in literature and by us in order to discuss the influence of Ar in plasmas generated in different reactor geometry types such as microplasmas, magnetron sputtering, and reactive ion etching. Finally, section 5 is a brief discussion about the influence of Ar incorporation on the properties of some grown thin films of interest in microelectronics and nanotechnology.

2. ARGON PLASMA CHEMISTRY

A reason to Ar was chosen as plasma forming gas in earlier studies in electric gas discharges was its chemical inertness. As no ordinary chemical reaction occurs in the discharge, it is easier to study the phenomena related to the "Plasma Physics" separating them from the phenomena of "Plasma Chemistry" inherent to reactive gas discharges. Because of this, Ar is often referred to as a background gas in reactive processes, used only to facilitate the plasma generation, without being important in chemical processes. However, this gas shows very interesting features not only in "Plasma Physics" but also in "Plasma Chemistry" due to the generation of excited species and ions in gas discharges. Among the tens of excited levels, there are many metastables states. These metastables affect considerably the electron energy distribution function (EEDF) [21-23]. A change in EEDF, by consequence, affects all the reactions involving electrons that, in general, are the dominant mechanisms in low-pressure gas discharges. The high concentration of metastable species facilitates the step ionization, reducing the energy cost to sustain the discharge [24], leading to an increase in electron density and consequent reduction of electron temperature, which affects the contraction of the positive column in high pressure [25]. In the presence of another gas species, the collisions with Ar metastables can be an important ionization mechanism through

Penning ionization [26]. In pure Ar discharges, if the concentration of metastable is high, the Penning ionization can be also relevant, due to collisions between metastables [14, 26].

The amount of energy that the metastables contain makes them chemically different from the atoms in ground state. This difference in chemical reactivity is more pronounced in the case of positive ions of Ar that have an incomplete valence shell unlike the neutral Ar atom. This unexpected reactivity of Ar can lead to interesting situations, for example a considerable density of Ar_2^+ ions in a discharge [25]. The presence of Ar ions can also affect processes like film deposition, due to incorporation of Ar in the film bulk [27, 28]. Despite the fact that these "reactive Ar atoms and ions" are not as reactive as a molecular gas like O_2, it is important to notice that the presence of Ar in reactive plasma is always accompanied by chemical effects, generally due to metastables. In the presence of molecular gases, the excitation energy of metastable levels is an efficient way to dissociate molecules like O_2, generating atomic species with greater reactivity [29]. Ar ions strongly influence the physical and chemical mechanism of etching processes in discharges with fluorine gases like SF_6 and CF_4 [15]. In film deposition and plasma nitriding, the bombardment of the samples with Ar ions is a known technique to transfer energy, due to high mass of Ar, affecting the kinects mechanisms of film growth or metal nitriding.

2.1. Argon Species and Collisional-Radiative Processes

In general, the low-pressure Ar gas/plasma phase chemistry is composed of five species, namely thermal argon atom in the ground state (Ar_g), different argon excited atoms (Ar_i, Ar_j), atomic ion (Ar^+), molecular ion (Ar_2^+), and electron (e). The main collisional and radiative processes involved between them are as follows: elastic scattering,

$$e + Ar_g \rightarrow Ar_g + e \qquad \text{(R0)}$$

electron impact excitation from the ground state Ar_g and de-excitation,

$$e + Ar_g \rightarrow Ar_i + e \qquad \text{(R1)}$$

$$e + Ar_i \rightarrow Ar_g + e$$

electron impact transition between two excited levels (Ar_i and Ar_j are the ith and j^{th} excited levels here),

$$e + Ar_i \rightarrow Ar_j + e \qquad \text{(R2)}$$

direct electron impact ionization,

$$e + Ar_g \rightarrow Ar^+ + 2e \qquad \text{(R3)}$$

stepwise ionization,

$$e + Ar_i \rightarrow Ar^+ + 2e \quad \text{(R4)}$$

Penning ionization [14, 26, 27],

$$Ar_i + Ar_i \rightarrow Ar^+ + Ar_g + e \quad \text{(R5)}$$

molecular ion formation [30],

$$Ar_g + Ar_g + Ar^+ \rightarrow Ar_g + Ar_2^+ \quad \text{(R6)}$$

dissociative recombination [30],

$$Ar_2^+ + e \rightarrow Ar_g + Ar_i \quad \text{(R7)}$$

spontaneous radiation and radiation trapping [29],

$$Ar_i \rightarrow Ar_g + hv$$

$$hv + Ar_g \rightarrow Ar_i \quad \text{(R8)}$$

$$Ar_j \rightarrow Ar_i + hv$$

and wall deactivation of metastable argon atoms Ar_m and argon ion Ar^+ [14, 30],

$$Ar_m \xrightarrow{wall} Ar_g \quad \text{(R9)}$$

$$Ar^+, Ar_2^+ \xrightarrow{wall} Ar_g$$

Figure 1 presents a simplified energy level diagram of the Ar atom illustrating all reactions listed above. The ground state of Ar is characterized by the closed layer $3p$ and for the spin momentum equal to zero, i.e., $\vec{S} = 0$ [31]. Individually, each excited state of Ar I (neutral atom) has the configuration $3s^2 3p^5 nls$, where, n, l, and s are, respectively, the quantum number principal, angular, and the spin of the outer electron.

The atomic core electrons, whose electronic distribution is $\{Mg\}3p^5$, couple first resulting in the vectors of spin $\vec{s}_c$ and angular momentum $\vec{l}_c$, which then are coupled to yield the total angular momentum vector of the core $\vec{j}_c$. In the case of Ar, the two possible values for $\vec{j}_c$ are $(1/2)$ and $(3/2)$. Taking into account the excited electron with orbital momentum $\vec{l}$, an intermediate vector $\vec{\mathrm{K}}$ is given by:

$$\vec{\mathrm{K}} = \vec{j}_c + \vec{l} \quad (1)$$

which in turn couples with the vector of spin $\vec{s}$ to form the total angular momentum vector $\vec{\mathrm{J}}$ of the complete atom, i.e.,

$$\vec{J} = \vec{K} + \vec{s}. \quad (2)$$

Due to the coupling $\vec{j}_c\vec{K}$, the energy levels of Ar I are divided into two sub-systems with different ionization limits [32]: "primed system" with $\vec{j}_c = 1/2$, ionization threshold equal to $15.937eV$, and the "unprimed system" with $\vec{j}_c = 3/2$, ionization threshold equal to $15.759eV$. Figure 2 shows a diagram of the energy levels of Ar. In total, there are 96 discrete effective levels, and all are below the first threshold (numerical value) of ionization. In Racah notation, the two systems of terms of the Ar I atoms are characterized by [31]:

$$[^2P_{1/2}]nl[\mathrm{K}]_\mathrm{J} \text{ para } J_c' = 1/2$$

$$[^2P_{3/2}]nl[\mathrm{K}]_\mathrm{J} \text{ para } J_c = 3/2$$

with,

$$K = J_c + l, J_c + l - 1, \ldots., |J_c - l|$$

$$\mathrm{J} = K \pm s = K \pm 1/2$$

For the designations in Paschen and Racah notation, the excitation energies and statistical weights of the effective energy levels are listed in Table 1.

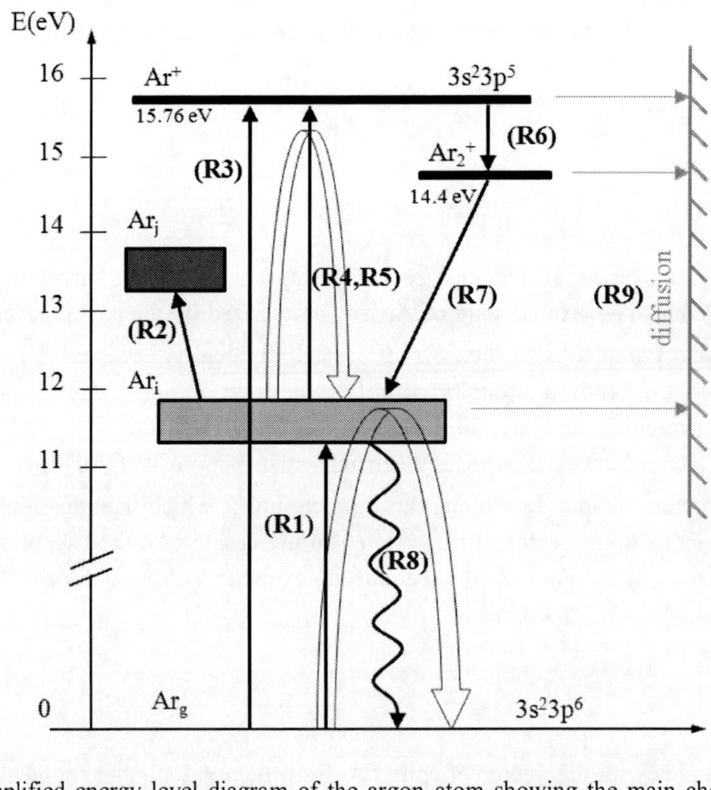

Figure 1. Simplified energy level diagram of the argon atom showing the main chemistry generated during the plasma. The excited states Ari and Arj are treated as a lumped level.

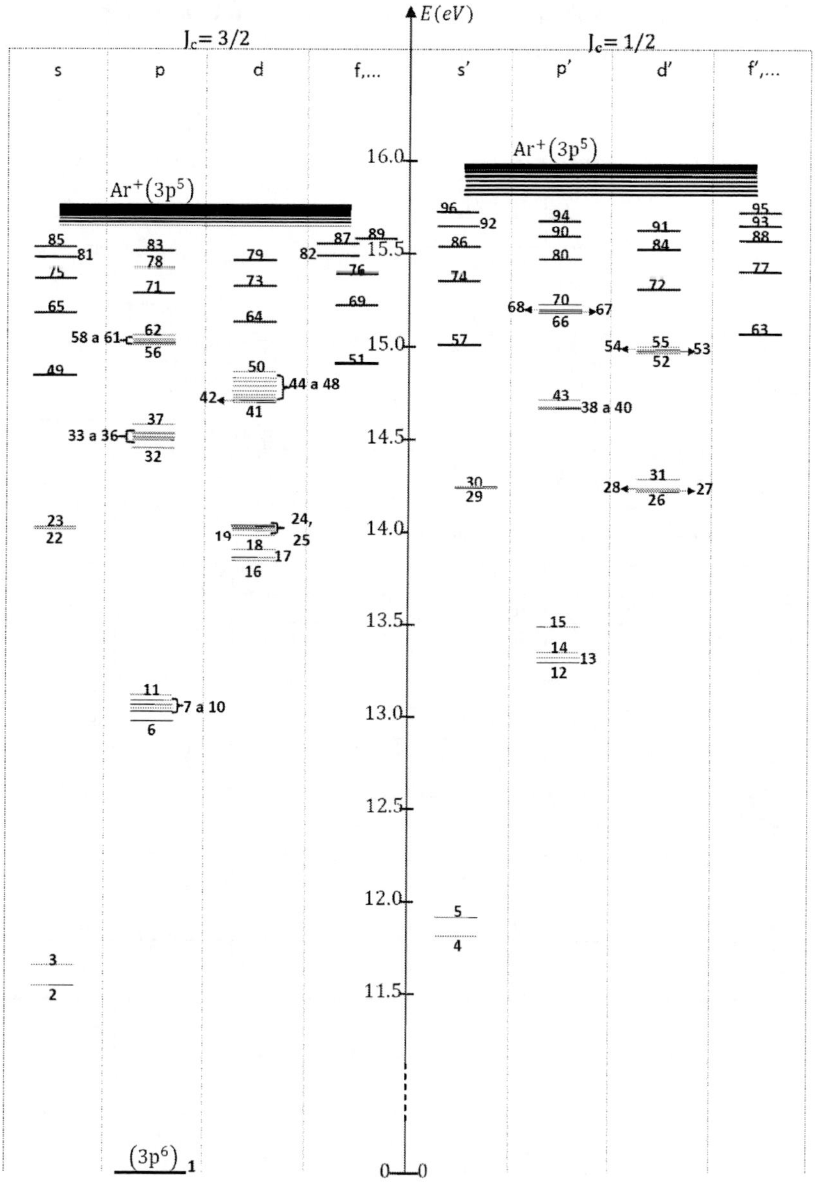

Figure 2. Diagram of energy levels of the argon atom [40].

These are indicated with the number S.N. (state number) of the level. It is important to notice that, to work with the present number of energy levels of Ar, Table 1, some levels were grouped (6s, 4f, 6s', 4f',....,10s'). The levels n(s, p, d, f, ...) are related with $J_c = 3/2$, whereas n(s', p', d', f', ...) are related with $J_c = 1/2$, and n is equal to 4, 5,...., 10. For this notation of Racah, where the energy levels are grouped, there is no corresponding Paschen notation. In the next sections of this chapter, we try to use the Racah notation; however, sometimes the Paschen notation will be used due to writing simplification. The chemistry of Ar excited species presented in this section was based on a detailed bibliographic search where the main references are cited below:

Table 1. List of energy levels of Ar I. S.N. refers to the state number, E and g are the statistical weight and energy level, respectively [40]

S.N.	Paschen Notation	Racah Notation $nl[K]_J$	E (cm^{-1})	E (eV)	g
1	$1p_0$	$3p_6$	0	0	1
2	$1s_5$	$4s[3/2]_2$	93144	11.548	5
3	$1s_4$	$4s[3/2]_1$	93751	11.624	3
4	$1s_3$	$4s'[1/2]_0$	94554	11.723	1
5	$1s_2$	$4s'[1/2]_1$	95400	11.828	3
6	$2p_{10}$	$4p[1/2]_1$	104102	12.907	3
7	$2p_9$	$4p[5/2]_3$	105463	13.076	7
8	$2p_8$	$4p[5/2]_2$	105617	13.095	5
9	$2p_7$	$4p[3/2]_1$	106087	13.153	3
10	$2p_6$	$4p[3/2]_2$	106238	13.172	5
11	$2p_5$	$4p[1/2]_0$	107054	13.273	1
12	$2p_4$	$4p'[3/2]_1$	107132	13.283	3
13	$2p_3$	$4p'[3/2]_2$	107290	13.302	5
14	$2p_2$	$4p'[1/2]_1$	107496	13.328	3
15	$2p_1$	$4p'[1/2]_0$	108723	13.480	1
16	$3d_6$	$3d[1/2]_0$	111668	13.845	1
17	$3d_5$	$3d[1/2]_1$	111818	13.864	3
18	$3d'_4$	$3d[3/2]_2$	112139	13.903	5
19	$3d_4$	$3d[7/2]_4$	112750	13.979	9
20	$3d_3$	$3d[7/2]_3$	113020	14.013	7
21	$3d_2$	$3d[5/2]_2$	113426	14.063	5
22	$2s_5$	$5s[3/2]_2$	113469	14.068	5
23	$2s_4$	$5s[3/2]_1$	113643	14.0899	3
24	$3d''_1$	$3d[5/2]_3$	113717	14.099	7
25	$3d'_1$	$3d[3/2]_1$	114148	14.153	3
26	$3s''''_1$	$3d[5/2]_2$	114641	14.214	5
27	$3s'''_1$	$3d[3/2]_2$	114805	14.234	5
28	$3s''_1$	$3d[5/2]_3$	114822	14.236	7
29	$2s_3$	$5s[1/2]_0$	114862	14.241	1
30	$2s_2$	$5s[1/2]_1$	114975	14.255	3
31	$3s'_1$	$3d[3/2]_1$	115367	14.304	3
32	$3p_{10}$	$5p[1/2]_1$	116660	14.464	3
33	$3p_9$	$5p[5/2]_3$	116943	14.499	7
34	$3p_8$	$5p[5/2]_2$	116999	14.506	5
35	$3p_7$	$5p[3/2]_1$	117151	14.525	3
36	$3p_6$	$5p[3/2]_2$	117184	14.529	5
37	$3p_5$	$5p[1/2]_0$	117563	14.576	1
38	$3p_4$	$5p'[3/2]_1$	118407	14.681	3
39	$3p_3$	$5p'[3/2]_1$	118460	14.687	3
40	$3p_2$	$5p'[1/2]_2$	118469	14.688	5
41	$4d_6$	$4d[1/2]_0$	118512	14.694	1
42	$4d_5$	$4d[1/2]_1$	118651	14.711	3
43	$3p_1$	$5p'[1/2]_0$	118871	14.738	1
44	$4d'_4$	$4d[3/2]_2$	118907	14.743	5
45	$4d_4$	$4d[7/2]_4$	119024	14.757	9
46	$4d_3$	$4d[7/2]_3$	119213	14.781	7
47	$4d_2$	$4d[5/2]_2$	119445	14.809	5
48	$4d''_1$	$4d[5/2]_3$	119645	14.834	7
49		6s	119711.8	14.842	8
50	$4d_1$	$4d[1/2]_0$	119848	14.859	3
51		4f	120225	14.906	56

S.N.	Paschen Notation	Racah Notation $nl[K]_J$	E (cm^{-1})	E (eV)	g
52	$4s''''_1$	$4d[3/2]_2$	120601	14.953	5
53	$4s'''_1$	$4d[5/2]_2$	120619	14.955	5
54	$4s''_1$	$4d[5/2]_3$	120754	14.972	7
55	$4s'_1$	$4d[3/2]_1$	121012	15.004	3
56	$4p_{10}$	$6p[1/2]_1$	121069	15.011	3
57		6s'	121145	15.020	4
58	$4p_9$	$6p[5/2]_3$	121165	15.023	7
59	$4p_8$	$6p[5/2]_2$	121192	15.026	5
60	$4p_7$	$6p[3/2]_1$	121257	15.034	3
61	$4p_6$	$6p[3/2]_2$	121271	15.036	5
62	$4p_5$	$6p[1/2]_0$	121470	15.060	1
63		4f'	121652.6	15.083	28
64		5d	122168	15.147	40
65		7s	122454.6	15.182	8
66	$4p_4$	$6p'[1/2]_1$	122601	15.201	3
67	$4p_3$	$6p'[3/2]_1$	122610	15.202	3
68	$4p_2$	$6p'[3/2]_2$	122635	15.205	5
69		5f	122708	15.214	56
70	$4p_1$	$6p'[1/2]_0$	122791	15.224	1
71		7p	123256.6	15.282	24
72		5d'	123537	15.317	20
73		6d	123753.7	15.344	40
74		7s'	123879.7	15.359	4
75		8s	123915	15.364	8
76		6f	124055	15.381	56
77		5f'	124140	15.391	28
78		8p	124390	15.422	24
79		7d	124652	15.455	40
80		7p'	124698.7	15.461	12
81		9s	124776	15.470	8
82		7f	124868	15.482	56
83		9p	125090	15.509	24
84		6d'	125142	15.516	20
85		10s	125330.7	15.539	8
86		8s'	125348.5	15.541	4
87		8f	125395	15.547	56
88		6f'	125487	15.558	28
89		9f	125756	15.592	56
90		8p'	125823.3	15.600	12
91		7d'	126096.7	15.634	20
92		9s'	126280	15.657	4
93		7f'	126300	15.659	28
94		9p'	126520	15.686	12
95		8f'	126827	15.724	28
96		10s'	126831	15.725	4
97		ion	127105	15.759	6

From 1980s, the works of Bretagne et al. [33] and J. Vlcek [34] propose a complete energy level diagram with 65 discrete effective levels. These works were followed by other researchers such as Guimarães and Bretagne (1993), Bogaerts et al. (1998), Kelkar et al. (1999) and Bultel et al. (2002), but most of them still make use of experimental electron impact cross sections obtained well before the 1990s [32, 35-37]. Then in 2005, A. Yanguas-Gil et al. presented an update of argon inelastic cross sections for plasma discharges and, consequently, a reviewed list of Ar excited levels [38]. This work considers 39 individual and lumped Ar excited states, ranging from the $3s^2p^6$ neutral ground state at 0 eV up to the $3s^2p^5$ singly ionized state at 15.76 eV. In 2007, A. Palmero et al. presented a simplified version of the energy level diagram of Ar presented by J. Vlcek justifying its complexity and the time required to find a stable solution when this is used in plasma models [39]. In a recent work, M. P. Gomes [40] developed a more complete set of reactions for Ar excited species. In many Ar plasmas, taking into account the experimental conditions, the most intense emission lines in the visible range of the electromagnetic spectrum, are associated with transitions from 4p to 4s levels (emission in red) and 5p to 4s (emission in the blue) [41].

Due to this fact, the radiative transitions considered in the input of the collisional-radiative (CR) models (see section 3) take mostly into account the transitions of the 4p and 5p levels, both for the level 4s as shown in Figure 3. Tables 2 and 3 present the values of the probabilities of radiative transitions for two of the most intense emission lines. Within the 4s levels, the levels $4s[3/2]_2$ (second level in Table 1) and $4s'[1/2]_0$ (fourth level in Table 1) have radiative lifetimes in the order of seconds.

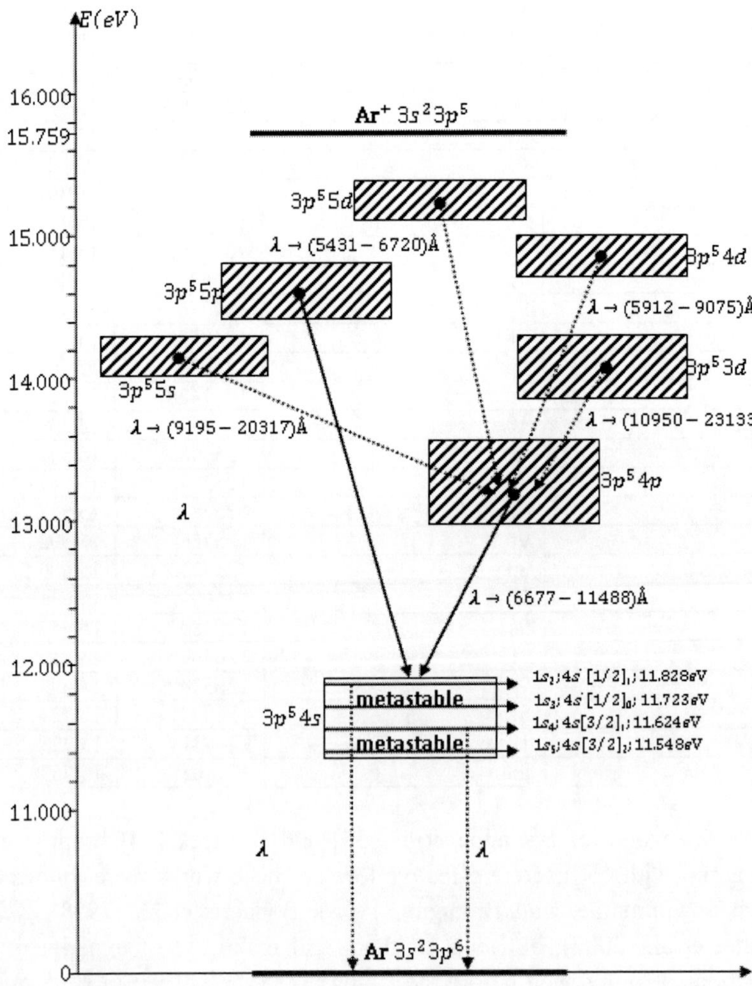

Figure 3. Emission lines of Ar I due to transitions between energy levels [40].

This lifetime is long enough to make other competing depopulating processes, such as electron collisional excitation or deexcitation processes, or diffusion towards and deexcitation at the walls, dominate over the radiative decay. This is why these two levels are known as metastable, and they do not live long enough to radiatively decay in the plasma under normal conditions.

Conversely, for the levels $4s[3/2]_1$ (third level in Table 1) and $4s'[1/2]_1$ (fifth level in Table 1), the radiative decay processes are the most important deexcitation processes, being responsible for emitting a large number of photons under typical plasma conditions. These are called resonance levels [22, 39]. The next block of levels in the energy scheme corresponds to the so-called 4p configuration of Ar, ranging from level 5 to 15. Each 4p level is coupled to the rest of the Ar levels through different optically allowed transitions, or different electron impact excitation cross sections.

Table 2. Atomic transition probabilities between the 4*p* and 4*s* levels of Ar atom [40]

$5p$		$4p$	data		
Racah notation	Paschen notation	Racah and Paschen notation	λ(nm)	$A(10^6 s^{-1})$	Uncertainty (%)
$4p'[1/2]_0$	$2p1$		750.358	44.5	±8
$4p'[1/2]_1$	$2p2$		826.452	13.3	±6
$4p'[3/2]_2$	$2p3$		840.821	22.3	±5
$4p'[3/2]_1$	$2p4$		852.144	13.9	±8
$4p[1/2]_0$	$2p5$	$4s'[1/2]_1, 1s2$	857.806	< 0.001	-
$4p[3/2]_2$	$2p6$		922.450	5.03	±10
$4p[3/2]_1$	$2p7$		935.422	1.06	±8
$4p[5/2]_2$	$2p8$		978.450	1.47	±8
$4p[1/2]_1$	$2p10$		1148.811	0.19	±30
$4p'[1/2]_1$	$2p3$		772.421	5.03	±5
$4p'[3/2]_1$	$2p4$	$4s'[1/2]_0, 1s3$	794.818	1.06	±8
$4p[3/2]_1$	$2p7$		866.794	1.47	±8
$4p[1/2]_1$	$2p10$		1047.005	0.19	±15
$4p'[1/2]_0$	$2p1$		667.728	0.236	±15
$4p'[1/2]_1$	$2p2$		727.293	1.83	±5
$4p'[3/2]_2$	$2p3$		738.398	8.47	±8
$4p'[3/2]_1$	$2p4$		747.117	0.022	±8
$4p[1/2]_0$	$2p5$	$4s[3/2]_1, 1s4$	751.465	40.2	±8
$4p[3/2]_2$	$2p6$		800.616	4.90	±15
$4p[3/2]_1$	$2p7$		810.369	25.0	±5
$4p[5/2]_2$	$2p8$		842.465	21.5	±5
$4p[1/2]_1$	$2p10$		965.778	5.43	±8
$2p'[1/2]_1$	$2p10$		694.543	6.39	±5
$2p'[3/2]_2$	$2p2$		706.722	3.80	±8
$2p'[3/2]_1$	$2p4$		714.704	0.625	±8
$2p[3/2]_2$	$2p6$	$4s[3/2]_2, 1s5$	763.511	24.5	±8
$2p[3/2]_1$	$2p7$		772.376	5.18	±5
$2p[5/2]_2$	$2p8$		801.479	9.28	±8
$2p[5/2]_3$	$2p9$		811.531	33.1	±8
$2p[1/2]_1$	$2p10$		912.297	18.9	±8

Table 3. Atomic transition probabilities between the levels 5*p* and 4*s* of Ar atom [40]

$5p$		$4p$	dados		
Racah notation	Paschen notation	Racah and Paschen notation	λ(nm)	$A(s^{-1})$	Uncertainty (%)
$5p'[1/2]_0$	$3p1$		425.936	3.98	± 6
$5p'[1/2]_1$	$3p2$		433.534	0.387	± 12
$5p'[3/2]_2$	$3p3$		433.356	0.568	± 7
$5p'[3/2]_1$	$3p4$		434.517	0.297	± 10
$5p[1/2]_0$	$3p5$	$4s'[1/2]_1, 1s2$	451.073	1.18	± 7
$5p[3/2]_2$	$3p6$		458.929	0.0062	± 20
$5p[3/2]_1$	$3p7$		459.610	0.0947	± 9
$5p[5/2]_2$	$3p8$		462.844	0.0383	± 6
$5p[1/2]_1$	$3p10$		470.232	0.109	± 9
$5p'[1/2]_1$	$3p2$		418.188	0.561	± 9
$5p'[3/2]_1$	$3p4$	$4s'[1/2]_0, 1s3$	419.103	0.539	± 15
$5p[3/2]_1$	$3p7$		442.400	0.0073	± 15
$5p[1/2]_1$	$3p10$		452.232	0.0898	± 8
$5p'[1/2]_0$	$3p1$		397.972	-	-
$5p'[1/2]_1$	$3p2$		404.596	0.041	± 10
$5p'[3/2]_2$	$3p3$		404.442	0.333	± 9
$5p'[3/2]_1$	$3p4$		405.453	0.027	± 15
$5p[1/2]_0$	$3p5$	$4s[3/2]_1, 1s4$	419.832	2.57	± 8
$5p[3/2]_2$	$3p6$		426.629	0.312	± 12
$5p[3/2]_1$	$3p7$		427.217	0.797	± 8
$5p[5/2]_2$	$3p8$		430.010	0.377	± 5
$5p[1/2]_1$	$3p10$		436.379	0.012	± 10
$5p[1/2]_1$	$3p10$		394.898	0.455	± 9
$5p'[3/2]_2$	$3p3$		394.750	0.056	± 15
$5p'[3/2]_1$	$3p4$		395.7	-	-
$5p[3/2]_2$	$3p6$	$4s[3/2]_2, 1s5$	415.859	1.40	± 9
$5p[3/2]_1$	$3p7$		416.418	0.288	± 9
$5p[5/2]_2$	$3p8$		419.071	0.280	± 8
$5p[5/2]_3$	$3p9$		420.067	0.967	± 7
$5p[1/2]_1$	$3p10$		425.118	0.111	± 15

The scheme of effective levels presented in [32, 42] were grouped into four levels according to their rate coefficient. The first one includes the 4p'$[1/2]_0$ and 4p$[1/2]_0$ levels. The second group consists of the 4p'$[3/2]_2$, 4p$[3/2]_1$ and 4p$[5/2]_2$ levels, having a large excitation rate coefficient from a resonance level. The third group has 4p'$[3/2]_1$ and 4p$[5/2]_3$, both of which have a large excitation rate coefficient from a metastable level. They have large excitation rate coefficients from the ground state (per degeneracy degree, see ref. [42]). The last group of 4p levels includes the 4p'$[1/2]_1$, 4p$[3/2]_2$, and 4p$[1/2]_1$. These levels have a relatively small rate coefficient of atom-collision population transfer [42]. Above the 4p levels, we have included levels corresponding to those with so-called 3d and 5s configurations. The reason to include these excited levels is that the levels 3d and 5s are strongly coupled to Ar upper excited states through radiative cascades, as well as to the 4p levels through optical allowed radiative transitions [32, 39]. Finally, above the 3d-5s levels, there are the 5p, 4d, 5d, and other high levels, and above then corresponds to the ionization channel where one electron is in the continuum (see Figure 2 and Table 1).

2.2. Kinetic Coefficients of the Reactions

The kinetic coefficients relating to electron impact excitations from level i to level j, with i < j, namely, k_{ij}, can be calculated using the expression

$$k_{ij} = \sqrt{\frac{2}{m_e}} \int_{\Delta\varepsilon_{ij}}^{+\infty} dE f(E) \sigma_{ij}(E) E \tag{3}$$

where E is the kinetic energy of the electron, $f(E)$ is the EEDF, normalized as $\int_0^{+\infty} dE f(E)\sqrt{E} = 1$, and $\Delta\varepsilon_{ij} = \varepsilon_j - \varepsilon_i$ the threshold energy of the process, ε_i being the energy of the level i, m_e the electron mass, and $\sigma_{ij}(u)$ the energy dependent angle integrated cross section for the process, defined as

$$\sigma_{ij}(E) = \sigma_{ij}^{A}(E) + \sigma_{ij}^{S}(E) + \sigma_{ij}^{P}(E) \tag{4}$$

where $\sigma_{ij}^{A}(E)$ is the cross section for optically allowed transitions, and $\sigma_{ij}^{S}(E)$ and $\sigma_{ij}^{P}(E)$ are the cross sections for optically spin forbidden and parity forbidden transitions, respectively. Under this description, only the corresponding cross section associated with the applicable transition between Ar levels (optically allowed, spin forbidden, or parity forbidden) is nonzero. For Ar gas, the cross sections can be taken in references [34, 42], which was found appropriate to describe the electron impact excitations in Ar [38]. More information about the cross sections used and the values of the parameters used in their semiempirical formulas can be found in [34, 42] and references therein.

For the electron collisional deexcitation process, from level j to level i, with i < j, the cross section of the process, $\sigma_{ji}(E)$, can be obtained using the detailed balance principle [43]. Thus, the kinetic coefficient for the electron collisional deexcitation process is

$$k_{ji} = \sqrt{\frac{2}{m_e}} \int_{0}^{+\infty} dE f(E) \sigma_{ji}(E) E \tag{5}$$

The electron impact ionization process from level j is characterized by a kinetic coefficient, namely, $k_{j\lambda}$, calculated as

$$k_{j\lambda} = \sqrt{\frac{2}{m_e}} \int_{\varepsilon_{j\lambda}}^{+\infty} dE f(E) \sigma_{j\lambda}(E) E \tag{6}$$

where $\sigma_{j\lambda}(E)$ is the ionization cross section, and $\Delta\varepsilon_{j\lambda} = \varepsilon_\lambda - \varepsilon_j$, with ε_λ being the ionization energy of argon (ε_λ = 15.76 eV) taken from [34]. In all the calculations of kinetic coefficients, it generally is assumed that the EEDF is Maxwellian; and therefore, $k \equiv k(T_e)$, where T_e is the electron temperature. The assumption of a Maxwellian distribution function is not always correct for low-pressure discharges; however, it can be used as a first approximation. In several cases, bi-Maxwellian or Druyvestein distribution functions are more appropriate to describe the electrons in the discharge.

The coefficient for the radiative decay from level j to level i, with j > i, A_{ji}^{eff}, is given by

$$A_{ji}^{eff} = \Lambda_{ji} A_{ji} \tag{7}$$

where A_{ji} is the Einstein coefficient for the transition, and Λ_{ji} the escape factor, which takes into account that an emitted photon can be reabsorbed by the plasma, via self-absorption. The probability of self-absorption increases with the population density of the lower level in the radiative transition, but in low-pressure Ar plasmas, the population density of the excited states is usually low enough to neglect this process.

In addition, the diffusive losses of the 4s metastable levels are given by the quantity $D_j \nabla^2 n_j$, where n_j denotes the population density of the excited metastable Ar atoms in level j and D_j the diffusion coefficient. All these reactions and respective kinetic coefficients can be used in computational plasma models that allow investigating the behavior of all relevant species in the plasma as well as their energy distribution as a function of input electromagnetic power or the gas pressure.

3. Simulation of Argon Plasmas

The kinetics of the plasmas species, in special excited species, is an important issue in low-pressure discharges. Detailed investigation of this kinetics can lead to better understanding of the electron dynamics, the radiative and chemical properties of plasmas. Several experimental and theoretical methods are used to investigate the plasma kinetics, and among these, plasma simulation stands out because it allows an overview of input and output parameters that govern the physics and chemistry of the plasma. The main simulation methods used are: collisional-radiative model [29, 34, 42, 44], global model (or zero-dimensional model) [18, 39], fluid model [45-47], and particle-in-cell [14, 47, 48].

3.1. Collisional-Radiative Model: Results and Advances

Collisional-radiative (CR) models are widely used in analyzing Ar discharges [34, 42, 44]. In CR model, generally, the plasma is characterized by a set of parameters, including the electron temperature, gas temperature, electron density, gas density, and the discharge geometry, such as tube radius. Tens of effective levels or hundreds of individual levels ranging from the Ar neutral ground state to the Ar singly ionized state (Table 1) are included. By considering the processes by which the levels are populated or depopulated, the rate balance equation of each level is obtained and solved. Thus, from a CR model and under certain discharge conditions, one can determine the most important processes for each level or calculate the concentration of a level, if the basic data involved with this level, such as electron impact cross sections and Einstein coefficient, are measured or calculated accurately.

Moreover, those CR models can be used to determine plasma parameters by optical emission spectroscopy (OES) in discharges containing Ar [39, 40, 42]. Several works noticed that the study of the excitation and deexcitation mechanisms of the 4p levels are important to the radiative properties of gas discharges due to the strong optical transitions between 4p and 4s levels. A. Palmero et al. developed a simple CR model based in effective 4p levels and some high levels (3d, 5s) to correlate optical emission lines from the radiofrequency (rf) magnetron sputtering plasma with the corresponding electron density and temperature [39]. Through comparisons with Langmuir probe measurements, they observed good agreement with model results and evidenced that the 4p levels play an important role in the excitation at an electron density above 10^{16} m^{-3} and pressures of order of mTorr. In addition, they found that the higher Ar levels have little influence in the investigated processes. In a recent work, X.-M. Zhu and Y.-K. Pu developed a simple CR model, in which individual 4s and 4p levels are used instead of the effective levels, to study low-temperature Ar discharges [42]. By using this model, the dominant kinetic processes of 4s and 4p levels are investigated for different reactor types, such as inductively coupled plasma (ICP) and capacitively coupled plasma (CCP).

Figure 4 illustrates some of their results. The population distribution of 1s and 2p levels (4s and 4p in Racah notation) is very different between the two types of low-pressure discharges. In ICP, the concentration of metastable $1s_5$ and $1s_3$ levels is much larger than resonance $1s_4$ and $1s_2$ levels; however, these levels have an equilibrium-like distribution for the medium pressure CCP (Figure 4a).

Relative to the 2p species (Figure 4b), the $2p_9$ level has the largest population for the ICP. For the medium-pressure CCP, the 2p levels have an equilibrium-like distribution with large population of the lowest $2p_8$, $2p_9$, and $2p_{10}$ levels. In particular, long-lived excited states "metastables" play a major role in maintaining the ionization balance due to their extremely low excitation threshold energy and large ionization cross sections. Consequently, the metastables can become an important source of ionization [20].

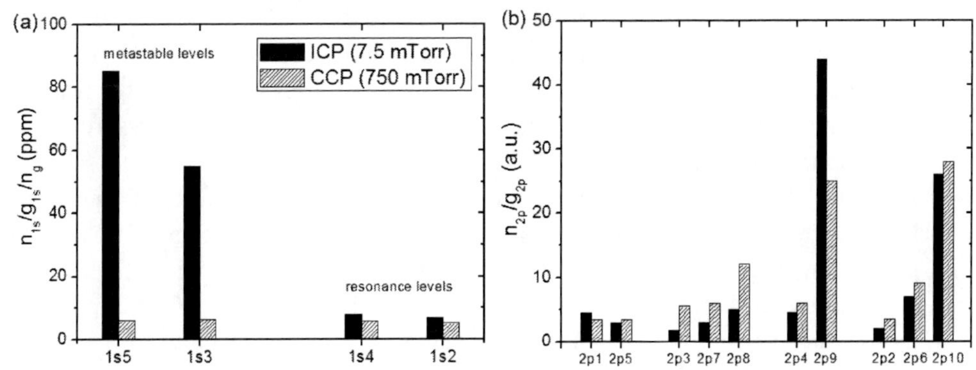

Figure 4. (a) Population density of 1s levels obtained from CR model for ICP and CCP Ar discharges. Here, n1s refers to the population density of 1s levels (Paschen notation), ng refers to the gas density, and g1s is the degeneracy degree. (b) Population distribution of 2p levels obtained from CR model for ICP and CCP Ar discharges. Here, n2p means the population density of 2p levels (Paschen notation), and g2p is the degeneracy degree. The values of n2p/g2p are normalized to one hundred (adapted from ref. [42]).

If the metastable density is high enough (ICP case) and the electron impact ionization kinetics is dominant, the two-step via metastables or multistep ionizations contribute significantly to the ionization balance of the plasma; however, at low electron density (order of 10^{15} m^{-3}, CCP case) direct ionization from the ground state mainly takes place. Therefore, the effect of stepwise ionization is expected to increase as the electron density increases. Several works have measured, using CR and global models, the behavior of Ar metastable species in ICP as a function of discharge parameters such as rf applied power and pressure; among them are references [20, 21, 42]. In general, for ICP, the density of the metastable species and, consequently, resonant and 4p species increase with pressure and power [20, 21, 51]. Depending on the operating condition, the metastable density can be equal or exceed the density of electrons and/or Ar^+ [21]. On the other hand, the density of resonance and 4p species are generally one to two orders of magnitude lower than metastable species [19, 52]. In a practice example, Y.-K. Lee and C.-W. Chung showed that the stepwise ionization accounts for about 3% at 3 mTorr and 134 W, but it reaches up to about 80% at 50 mTorr and 406 W for an ICP discharge [19]. However, for pressure values higher than 30 mTorr, the contribution of two-step ionization via metastables slightly decreases with increasing absorbed power. This indicates that the considerable increase in ionization occurs via other excited levels, especially the 4p level, which was supported by the higher 4p level density as well as its large ionization rate coefficient [19]. In order for a CR model to be valid for a wide range of pressures, the transitions between 4p and 4s levels as well as the population transfer processes between 4p levels and higher excited levels, such as 3d, 5s, and 5p levels, must be taken into account [42]. By including the processes for those higher levels, some other works developed a CR model for discharges at medium pressures, which was adopted for the investigation of discharges at low pressures, high pressures (75 Torr $< p <$ 750 Torr), or atmospheric pressure [42, 50, 51].

Additionally, from the studies of X.-M. Zhu and Y.-K. Pu, some kinetic regimes were defined by considering the dominant kinetic processes of 4s and 4p levels. In this chapter, we highlight four main regimes [42, 53, 54]:

- *C-regime*: the corona regime. The dominant processes are the ground-state excitation, the radiation to 4s levels, the cascade transition, and the resonance radiation to the ground state. It is suitable for discharges with low pressure and low ionization rate (order of 10^{-5});
- *L-regime*: the low-pressure nonequilibrium regime. In this regime, the metastable-level excitation becomes more important than the ground state excitation, i.e. the population transfer between 4s and 4p levels dominates. Reference [53] reports a nonequilibrium distribution with large $2p_9$ ($4p[5/2]_3$ in Racah notation) population due to the nonequilibrium distribution among 4s levels (see Fig. 4b). Here, the radiative processes are important;
- *H-regime*: the high-pressure nonequilibrium regime. The dominant processes are the electron-impact excitation from 4s levels and the atom-collision population transfer from 4p levels;
- *S-regime*: the partial local Saha equilibrium regime. The electron-impact population transfer is the only dominant process. It occurs at high pressure and high ionization

rate, in which the excited particles are in equilibrium and have the Boltzmann distribution at an excitation temperature equal to the electron temperature

It is important to underscore that other sub-regimes that depend on collisional reactions, such as Penning ionization, three-body collisional association, and electron impact ionization, can superpose these regimes. For more information, please see the references [42, 53]. Figure 5 shows a graph of these different kinetic regimes for 4p levels. The discharge conditions are pressure 10^{-4}–10^{3} Torr, ionization rate 10^{-6}–10^{-3}, and gas temperature 400 K. The electron temperature is assumed to be decreasing as pressure increases, which more or less represents the relationship between the electron temperature and the pressure for most typical low-temperature nonequilibrium discharges. The plasma dimension is also assumed to be decreasing with pressure, from 10 cm to 100μm, in agreement with the operating conditions of several discharges considered in this work: ICP, CCP, magnetron sputtering, and micro plasma.

In the transitional regime between L-regime and S-regime, it is generally believed that the excited levels have a Boltzmann-like distribution with an excitation temperature between gas temperature and electron temperature, as reported in the literature on low-temperature atmospheric-pressure microplasmas. However, X.-M. Zhu and Y.-K. Pu found that the population distribution of 4p levels significantly deviates from the expected Boltzmann-like distribution in the transitional regime, over a wide pressure range from mTorr to atmospheric pressure [53], which compromises the results obtained with the Boltzmann plot method. This fact was observed by M. P. Gomes where the S-regime was observed only for micro plasmas operating at pressures above 400 Torr [40].

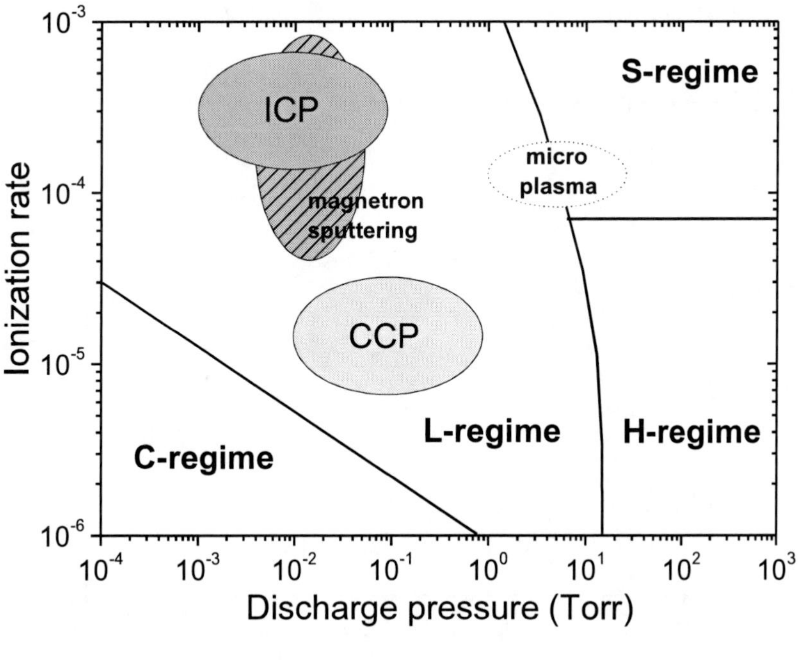

Figure 5. Kinetic regimes of 4p levels. The ICP, CCP, magnetron sputtering, and micro plasmas are denoted by the ionization rate and the pressure (adapted from [42, 53]).

3.2. Global Model of Ar+CF_4 ICP Plasma

Mixtures of Ar with other gases are often used in deposition or etching processes applied in microelectronics [14, 26]. In etching processes, the use of Ar is important to increase the ionization efficiency of the discharge and hence the density of radicals. Here we present an example of the mixture of Ar+CF_4 mixture in low-pressure plasma. Carbon tetrafluoride (CF_4) is a common etchant in the microelectronics industry. It is commonly used to etch silicon, silicon dioxide, and silicon nitride. The presence of carbon in its molecular structure allows, depending on the process parameters, highly anisotropic profiles to be obtained due to polymerization of the walls of etched structures. For the study of this discharge chemistry, simulation is common. This topic presents some results obtained by global model [14, 55, 56]. Although this plasma mixture is well investigated in the literature, few works focus on the study of Ar fraction (Ar%) in Ar+CF_4 mixture. A global model is formulated for the analysis of inductively coupled Ar+CF_4 plasma generated in a cylindrical aluminum chamber of radius R = 0.115 m and length L = 0.05 m. The governing equations are derived from mass and energy conservation arguments and may be expressed as a set of nonlinear ordinary differential equations. The global model was developed considering the most recently theory proposed by Kim et al. [56], which takes into account the modifications in electronegative region of plasma when the pressure (or electronegativity) is widely varied.

Moreover, the reactions examined in the model consider the presence of metastable, resonance and 4p state species for Ar chemistry [18], and for CF_4 chemistry, the most recent set of reactions were published by Kimura et al. [57]. Figure 6 shows the dependence of calculated atomic fluorine density near substrate surface, $n_{F,s}$, and atomic fluorine flux to substrate surface, Γ_F, as a function of Ar% in CF_4+Ar plasma. Two different total gas flows were investigated: 10 and 50 sccm, for a rf power fixed at 50 W.

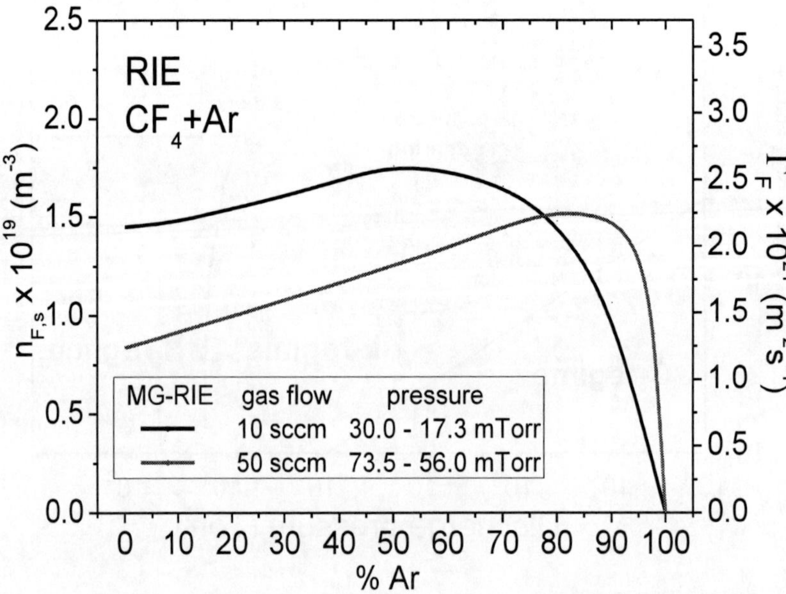

Figure 6. Effect of concentration of Ar over the $n_{F,s}$ and Γ_F for Ar+CF_4 plasma.

The results indicate an increase in $n_{F,s}$ with Ar% in up to 1.2 times for 10 sccm and up to 1.7 times for 50 sccm. Moreover, this increase in $n_{F,s}$ reaches a maximum value and is related to the elevation of electron density with Ar% that enhances the dissociative processes. It was also verified through the global model that charge exchange reactions contribute significantly to the creation of atomic fluorine [55]. However, for Ar% above the maximum value of $n_{F,s}$, the concentration of fluorinated gas starts to be limited, consequently reducing the production of F. Finally, the maximum value of $n_{F,s}$ shifts to higher Ar% as the flow (or pressure) is high. Similar dependence with Ar concentration was observed by Gudmundsson et al. for the product of the dissociation of O_2 gas (i.e., oxygen atom) in an Ar+O_2 ICP plasma [58].

4. Experimental Observations of Argon Plasmas

This section presents some of our main experimental results with Ar plasmas generated in the reactors: micro hollow cathode, magnetron sputtering, high power pulsed magnetron sputtering, and reactive ion etching.

4.1. Optical Spectrum of Micro Hollow Cathode Ar Plasma

One of the more stable configurations of micro plasma is hollow cathode geometry, where a hole is made through two parallel flat plates and a dielectric is placed between them [47, 59]. This hole should leak the entire assembly as shown in Figure 7. This configuration was used to generate Ar micro plasmas at pressures ranging from 10 to 760 Torr [40, 60]. For analysis of micro plasmas, the OES technique is often used. For this expirement, the radiation emitted by the Ar plasma was focused by means of an optical fiber of 1 mm i.d. into the entrance slit (width of 60 μm) of 1 m Czerny-Turner monochromator (THR1000) made by Jobin-Yvon. The monochromator was equipped with a holographic diffraction grating of 1800 lines mm^{-1}, blazed at 580 nm. The radiation was collimated at the exit slit where a photomultiplier tube, converted photons into an electric signal.

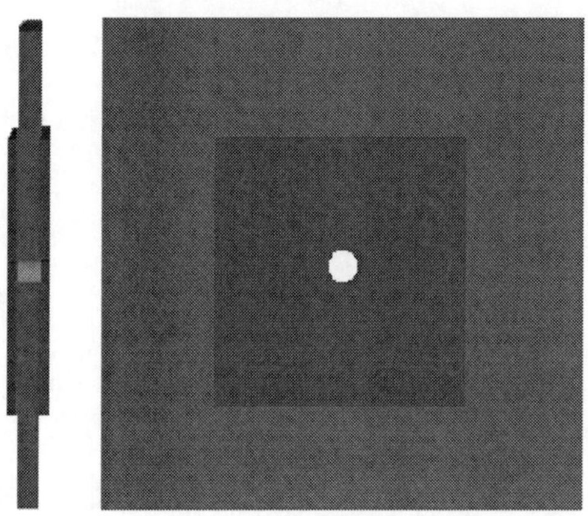

Figure 7. Experimental setup of open micro hollow cathode used in the production of Ar plasmas [40].

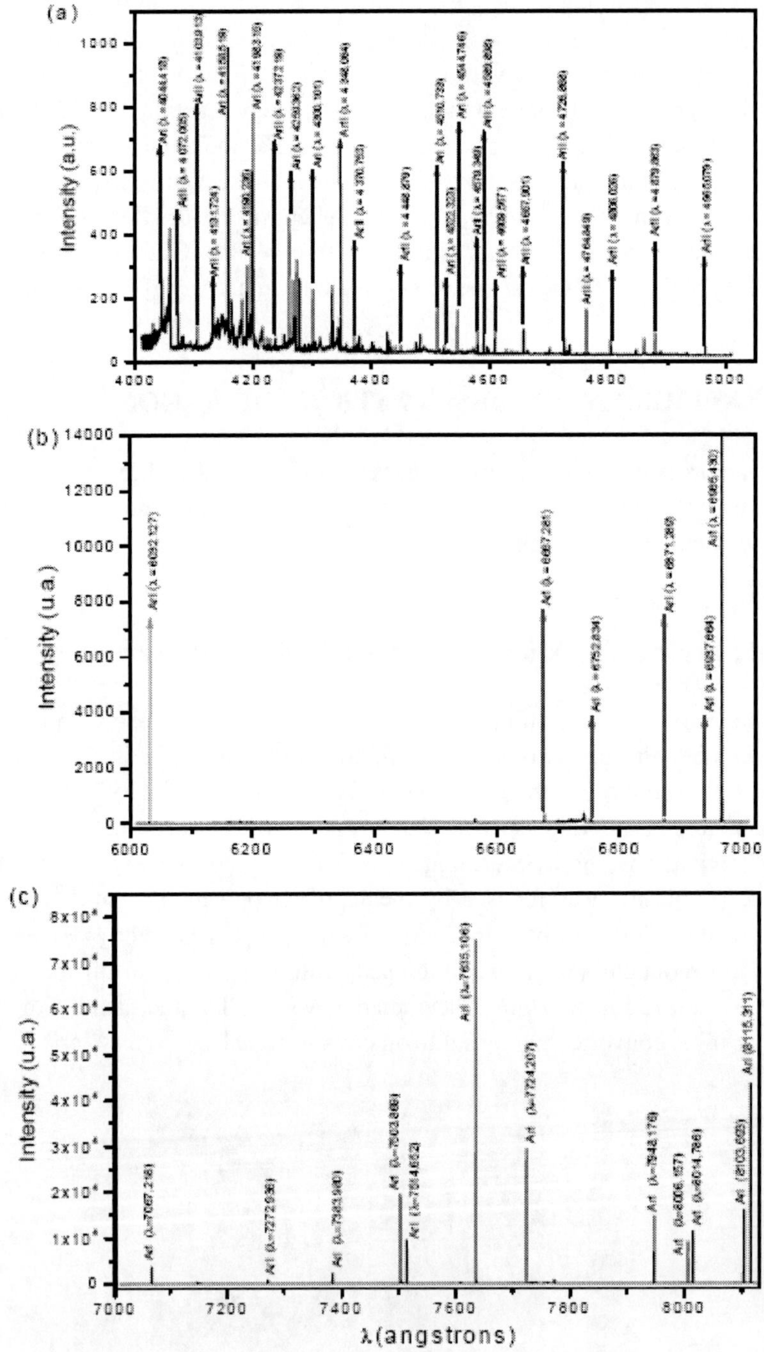

Figure 8. Emission spectra of Ar micro hollow cathode discharge obtained from a hole with a diameter of 250 μm, pressure of 60 Torr, and discharge current of 2 mA [40].

The current generated by the PMT was sent to data acquisition electronics, where the signal was processed by software. The optical system was absolutely calibrated with the aid of a tungsten lamp [60]. Figure 8 shows the spectra obtained for an Ar plasma generated in the micro hollow cathode reactor.

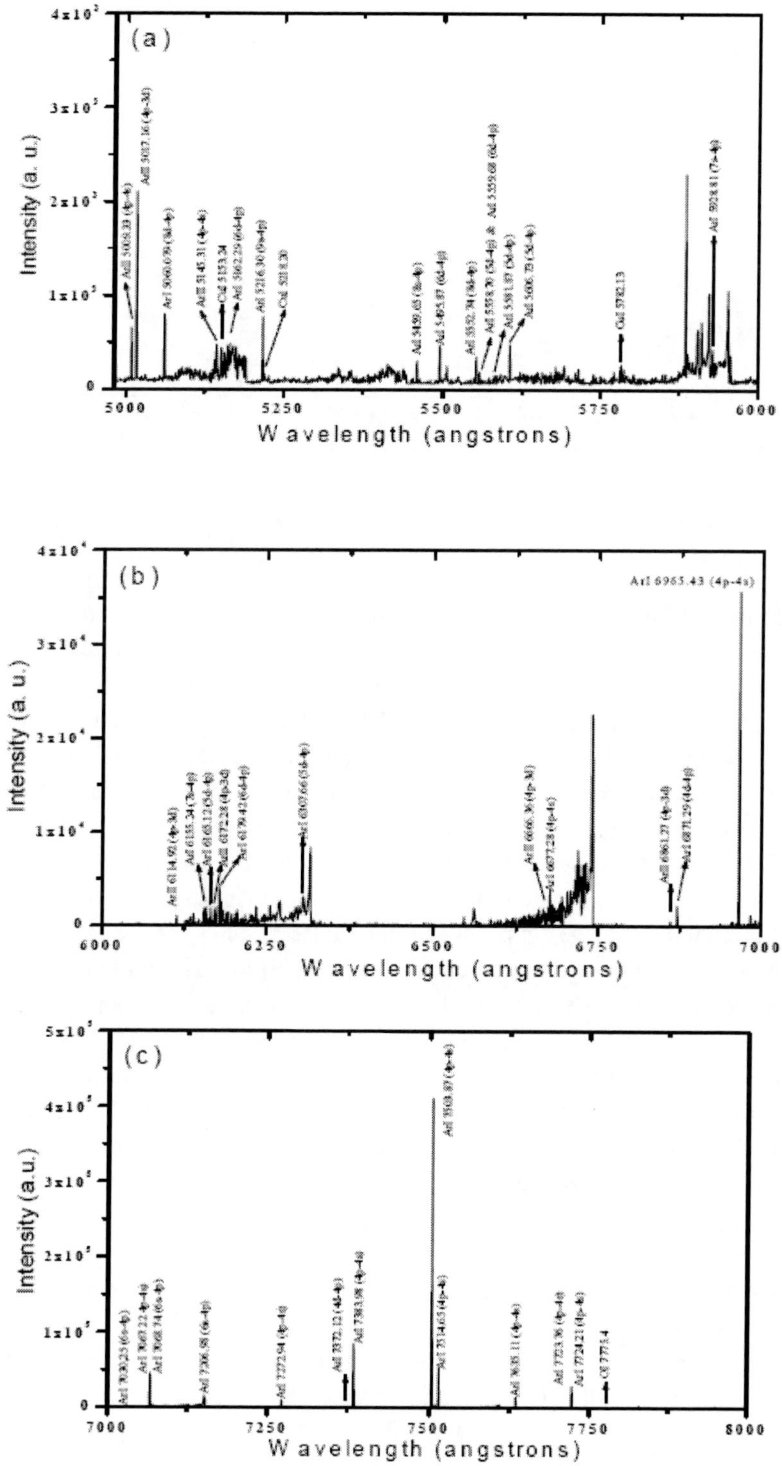

Figure 9. (Continued).

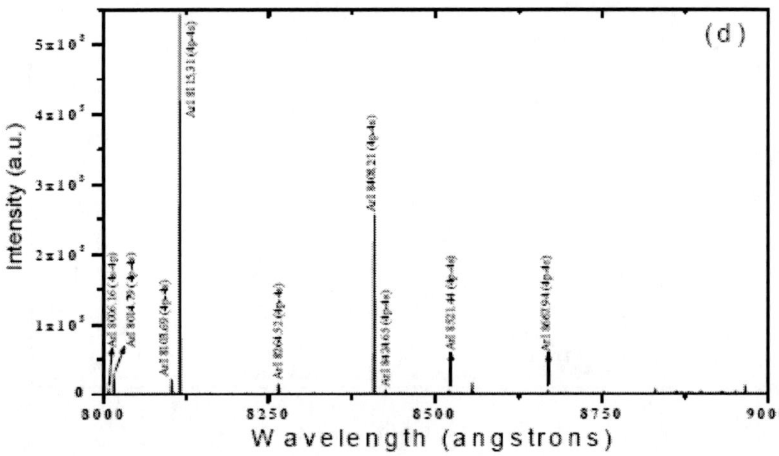

Figure 9. Optical spectra obtained from cathode boundary layer discharge, with hole diameter of $D = 600\ \mu m$, pressure of $P = 40\ Torr$, and discharge current of $I_d = 8.5mA$.

The spectrum in Figure 8a shows the emission line of the Ar atoms in the blue region of the electromagnetic spectrum, whereas Figures. 8b and 8c show the lines emitted in the red region. In the three figures, the lines are identified with their respective wavelengths. Generally, a glow discharge generated by an Ar gas has a greater radiation emission in the visible region because, as commented earlier, the principal transitions between Ar I states occur between the levels $4p \rightarrow 4s$ and $5p \rightarrow 4s$, resulting in emissions in the wavelength range of 6900 to 8115Å, red emission, and from 4000 to 5000Å, blue emission, respectively. In region of 4000 to 5000Å, several lines of ArII that correspond to higher states of ionized argon are generally observed. In addition, depending on the experimental conditions in which an electrical discharge in Ar gas is produced, the emission lines can be found with relatively low intensity for the wavelength range of 5000 a 6900Å [40]. The presence of these lines in the spectrum indicates that the microdischarge contains a significant population of electrons with energies between 14 and 24eV, meaning that the plasma does not contain large number of electrons with energy greater than 24eV, according to the energy values listed in Table 1. These electrons are not necessarily the average temperature on the EEDF. Electrons that are more energetic are usually found in the tail of the EEDF, and the ionization processes are possible because of this distribution [14, 60].

4.2. Radiative Emission from Cathode Boundary Layer Discharge

Figure 9 presents the spectral lines obtained experimentally from cathode boundary layer discharge. The gas used in this experiment is a mixture of 98%Ar - 2%H_2, both of 99.99% purity.

The Ar flow was 0.7 l min^{-1} and H_2 was 15 ml min^{-1}. A series ballast resistor of 47 kΩ was connected to the cathode to limit the current in the circuit. A high voltage dc power supply used provides a 5 kV voltage and maximum current of 200 mA.

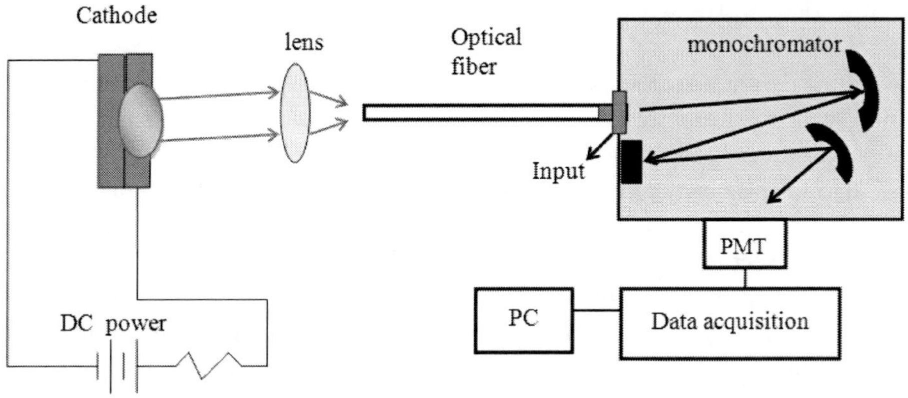

Figure 10. Experimental set-up showing the radiation acquisition.

The experimental setup to obtain the spectra is the same as topic 4.1, except that the radiation collected by the lens was focused onto the optical fiber aperture by means of a convergent lens (see Figure 10). Our plasma is considered optically thin and the self-absorption of the used lines can be neglected. The Abel inversion technique can be performed for the radial distribution of the OES diagnostics [60]. This studied region is far enough from cathode fall that there are intense electric fields. Thereby, the Stark effect in the line shapes in our focusing point is mainly due to the surrounding microfield of electrons and ions near the emitter. The apparatus function was obtained from Ar low-pressure lamp and found to be 0.0213 nm. Both electrical and optical emission spectroscopy measurements of electron number density, excitation and gas temperature, and hydrogen atom temperature could be obtained from the emitted radiation.

Optical emission spectroscopy (OES) measurements could be done in the region close to the cathode surface, where intense radiation could be seen in the negative glow region. Plasma parameters could be determined from spectral line intensity and broadening specifically for a specific cathode material. The optical measurements could be done in 98%Ar-2%H_2 gas mixture, using OH (from residual water vapor), H_2 (added to discharge), atoms (sputtered from cathode), and Ar lines. The main mechanisms responsible for H_α, H_β, Ar I 603.213, and Ar I 565.070 nm line broadenings could be investigated. Through H_α and H_β lines, the hydrogen atom temperature and electron number density could be determined. In the experimental conditions, these two argon lines are substantially broadened by Van der Waals and Stark effects, also enabling the estimation of gas and electron temperatures. The excitation temperature could be determined from the methods of two Cu I lines and Boltzmann-plot, using Ar I 4p – 4s and 5p - 4s transition lines. Cuprum cathode boundary layer microplasmas, with inner diameter D ~ 400 μm and pressure P = 40 Torr, were used to measure these parameters, for currents ranging from 3 to 30 mA. From two Cu I lines, the excitation temperature ranged from 5000 to 8000 K. From Boltzmann-plot method, this temperature, ranging from 6000 to 8500 K, was calculated from the spectral lines corresponding to the main Ar I emissions. From Balmer and Ar I 603.213 and 565.070 nm line broadenings, the electron number densities (n_e) were in good agreement, indicated that n_e has increased with current as required for glow discharges. For these currents, the densities ranges from 3.0 to 6.0×10^{14} cm^{-3}.

4.3. Magnetron Sputtering Discharges

One of the most used techniques for film deposition is the so-called magnetron sputtering deposition, which is a physical vapor deposition (PVD) technique based in a magnetically confined low-pressure discharge. The term "magnetron" arises from the utilization of a magnetic trap to enhance the ionization and, as a consequence, the plasma density. The term "sputtering" refers to the sublimation method of the material to be deposited. Sputtering consists in the ejection of atoms from a solid target due to ion bombardment. In the magnetron sputtering, the ions are created in the discharge and the solid target is the cathode, thus the ions are accelerated towards the target. The ejected atoms condense in the internal surfaces of the deposition chamber forming a thin film. The typical example of the utilization of the magnetron sputtering technique is the deposition of a metallic film. Thus, a metal (Cu, Ti, Al, ...) is used as the discharge cathode. To avoid contamination of the thin film, the plasma forming gas should be an inert gas, reducing the options to the family of noble gases. In addition, the sputtering process is highly dependent on the masses involved (ion and target atom). For a specific target, the higher the ion mass is, the higher the sputtering yield is (the amount of ejected atoms per incident ion), which is a key parameter for efficient film deposition. This mass dependence makes Helium and Neon a bad choice for this application. On the other hand, some of the more massive noble gases are radiative and others are considerably more expensive than Ar [61]. Thus, Ar is the standard gas for film deposition by magnetron sputtering. This choice has consequences in the characteristics of the discharge.

The existence of metastable species with energies greater than the typical ionization potential of the sputtered atoms (generally metals) makes the Penning ionization an important mechanism, because the presence of metal ions affects both the sputtering process and the energy transfer to the growing film. A more obvious effect of choosing Ar is that all the processes in the plasma are dependent on the characteristics of this gas, in particular, the excitation and ionization potentials as well as the cross-section. The comparative study of West et al. [62] for magnetron sputtering discharges generated in Ne, Ar, Kr, and Xe shows a difference in plasma parameters such as the floating potential and ion saturation current. More interesting for film deposition is the dependence of film parameters like hardness, roughness, and texture coefficient with the inert gas used. These differences arise principally due to the difference in the masses of the gases, thus in energy transfer to the film, both by atoms and ions of noble gas. Other important effect is the incorporation of the noble gas into the film. This feature for Ar discharges will be discussed in details in section 5. In typical magnetron sputtering discharges with Ar, the emission lines do not differ considerably from other low-pressure discharges, in which the most intense lines refer to Ar I due to transitions from $4p \rightarrow 4s$ and $5p \rightarrow 4s$.

However, in the recent years, new types of power supplies allow of ultra-dense discharges (electron density of the order of 10^{18} - 10^{19} m^{-3}) named High Power Impulse Magnetron Sputtering (HiPIMS), characterized by short pulses (from 5 to 5000 μs) of high power and current density (typically a few kW/cm^2 and several A/cm^2, respectively) with repetition frequencies from 10 Hz to 10 kHz [63]. In these ultra-dense discharges, the optical emission spectrum is completely altered, as can be seen in Figure 11, for a HiPIMS discharge in Ar using a graphite target.

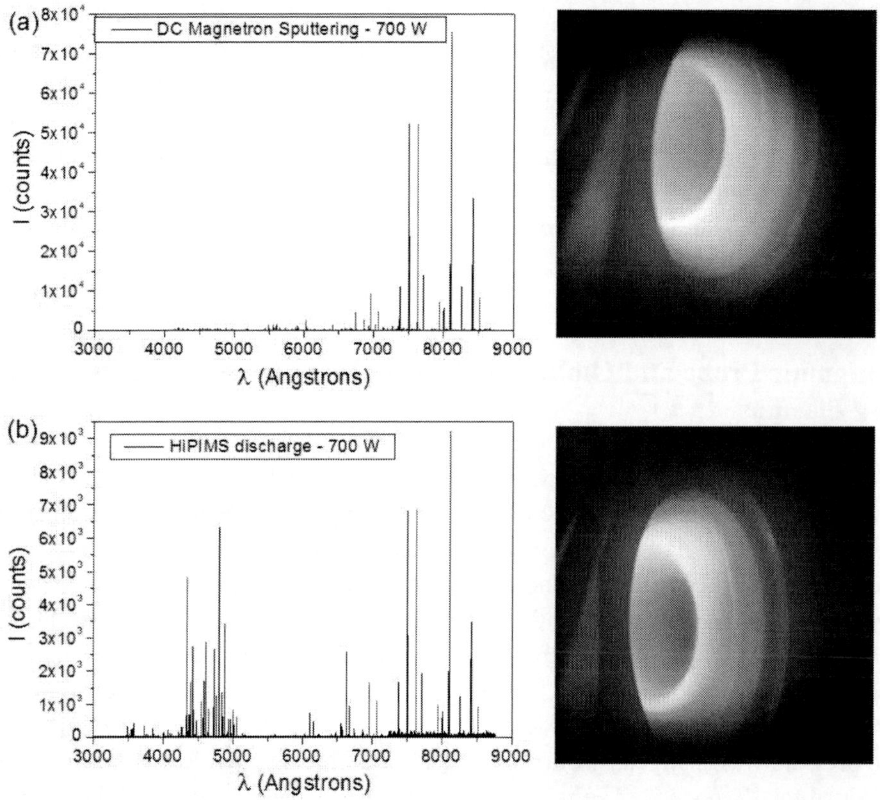

Figure 11. (a) Optical emission spectrum of an Ar discharge in a DC magnetron sputtering system using a carbon target. (b) Optical emission spectrum of an Ar discharge in a HiPIMS system also using a carbon target for the same average power (700 W). A repetition frequency of 250 Hz and a pulse width of 80 μs were used in the HiPIMS discharge. A change of a "purple" discharge in DC magnetron for a blue discharge in HiPIMS is observed due to the transitions between high energy levels of Ar I and transitions between levels of Ar II, as can be seen in the emission in the region 400 and 500 nm.

The spectrum is dominated by transitions of Ar II (due to increased ionization in the plasma) and a strong presence of Ar III is observed. In discharges with metal target, a strong presence of metal ions are also observed [63, 64]. The increased concentration of Ar ions, in particular of multiply ionized ions (like Ar^{2+}), has a strong influence on the main mechanism of discharge sustaining.

Besides sputtering, the collision of Ar ions on the target surface leads to the emission of secondary electrons that are accelerated in the cathode sheath and are the responsible by the great number of ionization in the discharge. The emission of secondary electron can be caused by two different mechanisms: the kinetic emission and the potential emission [63]. Below ion energies of 1 keV, which correspond to most of the magnetron sputtering processes, the potential emission is predominant according to the equation:

$$\gamma = 0.032(0.78E_{pot} - 2\phi) \tag{8}$$

where E_{pot} is the ionization potential of the incoming ion and ϕ is the work function of the target material. A doubled ionized ion has a greater ionization potential than a singly ionized ion; thus, the presence of Ar^{2+} in the HiPIMS discharge increases the potential emission of secondary electrons [63]. In addition, due to the greater electric charge, Ar^{2+} acquires a kinetic energy two times greater than Ar^{+} for a same potential drop; therefore, energies greater than 1 keV are achieved, improving the kinetic emission. This example of HiPIMS discharges shows how changes in the plasma characteristics affect the population of Ar species and how these modifications in the populations affects the plasma characteristics. This synergic effect is fundamental to understand the complex mechanisms of film formation by magnetron sputtering deposition.

4.4. Langmuir Probe and Global Model Studies of Pure Ar Reactive Ion Etching Plasmas

In this topic, we present the experimental and global model analysis of Ar plasmas generated in CCP reactors with single (RIE) and hollow cathode geometry (HCRIE) [55, 65]. Additionally, CCP and ICP global model results (for details about the global model see references. [14, 55-57]) are compared with experimental Langmuir probe and optical spectrometry results of planar RIE and HCRIE plasmas, respectively. Figure 12 illustrates the plasma density ($n_p = n_{e,total} = n_+$, where $n_{e,total} = n_{e,c} + n_{e,h}$) as a function of rf power applied in discharge, which is generated in the RIE and HCRIE reactors for different conditions of gas pressure.

The pressure was changed by only varying the gas flow rate while the vacuum gate valve is maintained totally opened. The n_p was determined experimentally by the Langmuir probe technique (the probe was positioned at the center of the discharge) for both reactors, and estimated by global model for CCP and ICP reactors. For this and the next graphs of this section, we use the following symbols: LP = data obtained by Langmuir probe; MG-X = data obtained by the global model for a reactor X; n_{eh}/n_{ec} = ratio of the hot electron density by the cold electron density measured by LP (the range of n_{eh}/n_{ec} presented in the graphs of Figures. 12 and 13 is given for discharge power increasing from 10 to 200 W). It is important to note that the value of n_{eh}/n_{ec} equaling zero corresponds to a Maxwellian EEDF, while other values are non-Maxwellian (Druyvesteyn) EEDF. It can be observed that n_p increases with rf power. With increasing of rf power, there is an increase in the energy supplied to ionizing collisions, thus increasing the n_p. Similarly, the increase in pressure (for a fixed value of rf power) promotes the increase of n_p. In this case, the elevation of the neutral gas density promotes the increase of collisions between the particles, which for the case of Ar gas favors the ionization processes.

This latter effect, however, happens in reverse for HCRIE plasma, where, due to the establishment of hollow cathode effect [47, 55], the n_p is higher with decreasing pressure. Under these conditions, n_p is of the order of 10^{17} m^{-3} for most values of rf power and pressure, these results are comparable with an ICP system operating at same experimental conditions, see Figure 12b. The inverse behavior of n_p with decreasing pressure for HCRIE discharge demonstrates the large capability of ionization of this geometry under low pressures and has a direct influence on electron temperature (T_e). Figure 13 shows the $T_{e,ef}$ (effective electron temperature) as a function of rf power for different values of pressure.

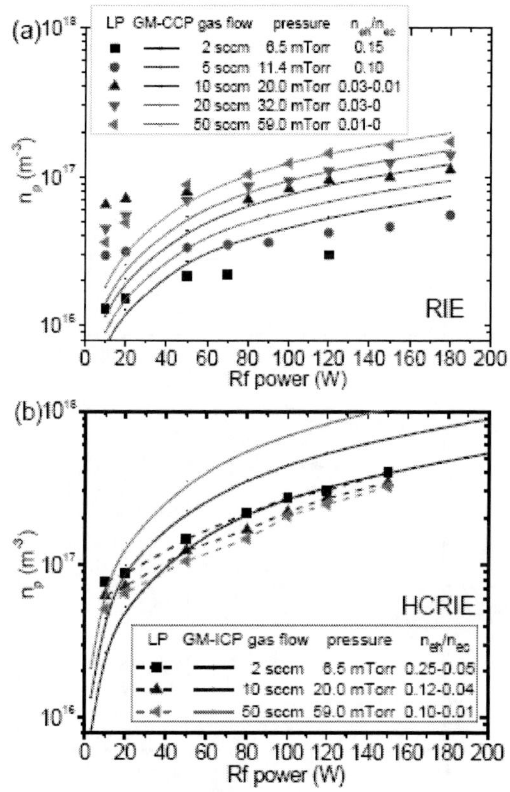

Figure 12. Plasma density as a function of applied rf power for Ar plasma reactors: RIE (a) and HCRIE (b). The solid lines correspond to global model results.

Figure 13 illistrates the following:

1) the increase of rf power and pressure generally promotes decreasing T_{eff};
2) for RIE plasmas (Figure 13a), there is a greater sensitivity of T_{eff} with pressure variation, which occurs due to large variation of n_p with pressure. For discharges operated at pressures $\geq$ 20 mTorr, the T_{eff} is of the order of 2-3 eV, assuming values close to Maxwellian plasmas (see curves MG-CCP). Additionally for pressures less than 20 mTorr, the EEDFs measured have a strongly non-Maxwellian behavior ($n_{eh}/n_{ec} > 0$, see table in Figure 13a), which results in a drastic increase of the T_{eff};
3) for plasma generated in HCRIE geometry (Figure 13b), the dependency of T_{eff} on rf power is higher. The values of T_{eff} are higher for low rf power, decreasing gradually up to 4 eV in the range of 10-150 W. These high values of T_{eff} are explained by strong non-Maxwellian EEDFs for lower rf power ($n_{eh}/n_{ec} > 0.1$). When the rf power is high, the distortion of the EEDF shape is minimized reducing n_{eh}/n_e and, consequently, T_{eff}. For the pressure effect, the values of T_{eff} is equal for all pressure values investigated, because n_p is close (Figure 12b).

Finally, we highlight the good agreement between the results of n_p and T_{eff} obtained by global model and Langmuir probe, which validates the set of reactions for the Ar gas and the energy balance theory proposed for the single CCP reactor in reference [55].

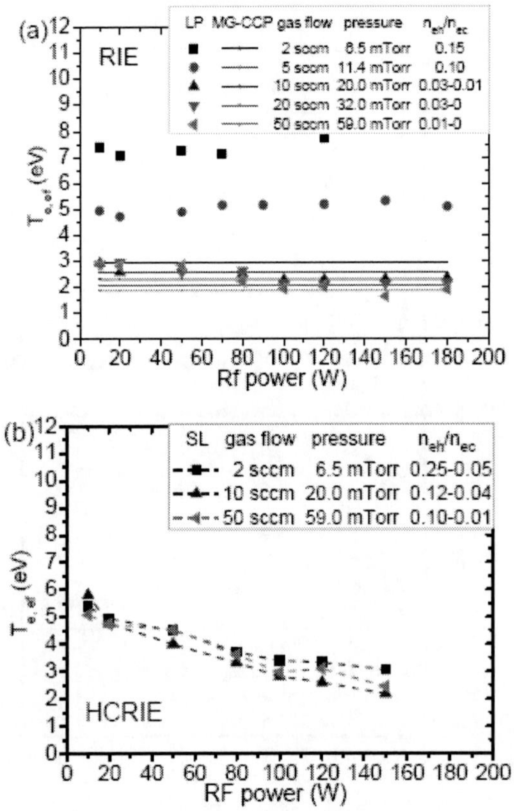

Figure 13. Effective electron temperature, T_{eff}, as a function of rf power applied to Ar plasma for the reactors: RIE (a) and HCRIE (b). The solid lines in (a) correspond to the results of global model. Here, T_{eff} calculated by GM to ICP system is not shown because it is similar to that obtained for the CCP system [55].

The discrepancies between the experimental data and calculated values were observed only for conditions where n_{eh}/n_{ec} is high, i.e. for low rf power and pressures lower than 20 mTorr. This is due to the assumption of a Maxwellian EEDF in the calculation of kinetics coefficients in global model [55-57].

5. Influence of Argon Incorporation in the Properties of the Grown Thin Films

The influence of contaminants incorporation on chemical, structural, optical, electrical, and mechanical properties of the grown thin films have been attracting many researchers' attention for a long time [66-70]. The most common and often contaminant during the deposition processes is oxygen. However, special attention to Ar, which is present in almost all plasma discharge processes, has been given in order to investigate how the Ar atom incorporation occurs and how it affects the properties of the grown thin film [68-72]. Due to the low binding energy of Ar with most surfaces of the substrates (~ 0.02 eV on silicon, for example [73]), the main explanation for Ar incorporation mechanism in the film bulk is the implantation of Ar ions, which are ionized in the plasma and accelerated towards the deposition surface with kinetic energy approximately equal to the plasma potential (30-50 eV) [74]. These implanted Ar atoms have been reported to be present in the metal films as single atoms as well as gas bubbles (a small content) [75, 76]. Evans et al. showed that the internal pressure in these bubbles can be sufficient for the gas to solidify [77]. Although Ar incorporation within the atomic network of thin films is recognized to occur when the deposition methods such as magnetron sputtering or ion beam assisted deposition are used, few studies show a correlation between the Ar incorporation and the film properties. Recent research notices that the presence of Ar in the film composition causes modifications in the structure and microstructure of the films, and hence in the film hardness (increasing or decreasing the film hardness depending of the deposition method) as well as in their optical properties. Paturaud et al. showed that Ar gas entrapment, originating from the growing film bombardment by both ions and reflected neutrals (backscattered charge neutralized sputtering ions) in magnetron sputter deposition, can be understood by calculating the net trapping probability of these two argon species as a function of their energy [78]. The net trapping probability Γ_T is given by [79]:

$$\Gamma_T = \frac{C_{Ar}}{1-C_{Ar}}\left(\frac{J_i}{J_s}\right)^{-1} \tag{9}$$

where C_{Ar} is the total Ar content in the film and is given by (where C_i and are the Ar content assigned to the ions and the reflected neutrals, respectively). J_i and J_s are the incident flux of Ar ions and the sputtered atoms from the target, respectively. Andriotis et al. provided a theoretical description about the interaction of Ar with the Al(100) surface. The main result was that in the absence of the van der Waal's interaction between the substrate and the adatom, the interaction between Ar and Al(100) is everywhere repulsive. This result and the theoretical evidence of a small attractive interaction between Ar and other surfaces ($A1_2O_3$ for example) [80] indicate the importance of the van der Waals term in a complete description of the physisorption of noble gases onto metallic surfaces [81].

The Ar incorporation appears to affect the structure and microstructure of the films. The effects on this property due to the incorporation of Ar atoms in the lattice of ZrO_2 thin films prepared by ion beam assisted deposition are reported by Holgado et al. [78]. They showed that Ar incorporation produces some amorphisation of the original thin films. Furthermore, they inferred that the influence of Ar incorporation on the film properties is directly affected by the temperature annealing of the samples, because for annealing of samples with embedded Ar at the cubic/tetragonal phases of ZrO_2 with a preferential growth of crystal planes are stabilized. However, for annealing at $T = 1023K$, a distorted cubic/tetragonal phase is detected by XRD, while FTIR reveals that the monoclinic structure of zirconia starts to form. The film structure modification by Ar incorporation leads to a film hardness modification. Paturaud et al. found a clear correlation between the microhardness of sputtered films and the amount of embedded Ar, in the case of the chromium films deposition.

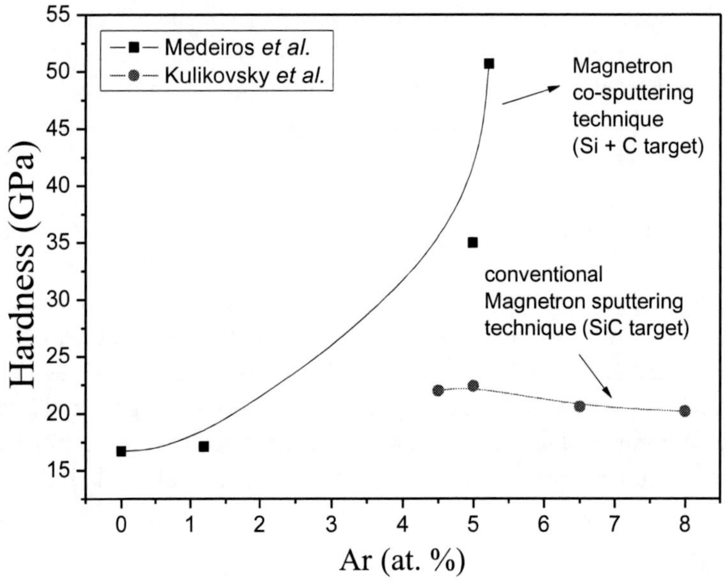

Figure 14. Comparison between the hardness of the SiC thin films deposited by magnetron co-sputtering and conventional magnetron sputtering technique as a function of the Ar content in the films.

For all parameters investigated in this work, it appears that the Ar entrapment in the chromium films deposited is the most unequivocal manifestation of microhardness changes induced by energetic bombardment during film growth [78]. In the same way, Medeiros et al. reported the hardness modification by Ar incorporation on silicon carbide (SiC) thin films deposited by bias co-sputtering technique simaltanoiusly using a graphite (C) and a silicon (Si) target [72]. They found that the hardness is strongly affected by the presence of Ar in the SiC thin films where the hardness increases from 16.7 to 50.7 GPa with increasing of the argon content. Kulikovsky et al. also observed the influence of the Ar incorporation on the hardness of the SiC films but this time the films were deposited by conventional magnetron sputtering technique using a SiC target [82]. However, the hardness modification in this case was not so great, as can be seen in the Figure 14, which shows a comparison between the hardness of the SiC films deposited by these two techniques as a function of the Ar content in the films. Another film property reported in this study is the electrical resistivity, which was changed from ~7.0 to 0.45 Ω.cm with the increase of the Ar content in film from 0 to 5.22%, respectively.

CONCLUSION

This chapter reviews some recent studies about the chemistry of low-pressure discharges generated with Ar gas. This noble gas is widely used in gas-discharge lamps, as sputtering gas for deposition and/or etching processes, arc welding, among others. Especially, in the semiconductor industry, this gas is present in most of the processing steps of a chip or a micro device. In summary, details of the chemistry that occurs in pure Ar plasma at sub-atmospheric pressure, particularly detailing the transitions between excited species, was presented. The use of cross sections and kinetic constants of these reactions are input for computational models such as collisional-radiative model and global model. Through these models, we can predict the density of each species considered in the model as well as its influence on the global balance of reactions. Additionally, some results were obtained through analysis of experimental reactors namely micro hollow cathode, magnetron sputtering, high power pulsed magnetron sputtering, and reactive ion etching. As a rule, the chemistry of Ar is very dependent on the type of reactor and, consequently, on the distributions of the population of electrons in the discharge. All this synergism allows enhanced plasmas to be obtained and enables the formation of species such as Ar^{2+}, as in the case of HiPIMS system. Allied to this, it is important to understand and analyze the optical emission of plasmas, because this experimental method permits the complete analysis of the collisional-radiative processes in Ar plasmas.

Acknowledgment

This work was supported by the following Brazilian agencies projects: CAPES/PNPD (process nº 02765/09-8), FAPESP/CNPq/PRONEX (process nº 2011/50773-0) and FAPESP (process nº 2012/13064-4).

References

[1] Argon: An Overview, WikiProject Wikipedia-Books, site: http://en.wikipedia.org/wiki/Book:argon, access: 07/20/2012.

[2] J. C. Maxwell, *The scientific papers of the honorable Henry Cavendish, F. R. S.*, Cambridge University Press (1921).

[3] *Lord Rayleigh - Biography*. Nobelprize.org. 28 Jul 2012 http://www.nobelprize.org/nobel_prizes/physics/laureates/1904/strutt-bio.html.

[4] Argon history, site: http://www.airliquide.com/en/company/our-businesses-our-products/air-gases/argon-5/the-history-of-argon.html, access: 07/20/2012.

[5] J. T. Bernstein, *Cryogenic argon Production*, presented in: Modern Air Separation Plant Technology Conference (1999).

[6] L. Khriachtchev, M. Pettersson, N. Runeberg, J. Lundell, M. Räsänen, *Nature* 406, 874–876 (2000).

[7] V. R. Belosludov, O. S. Subbotin, D. S. Krupskii, O. V. Prokuda, R. V. Belosludov, Y. Kawazoe, *Journal of Physics: Conference Series* 29, 1–7, (2006).

[8] D. Petsalakis, G. Theodorakopoulos, *J. Phys. B: At. Mol. Opt. Phys.* 27, 4483 (1994).

[9] H. Akashi, Y. Sakai, T. Tagashira, *J. Phys. D: Appl. Phys.* 27, 1097 (1994).

[10] Cohen, J. Lundell, R. B. Gerber, *Journal of Chemical Physics* 119(13), 6415-6417 (2003).

[11] *Argon applications*, site: http://www.airliquide.com/en/company/our-businesses-our-products/air-gases/argon-5/argon-applications.html, access: 07/20/2012.

[12] C.L. Heng, L.W. Teo, V. Ho, M.S. Tay, Y. Lei, W.K. Choi, W.K. Chim, *Microelectronic Engineering* 66, 218-223 (2003).

[13] S. A. Campbell, *The Science and Engineering of Microelectronic Fabrication*. New York: Oxford University Press (2001).

[14] M. A. Lieberman, A. J. Lichtenberg, *Principles of Plasma Discharges and Materials Processing*, 2th edition, New York: Wiley (2005).

[15] M. Mao, Y.N. Wang, A. Bogaerts, *J. Phys. D: Appl. Phys.* 44, 435202 (15pp) (2011).

[16] J. H. Moraes, A. S. da Silva Sobrinho, H. S. Maciel, J. C. N. Dutra, M. Massi, S. A. C. Mello, W. H. Schreiner, *J. Phys. D: Appl. Phys.* 40, 7747 (2007).

[17] X.-M. Zhu, Y.-K. Pu, *J. Phys. D: Appl. Phys.* 43, 403001 (24pp) (2010).

[18] M.-H. Lee, S.-H. Jang, C.-W. Chung, *Physics of Plasmas* 13, 053502 (2006).

[19] Y.-K. Lee, C.-W. Chung, *Journal of Applied Physics* 109, 013306 (2011).

[20] D.S. Scott, F.A.L. Dullien, *Amer. Inst. Chem. Eng. Journal*, 15, 293 (1962).

[21] D. Mariotti, Y. Shimizu, T. Sasaki, N. Koshizaki, *Applied Physics Letters* 89, 201502 (2006).

[22] M.-H. Lee, C-W. Chung, *Applied Physics Letters* 87, 131502 (2005).

[23] M. Roberto, H.B. Smith, J.P. Verboncoeur, *IEEE Transactions on Plasma Science* 31, 6 (2003).

[24] Gutsol, A. Rabinovich, A. Fridman, *J. Phys. D: Appl. Phys.* 44, 274001 (14pp) (2011).

[25] N. A. Dyatko, Y. Z. Ionikh, I. V. Kochetov, D. L. Marinov, A. V. Meshchanov, A. P. Napartovich, F. B Petrov, S A. Starostin, *J. Phys. D: Appl. Phys.* 41, 055204 (2008).

[26] Grill, Cold Plasma Materials Fabrication: From Fundamentals to Applications, Wiley, John and Sons, Incorporated (1994).

[27] M. A. Fraga, R. S. Pessoa, M. Massi, H. S. Maciel, *Applications of SiC-based thin films in electronic and MEMS devices*. In: Y. Hijikata. (Org.). Physics and Technology of Silicon Carbide Devices: Intech, (2012).

[28] R. S. Pessoa, H. S. Medeiros, M. A. Fraga, L. V. Santos, H. S. Maciel, M. Massi, A. S. da Silva Sobrinho, *Argon incorporation on silicon carbide thin films deposited by bias co-sputtering technique*. In: 2012 MRS Spring Meeting and Exhibit, 2012, San Francisco. Mater. Res. Soc. Symp. Proc., 2012. v. 1433. p. 1-6.

[29] X.-M. Zhu, Y.-K. Pu, *J. Phys. D: Appl. Phys.* 40, 5202 (2007).

[30] N. Balcon, G. J. M. Hagelaar, J. P. Boeuf, *IEEE Trans. Plasma Sci.* 36, 2782–7 (2008).

[31] K. Katsonis, H. W. Drawin, *Journal of Quantitative Spectroscopy and Radiative Trasnfer* 23(1), 1-55, (1980).

[32] Bogaerts, R. Gijbels, J. Vlcek, *J. Appl. Phys.* 84, 121-136 (1998).

[33] J. Bretagne, G. Callede, M. Legentil, V. Puech, *J. Phys. D: Appl. Phys.* 19, 761 (1986).

[34] J. Vlcek, *J. Phys. D: Appl. Phys.* 22, 623 (1989).

[35] F. Guimarães, F. Bretagne, *Plasma Sources Sci. Technol.* 2, 127 (1993).

[36] U. M. Kelkar, M H. Gordon, L. A. Roe, Y. Li, *J. Vac. Sci. Technol. A* 17, 125 (1999).

[37] Bultel, B. van Ootegem, A. Bourdon, P. Vervisch, *Phys. Rev. E* 65, 046406 (2002).

[38] Yanguas-Gil, J. Cotrino, L. L. Alves, *J. Phys. D: Appl. Phys.* 38, 1588–1598 (2005).

[39] Palmero, E. D. van Hattum, H. Rudolph, F. H. P. M. Habraken, *Journal of Applied Physics* 101, 053306 (2007).

[40] M. P. Gomes, *Microplasmas em Equilíbrio de Excitação*, PhD Thesis, Instituto Tecnológico de Aeronáutica, Brazil (2011).

[41] W. L. Wise, J. W. Brault, K. Danzmann, V. Helbig, M. Kock, *Physical Review A* 39(5), 2461-2471 (1989).

[42] X.-M. Zhu, Y.-K. Pu, *J. Phys. D: Appl. Phys.* 43, 015204 (17pp) (2010).

[43] M. Mitchner, C. H. Kruger, *Partially Ionized Gases*, Wiley, New York (1973).

[44] Yanguas-Gil, J. Cotrino, A. R. González-Elipe, *J. Appl. Phys.* 99, 033104 (2006).

[45] Ramamurthi, D. J. Economou, *Plasma Sources Sci. Technol.* 11, 324–332 (2002).

[46] P. Bartos, R. Hrach, P. Jelínek, *Vacuum* 82, 220–223 (2008).

[47] J. Amorim Filho, *Gas Discharges Fundamentals and Applications*. Kerala: Transworld Research Network, v. 1 (2007).

[48] Sainz, J. Margot, M. C. Garcia, M. D. Calzada, *J. Appl. Phys.* 97, 113305 (2005).

[49] Ph. Belenguer, J. Ph. Blondeau, L. Boufendi, M. Toogood, A. Plain, A. Bouchoule, C. Laure, J. P. Boeuf, *Phys. Rev. A* 46, 7923–7933 (1992).

[50] J. T. Gudmundsson, E. G. Thorsteinsson, *Plasma Sources Sci. Technol.* 16, 399-412 (2007).

[51] T. Sato, T. Makabe, *Journal of Applied Physics* 98, 113304 (2005).

[52] Yanguas-Gil, J. Cotrino, A. R. González-Elipe, *J. Appl. Phys.* 99, 033104 (2006).

[53] X.-M. Zhu, Y.-K. Pu, *J. Phys. D: Appl. Phys.* 42, 182002 (5pp) (2009).

[54] H. R. Griem, *Principles of Plasma Spectroscopy*, Cambridge: Cambridge University Press (1997).

[55] R.S. Pessoa, *Estudos de plasmas fluorados aplicados em corrosão de silício usando simulação de modelo global e diagnósticos experimentais*, PhD Thesis, Instituto Tecnológico de Aeronáutica, Brazil (2009).

[56] S. Kim, M.A. Lieberman, A.J. Lichtenberg, *J. Vac. Sci. Technol. A* 24(6), 2025-2040 (2006).

[57] T. Kimura, K. Hanaki, *Japanese Journal of Applied Physics* 47(11), 8537-8545 (2008).

[58] J. T. Gudmundsson, E. G. Thorsteinsson, *Plasma Sources Sci. Technol.* 16, 399-412, (2007).

[59] Rousseau, X. Aubert, *Journal Physics D: Applied Physics* 39(8), 1619-1622 (2006).

[60] B.N. Sismanoglu, *Estudo e caracterização de microplasmas luminescentes através da espectroscopia óptica de emissão,* PhD Thesis, Instituto Tecnológico de Aeronáutica, Brazil (2010).

[61] Chapman, *Glow Discharge Processes.* New York: John Wiley and Sons, (1980).

[62] G.T. West, P.J. Kelly, *Surf. Coat. Technol.* 206, 1648 (6pp) (2011)

[63] K. Sarakinos, J. Alami, S. Konstantinidis, *Surf. Coat. Technol.* 204, 1661 (24pp) (2010).

[64] J. Čapek, M. Hála, O. Zabeida, J. E. Klemberg-Sapieha, L. Martinu, *J. Appl. Phys.* 111, 023301(9pp) (2012).

[65] R.S. Pessoa, L.L. Tezani, H.S. Maciel, G. Petraconi, M. Massi, *Plasma Sources Sci. Technol.* 19 025013 (7pp), (2010).

[66] G. W. Wilson, B. P. Sinha, *Thin Solid Films* 8, 207-211, (1971).

[67] J. Vlcek, M. Kormunda, J. Cızek, V. Perina, J. Zemek, *Surface and Coatings Technology* 160, 74-81, (2002).

[68] J.P. Holgado, R. Escobar Galindo, A. van Veen, H. Schut, J.Th.M. de Hosson, A.R. González-Elipe, *Nuclear Instruments and Methods in Physics Research B* 194, 333–345, (2002).

[69] V. Stambouli, O. Burat, D. Bouchier, F. Meyer, J. Gilles, G. Gautherin, *Thin Solid Films* 193/194, 181 – 188, (1990).

[70] F. Yubero ,M. Ocana, A. Caballero, R. González-Elipe, *Acta Mater.* 48, 4555–4561, (2000).

[71] Window, *J. Vac. Sci. Technol. A* 11, 1522 (1993).

[72] H. S. Medeiros, R. S. Pessoa, M. A. Fraga, L. V. Santos, H. S. Maciel, M. Massi, A. S.da Silva Sobrinho, *Mater. Res. Soc. Symp. Proc.* 1433 (2012).

[73] G.N.A. van Veen, F.H.M. Sanders, J. Dieleman, A. van Veen, D.J. Oostra, A.E. de Vries, *Phys. Rev. Lett.* 57, 739, (1986).

[74] E.D. Van Hattum, A. Palmero, W.M. Arnoldbik, H. Rudolph, F.H.P.M. Habraken, *Appl. Phys. Lett.* 91, 171501, (2007).

[75] *Proceedings of the International Symposium on Fundamental Aspects of Helium in Metals*, Rad. Eff. 78 (1983).

[76] M. U. Scherzer, in *Sputtering by Particle Bombardment II*, Topics in Applied Physics Vol. 52, edited by R. Behrisch, Springer-Verlag, Berlin, (1983).

[77] J. H. Evans, D. J. Mazey, *Scr. Metall.* 19, 621, (1985).

[78] Paturaud, G. Farges, M.C. Sainte Catherine, J. Machet, *Thin Solid Films* 347, 46-55, (1999).

[79] Paturaud, G. Farges, M.C. Sainte Catherine, J. Machet, *Surf. Coat. Technol.* 86-87, 388-393, (1996).

[80] A.D. Zdetsis, A.B. Kunz, *Phys. Rev. B* 26, 4756, (1982).

[81] A.N. Andriotis, C.A. Nicolaides, *Solid State Communications* 51, 251-254, (1984).

[82] V. Kulikovsky, V. Vorlíček, P. Boháč, M. Stranyánek, R. Čtvrtlík, A. Kurdyumov, L. Jastrabik, *Surface and Coatings Technology* 202, 1738–1745, (2008).

In: Argon: Production, Characteristics and Applications ISBN: 978-1-62618-204-2
Editors: B.N. Sismanoglu, H.S. Maciel et al.

Chapter 10

EXPERIMENTAL AND THEORETICAL STUDIES OF THE CATHODE SHEATH OF AN ARGON LOW PRESSURE HOLLOW CATHODE DISCHARGE

***G. Petraconi*[1,*], *A. B. Guimarães Neto*[1], *H. S. Maciel*[1,2], *B. N. Sismanoglu*[1] *and R. S. Pessoa*[1,2]**

[1]Technological Institute of Aeronautics (ITA-DCTA), São José dos Campos/SP, Brazil
[2]IP and D, University of Paraiba Valley (UniVap), São José dos Campos/SP, Brazil

ABSTRACT

This chapter presents the investigations about the characteristics of the cathode sheath generated in an argon low-pressure hollow cathode discharge (HCD). The theoretical model for the cathode sheath is considered to be non-collisional. Secondary electrons emitted from the cathode surface are taken into account in the discharge model and their influence on the theoretical sheath potential profile is investigated. The plasma parameters and the floating potential profile along the discharge axis were inferred from the current-voltage characteristics of a single Langmuir probe positioned at the inter-cathode space of the HCD. For a low pressure HCD, typical values of the electron density and electron temperature are $n_e \approx 10^{16}$ m^{-3} and $T_e = 4$ eV, respectively. By using the probe data, the floating potential profile was determined to verify the position of the plasma-cathode sheath interface and to promote a qualitative discussion between theoretical and experimental results.

1. INTRODUCTION

The most important effects in a glow discharge occur next to the cathode, i.e., in the cathode sheath and negative-glow regions. The cathode sheath supports almost all the discharge voltage and is mainly responsible for the production processes of neutral and charged species. The electrons crossing the cathode fall and streaming into the negative-glow

[*] Email address: gilberto@ita.br.

acquire enough energy to promote more intense ionization and excitation rates as compared with the positive-column plasma. This makes the negative-glow based discharges a useful tool in laser, lighting, and surface-processing applications [1-3].

Hollow cathode discharges are the most common examples of discharges in which the negative-glow plasma is generated. The hollow cathode device consists of a cathode with hollow structure (hole, aperture, etc.) and an arbitrarily shaped anode [4]. Compared to a conventional glow discharge between parallel plates, the hollow cathode geometry is characterized by higher current densities and larger concentrations of high-energy electrons. These properties are consequences of the phenomenon known as *hollow cathode effect* (HCE) [3, 4].

In 1916, Friedrich Paschen used the first hollow cathode from an aluminum sheet shaped into a simple rectangular tube with a helium DC discharge to study helium spectral lines [4]. For quite a long time, hollow cathodes were studied and used mainly as intense light sources for spectrochemistry. Although several authors observed an increase in the current (1-3 orders of magnitude) flowing through the discharge in hollow cathode geometry, as compared with the planar cathodes, in 1923, Günterschulze was the first to evidence experimentally the HCE [5]. In 1954, Little and von Engel reported a theory for this effect [6]: *when the separation between the cathodes is reduced enough, a potential well is formed by the plasma with intensity next to the cathode potential. This leads to an oscillatory movement of the electrons between the opposing regions of the cathode fall known as "pendulum effect", and this effect causes a drastic increase in the number of ionization processes, consequently increasing the discharge efficiency.*

Theoretic [7, 8] and experimental [9, 10] studies, about the electron energy distribution function (EEDF) of the low-pressure HCDs, indicate that EEDF is strongly non-Maxwellian and contains a significant amount of high-energy electrons. These studies found a copious amount of electrons with energies well above 10 eV and a tail extending up to the plasma operating voltage. Furthermore, a fraction of high-energy "beam" electrons was measured with energies near the plasma voltage, which are electrons that were accelerated across the full potential of the cathode fall, crossed the negative glow, and were repelled by the negative fall in the opposing cathode sheath.

Based in this fact, detailed studies were conducted by Boeuf [11], Kolobov and Tsendin [12] that considered an analytic model based on a kinetic treatment of the high-energy electrons (named as fast electrons). Their results suggest that the enhancement of the ionization in the cathode fall by the oscillating fast electrons is responsible for the HCE. In addition, Arslanbekov et al. [13] reported that HCE is not only due to more efficient use of the fast electrons, but also by enhancement of the ion flux from the negative glow to the cathode fall.

A simplified theoretical model of non-collisional cathode sheath of a plane HCD is usually made by only considering the contribution of electrons and ions from the plasma volume. From this model, we conclude that the sheath region should have a net positive electrical charge [14]. However, in real situation, we must take into account the contribution of secondary electrons produced as a result of the impact of ions, photons, or neutral and metastable species against the cathode surface.

This chapter presents experimental and theoretical investigations about the characteristics of the cathode sheath generated in an argon low pressure HCD. The next section is a description of the model for a planar HCD. Furthermore, the experimental setup and the results and discussions are presented. Finally, the last section discusses the main points observed.

2. Description of the Model

To develop the model, a plane of abscissa $x = 0$ coincident with the flat surface of the cathode is considered. The sheath region is confined between the plane $x = 0$ and an abscissa $x_b > 0$. The electric potential in the sheath, as well as in the plasma, is described by the function $V = V(x)$. At any point in the plasma, $V(x) = 0$ is attributed to be consistent with the quasi-neutral conditions. The potential of the cathode in $x = 0$ is $V(0) = V_c < 0$ because, in this case, the plasma is the region of the highest potential [14]. According to the condition of quasi-neutrality, the plasma density n_o is equal to the ion density n_{oi} and to the electron density n_{oe}. The secondary electrons are supposed to travel as a beam, leaving the cathode at $x = 0$. The electron beam is accelerated through the cathode sheath towards the plasma. The ions from plasma are accelerated by the cathode field, and the electrons from the plasma are decelerated by this field.

The secondary electrons are considered to be generated by the impact of ions, photons, and neutral species - both stable and metastable - remembering that in the cathode region, the emission by electron impact is not very significant [14]. Thus, it can make use of a dimensionless coefficient denoted by the effective (or apparent) electron secondary emission coefficient [15, 16], because it includes the effect of all relevant mechanisms in electron emission. This effective secondary electron emission coefficient is given by:

$$\gamma = \left.\frac{J^-}{J^+}\right|_{cathode} \tag{1}$$

Based on equation (1), and with the knowledge of the effective coefficient, the current density of secondary electrons J^- is determined. Therefore, it is necessary to determine J^+, by the continuity equation, $J^+ = n_o e v_o$, so that the equation (1), results in:

$$J^- = \gamma e n_o v_o = e n_{ose} v_{se} \tag{2}$$

where v_o is the ion velocity at the sheath edge, the n_{ose} and v_{se} are the secondary electron density and secondary electron velocity, respectively. Thus, we have:

$$n_{ose} = \gamma n_o v_o \left(\frac{-2eV_C}{m}\right)^{-\frac{1}{2}} \tag{3}$$

The number density of secondary electrons in the plasma remains constant, but the same does not occur in the sheath. Such density can be denoted by n^- and calculated based on the equations of energy conservation and continuity:

$$n^- = \gamma n_0 v_0 \left[\frac{2e(V - V_C)}{m}\right]^{-\frac{1}{2}}, V \neq V_C \tag{4}$$

Furthermore, the electron density in the plasma volume is given by the difference between n_o and n_{ose}, so:

$$n_{oep} = n_o \left[1 - \gamma v_o \left(\frac{-2eV_C}{m}\right)^{-\frac{1}{2}}\right] \tag{5}$$

The electrons from the plasma are assumed to be in thermal equilibrium, called thermal electrons. The electrons density (n_e) in the sheath region satisfies the Boltzmann relation given by [17]:

$$n_e(x) = n_{oep} \exp\left(\frac{e\,V(x)}{k\,T_e}\right), \tag{6}$$

where n_{oep} is the electron density of plasma without the contribution from the electron beam, V is the electric potential in the sheath, e is electron charge, and kT_e is the electron temperature in eV. The ions are cold and monoenergetic. Using both the continuity equation and the energy conservation, the ion density n_i in the sheath can be obtained by:

$$n_i(x) = n_o \left(1 - \frac{2eV(x)}{Mv_o^2}\right)^{-1/2}, \tag{7}$$

where M is the ion mass. The potential in the sheath is given by Poisson's equation

$$\frac{d^2V}{dx^2} = -\frac{\rho}{\varepsilon_o} = -\frac{1}{\varepsilon_o}\left(-n_e e - n^- e + n_i e\right), \tag{8}$$

where ε_o is the vacuum permittivity and ρ is the total charge density. By using equations 4, 5, 6, 7, we obtain

$$\frac{\varepsilon_0}{en_0}\frac{d^2V}{dx^2} = \left[1 - \gamma v_0 \left(\frac{-2eV_C}{m}\right)^{-\frac{1}{2}}\right] \exp\left(\frac{e\,V}{k\,T_e}\right) + \gamma v_0 \left[\frac{2e(V - V_C)}{m}\right]^{-\frac{1}{2}} - \left(1 - \frac{2eV}{Mv_0^2}\right)^{-1/2} \tag{9}$$

Introducing the normalized quantities

$$\eta = -\frac{eV}{kT_e},\ \eta_C = -\frac{eV_C}{kT_e},\ \xi = \frac{x}{\lambda_D},\ \mu = \frac{M}{m},\ \nu = \frac{v_o}{c_s}, \tag{10}$$

where λ_D is the electron Debye length at the sheath edge and $c_s=(kT_e/M)^{1/2}$ is the ion sound speed [18, 19], we obtain

$$\frac{d^2\eta}{d\xi^2} = -\left[1-\gamma\nu\left(2\mu\eta_C\right)^{-\frac{1}{2}}\right]\exp(-\eta) - \gamma\nu\left(2\mu\eta_C\right)^{-\frac{1}{2}}\left(1-\frac{\eta}{\eta_C}\right)^{-\frac{1}{2}} + \left(1+2\eta\nu^{-2}\right)^{-\frac{1}{2}} . \tag{11}$$

The equation (12) has the inconvenient fact that ν is an unknown. To circumvent this, it is necessary to establish a relationship involving ν and other parameters of this equation, namely, γ, μ, and η_C. By imposing equality of charge density at $\eta \to 0$, we have ensured one of the matching conditions as ξ becomes large. In addition, if η is to evolve smoothly to zero as ξ increases, then it may vary asymptotically as [20]:

$$\eta = k\xi^{-n} \tag{12}$$

The substitution of equation (12) in equation (11), after performing the Taylor series expansion around $\eta = 0$, which is particularly valid in the limiting region between the sheath and the plasma, results in an equation which can be satisfied only if the terms that multiply ξ^n vanish. This condition leads to the following third-degree polynomial equation in ν:

$$k\nu^3 - \nu^2 + 1 = 0 \tag{13}$$

By solving equation (13) we obtain:

$$k = \gamma\left(2\mu\eta_C\right)^{-\frac{1}{2}}\left(1+\frac{1}{2\eta_C}\right) \tag{14}$$

The solution of equation (14) should be adopted, considering the fact that *k* is typically much smaller than unity, gives the value of ν such that the Bohm criterion is satisfied, and tends to unity when *k* approaches zero. It can be shown that ν is a positive real number if $0 < k < \frac{2}{3\sqrt{3}}$.

A wide range of typical values of γ, μ, and satisfies this condition. The electric field and potential in the sheath can be obtained by numerically integrating equation (12) using and $\frac{d\eta}{d\xi} = 0$, so:

$$\frac{1}{2}\left(\frac{d\eta}{d\xi}\right)^2 = \left[1 - \gamma\nu\left(2\mu\eta_C\right)^{-\frac{1}{2}}\right]\left[\exp(-\eta) - 1\right] + 2\gamma\nu\left(2\mu\eta_C\right)^{-\frac{1}{2}}\eta_C\left(1 - \frac{\eta}{\eta_C}\right)^{\frac{1}{2}} - $$
$$-2\gamma\nu\left(2\mu\eta_C\right)^{-\frac{1}{2}}\eta_C + \nu^2\left(1 + 2\eta\nu^{-2}\right)^{\frac{1}{2}} - \nu^2 \tag{15}$$

The boundary condition for determining where the sheath edge is located, at ξ=s, is

$$\eta\left(s/\lambda_D\right) = -\frac{eV_b}{kT_e} = \frac{\nu^2}{2} \tag{16}$$

because the sheath potential (V_b) in this position is given by:

$$V_b = -\frac{1}{2e}Mv_0^{\ 2} = -\frac{\nu^2 kT_e}{2e} \tag{17}$$

The mathematical problem constituted by equation (11) and condition must be solved numerically to determine the theoretical profile of the electrical potential in the cathode sheath.

The solution of the problem without the use of dimensionless variables depends on the knowledge of plasma parameters according to equations (10).

3. Reactor Chamber Details and Experimental Procedure

A schematic diagram of the plane hollow-cathode discharge apparatus is illustrated in Figure 1. The discharge resembles a glow discharge regime extending from the hollow anode (earth potential) up to flat cathodes made of aluminum and covered by Teflon. The hollow anode is made of titanium of 7 mm external diameter, which is covered by a ceramic tube terminated by an orifice of 2.0 mm diameter.

The internal diameter of the hollow anode cavity is 4 mm. The orifice constricts the anode discharge to form electrostatic layers, which help increase the ionization rate in the anodic region. The vacuum glass, chamber of 130 mm internal diameter and 300 mm length, was preliminary evacuated to a pressure below 10^{-5} Torr.

The plane hollow-cathode discharge was operated with argon gas for pressures (P) in the range of (10^{-2} – 10^{-1}) Torr. The gases were fed through the anode hole at a flow rate in the range of (1 – 40) sccm. Thus, the gas flows through the hollow anode and becomes partially ionized in the process before it reaches the vacuum chamber. The self-sustained cathode-anode voltage (V_d) during operation was in the range of (300 – 900) V with a discharge current (I_d) in the range of (10 – 800) mA.

A cylindrical tungsten probe, with 4.0 mm in length and 0.1 mm in diameter, was positioned at the axis of the discharge between the cathodes. The axis of the probe is perpendicular to the normal cathode surface. Standard technique [14, 21] was used to extract the argon plasma parameters from the current-voltage characteristic of the probe.

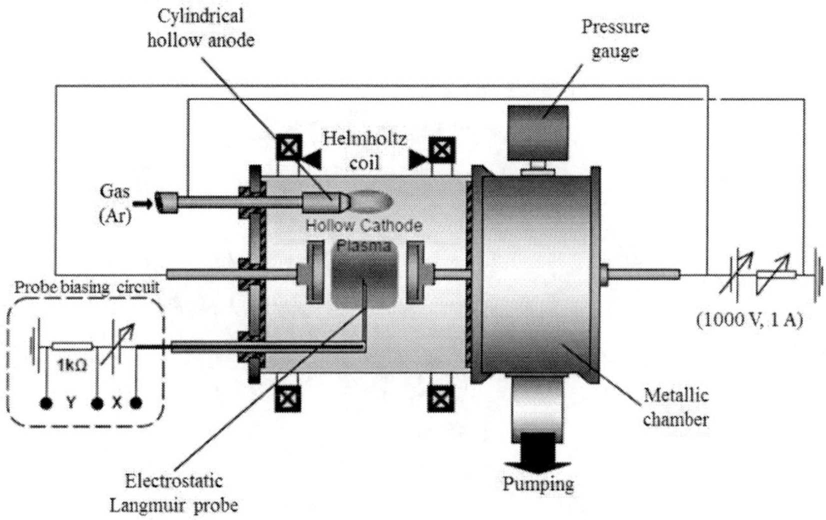

Figure 1. Simplified energy level diagram of the argon atom showing the main chemistry generated during the plasma. The excited states Ar_i and Ar_j are treated as a lumped level.

The plasma generated between the cathodes is characterized by the presence of impurity metal-ions and gas ions due to intensive ion sputtering of the cathode. A relatively high atom content of the cathode material in the plasma is explained by a high kinetic energy (1 – 10 eV) [14] of the sputtered atoms. This effect promotes the contamination of the chamber walls and of the electrostatic probe. The combined effect of the ion bombardment by polarizing the probe with the cathode potential promotes control over the incandescence intensity of the probe to clean it.

4. Results and Discussion

The resolution of several computational problems constituted by equation (15), with the variation of the parameters involved, shows that the model developed for the cathode sheath is not very sensitive to variations of the coefficient of secondary electron emission, γ, within the typical range of this parameter.

Figure 2 shows the result from numerical integration of the equation (15). For a certain set of parameters, variations of the γ coefficient have little influence on the dimensionless potential profile in the cathode sheath.

On the one hand, this low sensitivity may indicate that the same effect is valid in practice due to the relative low densities of secondary electrons. Only an analysis that includes values of the coefficient γ of the order of hundreds shows significant changes in the profiles of the electric potential for the same parameters adopted in Figure 2. A relevant theoretical analysis refers to the variation of thickness of the cathode sheath with the plasma density, which can be easily varied by changing some discharge parameters such as gas pressure, axial magnetic field, inter-cathode distance, and discharge voltage.

Figure 3 shows that the thickness of the cathode sheath decreases monotonically when the plasma density increases, which is a consequence of the direct dependence of such thickness on the Debye length.

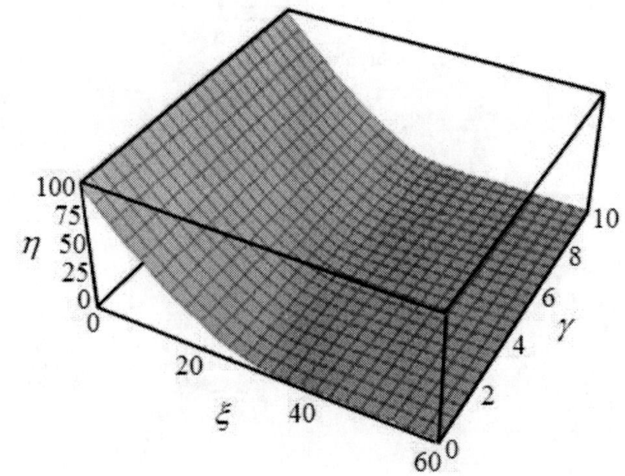

Figure 2. The normalized electric potential η as a function of distance normalized to the plasma Debye length and secondary electron coefficient, with $kT_e = 4\ eV$, $n_o = 1.0 \times 10^{16}\ m^{-3}$, $\eta_C = 104$.

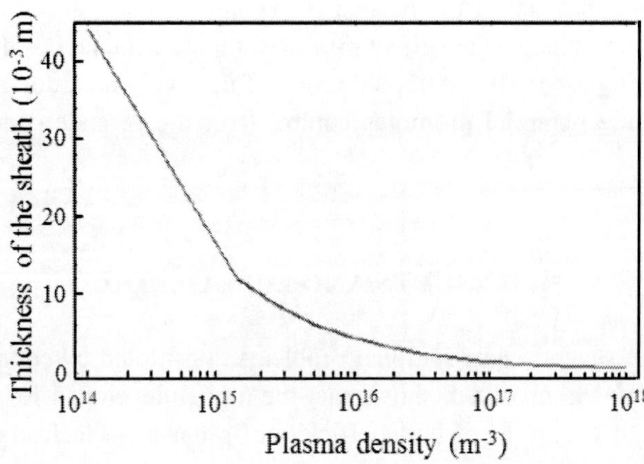

Figure 3. Thickness of the cathode sheath as function of the plasma density. The parameters used for making this simulation were $kT_e = 4$ eV, $\eta_C = 104$ and $\gamma = 0.07$.

If the electron temperature T_e is expressed in electron volts and the number density (n_o) in electrons per cubic meter, the Debye length (λ_D) is given by [17]:

$$\lambda_D \approx 7434\left(\frac{T_e}{n_o}\right)^{1/2} \tag{18}$$

The Debye length is the characteristic thickness of sheaths that form between a plasma and an electrode, or a plasma and a wall. Equation (18) predicts a Debye shielding distance of approximately 0.15 mm for plasmas with an electron temperature of 4 eV and a number density of 10^{16} electrons/m^3. The increase of the discharge current and the plasma density gives lower Debye lengths, consequently, the thickness of the cathode sheaths decreases. This effect generates a greater plasma volume, and consequently, it is possible to enhance the discharge current by lowering the inter-cathode distance. In our experimental apparatus, the inter-cathode distance was kept fixed at 19 mm. The discharge was operated by using argon as working gas. The mean values of the plasma parameters summarized in Table 1 were obtained at different positions inside the negative glow formed between the flat cathode surfaces for two operational conditions of the hollow cathode discharge:

- Condition I with $P = 15$ Pa, $V_d = 431$ V and $I_d=15$ mA.
- Condition II with $P = 25$ Pa, $V_d = 428$ V and $I_d=19$ mA.

The position of the interface region between the negative glow and the cathode sheath is visually identified by an intense difference in brightness between these two regions. This interface region can be evaluated experimentally by moving the cylindrical Langmuir probe with respect to the cathode surface along the axis of the discharge. The mean values of the interface positions also are presented in Table 1. The sheath-negative glow interface region can also be evaluated by determining the floating potential profile along the axis of the discharge, as shown in Figures. 4 and 5 for conditions I and II, respectively, with emphasis on the transition range of brightness in the sheath-negative glow interface region.

Table 1. Plasma parameters and the position of the sheath-negative glow interface for conditions I and II

Condition	kT_e (eV)	$n_o(10^{16}\text{m}^{-3})$	λ_D ($\times 10^{-3}$ m)	Position of interface (10^{-3} m)
I	7.7±0.9	1.9±0.3	0.15±0.08	3.3±0.5
II	5.9±0.6	2.3±0.2	0.12±0.06	2.8±0.5

The floating potential corresponds to the probe polarization potential for which current the probe current vanishes. It is supposed here that the floating potential profile has approximately the same qualitative behavior as the space potential profile, meaning that at the point where the electric potential of the medium begins to change, the same is expected to happen with the floating potential of the electrostatic probe.

The sheath-negative glow interface region can be identified from the numerical solution of equation 15 (see Figure 6) for comparisons with the experimental results determined from the floating potential profile along the axis of the discharge.

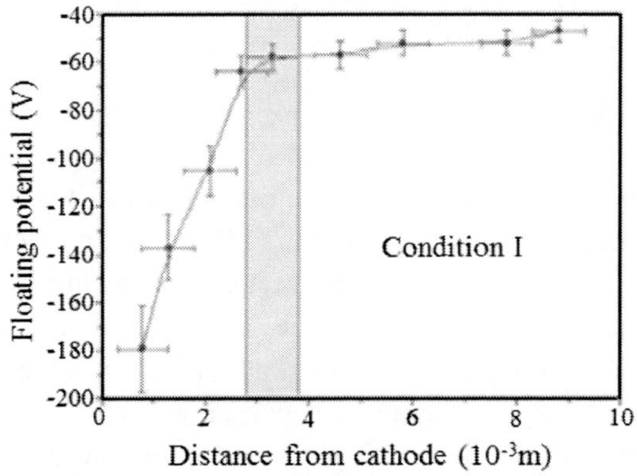

Figure 4. Floating potential as function of the distance from the cathode along the axis of the discharge; P=15 Pa, V_d= 431 V, I_d= 15 mA. The position of the sheath edge is within the range indicated by the red column.

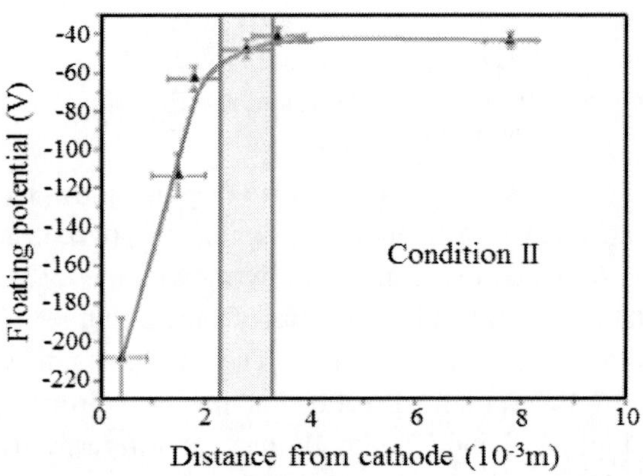

Figure 5. Floating potential as function of the distance from the cathode along the axis of the discharge; P = 25 Pa, V_d= 428 V, I_d= 19 mA. The position of the sheath edge is within the range indicated by the blue column.

The floating potential profile inferred from the probe characteristic may be qualitatively compared with the theoretical curve of electric potential of the medium (without the interference of the probe), obtained from the numerical solution of the equation (16), as shown in Figure 6 simulated for condition II. In this case, the electron secondary emission coefficient, which is already known not to influence significantly the results for the electrical potential, was kept fixed at γ =0.38 for argon and aluminum electrode [22]. For the argon ion, supposedly the only ionic species present in the discharge is $\mu \cong 72819.7$. The cathode potential fall, -V_C, is theoretically estimated as the sum of the discharge voltage (V_d), with the anode potential fall generally near the ionization potential of the gas [23]. In the case of argon, this potential is 15.76 V [17]. The electric potential along the cathode sheath region increases monotonically with the distance from cathode and tends to zero in the negative glow region where we have supposed the charge neutrality. The theoretical sheath thickness (s) was evaluated to be about 3.13 mm and, therefore, within the range of sheath edge position shown in Figure 5. In Figure 7, the electric field is relatively small at distances larger than the sheath thickness. This is in agreement with the physical concept of the pre-sheath region, where a small electric field must exist to ensure the acceleration of ions coming from the plasma at a speed at least equal to the ion sound speed c_s. This electric field, although of small intensity, is capable of producing significant modifications in charge densities of ions and electrons from plasma shortly before the entry in the cathode sheath region, as can be observed in Figures 8 and 9. Figure 9 also indicates that the electron number density in the sheath suffers a very drastic reduction governed by the ratio of Boltzmann, equation (6), and practically vanishes at a distance of 2.5 mm from the cathode. The net electrical charge in Figure 9 shows that for the parameters simulated, this density does not become negative at any point.

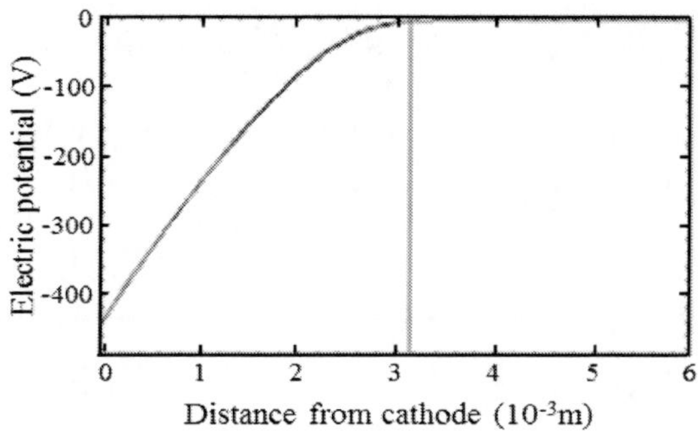

Figure 6. Electric potential profile along the x axis; kT_e = 5.9 eV, n_o = 2.3 x 10^{16} m^{-3}, V_C = -444 V, and γ = 0.38.

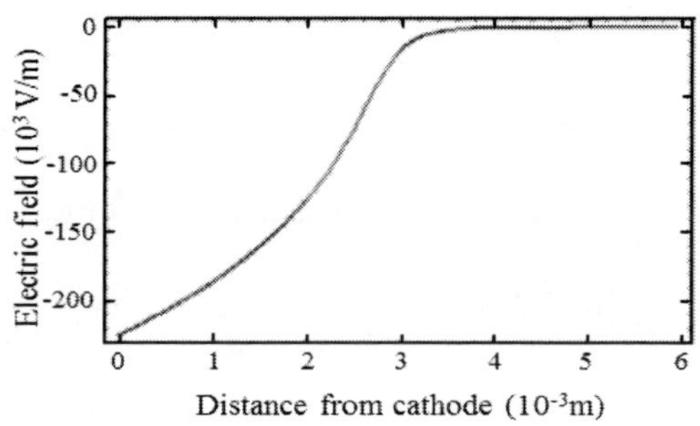

Figure 7. Electric field profile along the x axis; kT_e = 5.9 eV, n_o = 2.3 x 10^{16} m^{-3}, V_C = -444 V and, γ = 0.38.

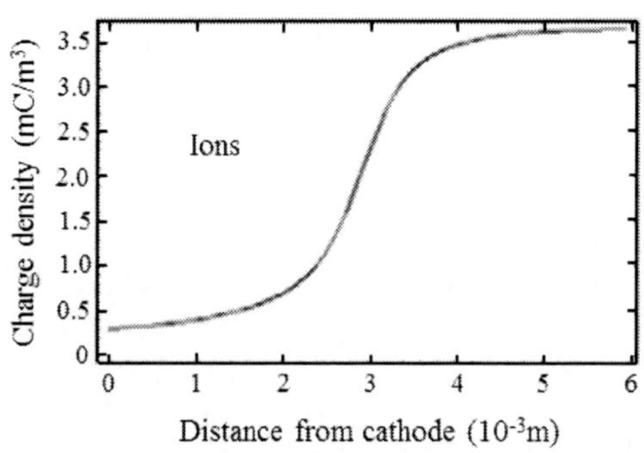

Figure 8. Variation of the ion charge density with the distance from the cathode along the x axis; kT_e = 5.9 eV, n_o = 2.3 x 10^{16} m^{-3}, V_C = -444 V, and γ = 0.38.

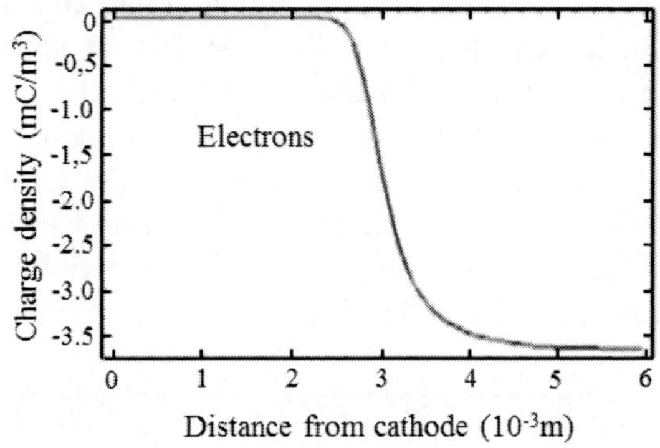

Figure 9. Variation of the electrons charge density with the distance from the cathode along the x axis; kTe = 5.9 eV, no = 2.3 x 1016 m-3, VC = -444 V, and γ = 0.38.

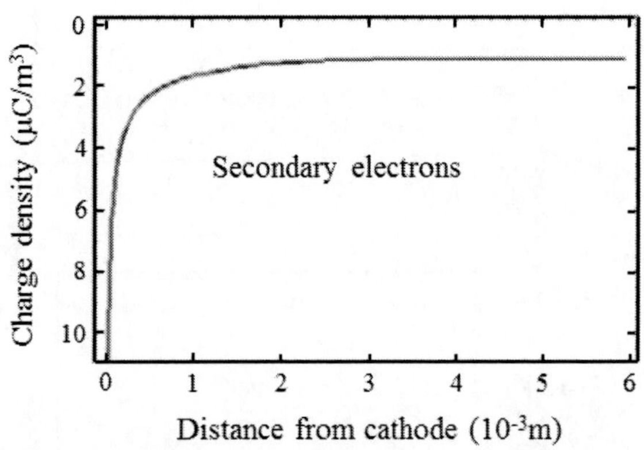

Figure 10. Variation of the secondary electrons charge density with the distance from the cathode along the x axis; kT_e= 5.9 eV, n_o= 2.3 x 10^{16} m^{-3}, V_C= -444 V, and γ = 0.38.

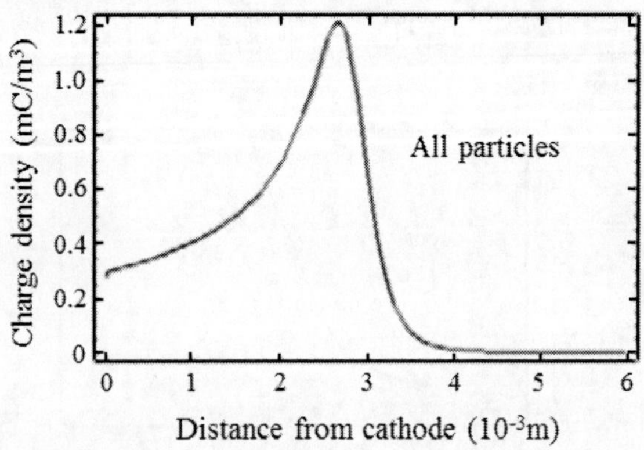

Figure 11. Variation of the total charge density with the distance from the cathode along the x axis; kT_e = 5.9 eV, n_o= 2.3 x 10^{16} m^{-3}, V_C= -444 V, and γ = 0.38.

Furthermore, the explanation for the behavior of this profile is that near the limiting plane between the sheath and negative glow, exists a maximum of net charge density.

As the ions coming from the plasma have not yet been sufficiently accelerated, the continuity equation establishes that their density must still be high at this point, which is confirmed in Figure 9. On the other hand, the secondary electrons released from the cathode surface are accelerated through the cathode sheath towards the plasma region followed by an expressive reduction of their charge density, as shown in the Figure 10.

This observation indicates that the electrons from the plasma region, decelerated by the sheath potential, are the ones that play the dominant role for the strong reduction of the net charge density at the sheath-plasma boundary (see Figure 11).

On the left side of the maximum of the net charge density profile, the ions density is much higher than the total electron density. The secondary electron charge density decreases with the distance from cathode, becoming three orders of magnitude lower than that of the ions in the plasma region. Overall, this means that the contribution of secondary electrons for the electrical potential profiles on the cathode sheath region is not significant, which is in agreement with the theoretical model.

Conclusion

The non-collisional model for the electric potential near the cathode of a HCD in conjunction with experimental data obtained from a Langmuir probe indicates a weak relationship of the cathode sheath potential profile with the secondary electron coefficient (γ). We include atypical values of γ to show that significant changes can occur in the profiles of electric potential only in the case of a net negative charge near the cathode sheath. This means that in these hypothetical cases, the emission of secondary electrons from the cathode is so intense that, in the cathode region, the number density of these electrons exceeds that of ions from the plasma.

The floating potential profiles inferred from the current-voltage characteristic of a single Langmuir probe are qualitatively similar to the theoretical electrical potential profiles. The position of the plasma-cathode sheath interface determined from the simulated model shows good agreement with the experimental data. This transition region allows a good estimation of the cathode sheath thickness.

Acknowledgments

This work was supported by the following Brazilian agencies projects: CAPES/PNPD (process nº 02765/09-8), FAPESP/CNPq/PRONEX (process nº 2011/50773-0) and FAPESP (process nº 2012/13064-4).

Appendix A - List of Physical Constants

The physical constants listed in Table A1 were used in this chapter:

Table A1. Physical Constants used

CONSTANT	SYMBOL	VALUE
Free space permittivity	ε_0	8.8542×10^{-12} F/m
Boltzmann constant	k	1.3807×10^{-23} J/K
Elementary charge	e	1.6022×10-19 C
Electron mass	m	9.1094×10^{-31} kg

APPENDIX B - DIMENSIONLESS VARIABLES AND DERIVED

In equation (10), they are defined with $\lambda_D = \left(\dfrac{\varepsilon_0 k T_e}{n_0 e^2}\right)^{1/2}$, the dimensionless variables, or normalized:

$$\eta = -\frac{e V}{k T_e}, \quad \xi = \frac{x}{\lambda_D} \tag{B.1}$$

which allow a more succinct and general representation of the equation (9).

Once such variables are defined from the calculation, the following relationships can be obtained:

$$\begin{aligned} \frac{d\xi}{dx} &= \frac{1}{\lambda_D} \\ \frac{dV}{dx} &= \frac{dV}{d\xi}\frac{d\xi}{dx} = \frac{1}{\lambda_D}\frac{dV}{d\xi} = -\frac{k T_e}{e\lambda_D}\frac{d\eta}{d\xi} \\ \frac{d^2V}{dx^2} &= \frac{d}{dx}\left(\frac{dV}{d\xi}\frac{d\xi}{dx}\right) = \frac{1}{\lambda_D{}^2}\frac{d^2V}{d\xi^2} = -\frac{k T_e}{e\lambda_D{}^2}\frac{d^2\eta}{d\xi^2} \end{aligned} \tag{B.2}$$

As is well known, the electric field in one dimension is given by $E = -\dfrac{dV}{dx}$. From the Poisson equation, also in one dimension, the space-charge density is $\rho = -\varepsilon_0 \dfrac{d^2V}{dx^2}$. Therefore, it follows that:

$$\begin{aligned} E &= \frac{k T_e}{e\lambda_D}\frac{d\eta}{d\xi} \\ \rho &= n_0 e \frac{d^2\eta}{d\xi^2} \end{aligned} \tag{B.3}$$

Another property involving derivatives of the function in relation to the independent variable is expressed by the equation:

$$\frac{1}{2}\frac{d}{d\xi}\left(\frac{d\eta}{d\xi}\right)^2=\frac{d\eta}{d\xi}\frac{d^2\eta}{d\xi^2} \tag{B.4}$$

so that a problem that contains an equation of the type:

$$\frac{d^2\eta}{d\xi^2}=f(\eta) \tag{B.5}$$

such as that described by the equation (11):

$$\frac{d^2\eta}{d\xi^2}=-\left[1-\gamma\nu(2\mu\eta_C)^{-\frac{1}{2}}\right]\exp(-\eta)-\gamma\nu(2\mu\eta_C)^{-\frac{1}{2}}\left(1-\frac{\eta}{\eta_C}\right)^{-\frac{1}{2}}+\left(1+2\eta\nu^{-2}\right)^{-\frac{1}{2}} \tag{B.6}$$

can be simplified to a first order differential equation using the following procedure:

$$\begin{gathered}\frac{d^2\eta}{d\xi^2}=f(\eta)\Rightarrow\frac{d\eta}{d\xi}\frac{d^2\eta}{d\xi^2}=f(\eta)\frac{d\eta}{d\xi}\Rightarrow\\ \Rightarrow\frac{1}{2}\frac{d}{d\xi}\left(\frac{d\eta}{d\xi}\right)^2=f(\eta)\frac{d\eta}{d\xi}\Rightarrow\frac{1}{2}\int d\left(\frac{d\eta}{d\xi}\right)^2=\int f(\eta)\,d\eta\end{gathered} \tag{B.7}$$

The equation (B.6) has, by definition, in the plane of abscissa $x = x_b$, null potential and electric field, or and $\frac{d\eta}{d\xi}=0$, with and, in a generic abscissa x — or —, within the sheath, and have determined values of and $\frac{d\eta}{d\xi}$. Thus, the integration performed in (B.7) should take, with the function f defined by (B.6), the boundaries:

$$\frac{1}{2}\int_0^{\left(\frac{d\eta}{d\xi}\right)^2}du=\int_0^{\eta}f(t)\,dt \tag{B.8}$$

i.e.:

$$\frac{1}{2}\left(\frac{d\eta}{d\xi}\right)^2=\int_0^{\eta}\left\{-\left[1-\gamma\nu(2\mu\eta_C)^{-\frac{1}{2}}\right]\exp(-t)-\gamma\nu(2\mu\eta_C)^{-\frac{1}{2}}\left(1-\frac{t}{\eta_C}\right)^{-\frac{1}{2}}+\left(1+2t\nu^{-2}\right)^{-\frac{1}{2}}\right\}dt \tag{B.9}$$

that with the second member integrated leads to the equation (15).

APPENDIX C - TAYLOR SERIES EXPANSIONS EMPLOYED

Substituting the expression (12) in (11), we obtain:

$$Kn(n+1)\xi^{-(n+2)} = -\left[1-\gamma\nu\left(2\mu\eta_C\right)^{-\frac{1}{2}}\right]\exp\left(-K\xi^{-n}\right)-\gamma\nu\left[2\mu\eta_C\left(1-\frac{K\xi^{-n}}{\eta_C}\right)\right]^{-\frac{1}{2}}+\left(1+2K\xi^{-n}\nu^{-2}\right)^{-\frac{1}{2}} \tag{C.1}$$

One can therefore make use of series expansions to obtain the conditions that the terms the second member of the equation generated are zero sum, which is equivalent to the expression:

$$\left[1-\gamma\nu\left(2\mu\eta_C\right)^{-\frac{1}{2}}\right]-\gamma\nu\left(2\mu\eta_C\right)^{-\frac{1}{2}}\cdot\frac{1}{2\eta_C}-\nu^{-2}=0 \tag{C.2}$$

Obtaining (C.2) from (C.1) is based on the expansions of the functions below in Taylor series, around $x=0$:

$$\exp(-x)=\sum_{n=0}^{\infty}\frac{(-1)^n x^n}{n!}=1-x+\frac{x^2}{2}-\frac{x^3}{6}+\ldots \tag{C.3}$$

$$(1-x)^{-1/2}=1+\frac{x}{2}+\frac{3x^2}{8}+\frac{5x^3}{16}+\frac{35x^4}{128}+\ldots \tag{C.4}$$

$$(1+x)^{-1/2}=1-\frac{x}{2}+\frac{3x^2}{8}-\frac{5x^3}{16}+\frac{35x^4}{128}-\ldots \tag{C.5}$$

Thus, considering only the first three terms of each expansion results in the following equalities:

$$\exp\left(-K\xi^{-n}\right)=1-K\xi^{-n}+\frac{K^2}{2}\xi^{-2n}+\ldots \tag{C.6}$$

$$\left(1-\frac{K\xi^{-n}}{\eta_C}\right)^{-\frac{1}{2}}=1+\frac{K\xi^{-n}}{2\eta_C}+\frac{3K^2\xi^{-2n}}{8\eta_C{}^2}+\ldots \tag{C.7}$$

$$\left(1+2K\xi^{-n}\nu^{-2}\right)^{-\frac{1}{2}}=1-K\xi^{-n}\nu^{-2}+\frac{3K^2\xi^{-2n}}{2}\nu^{-4}+\ldots \tag{C.8}$$

By substituting the equations (C.6) to (C.8) in (C.1), one obtains the equation which, among other potencies ξ, involves those of order $-n$. However, as it has certainly $n\neq n+2$, and is not a possible solution, the sum of terms in can only result nil. And this sum is precisely the one expressed in the equation (C.2), unless the multiplicative factor $K\xi^{-n}$, which is non-zero, by assumption.

APPENDIX D - DISCUSSION OF THE EQUATION $\kappa\nu^3 - \nu^2 + 1 = 0$

The determination of an consistent estimate for the ratio $\nu = \frac{v_0}{c_s}$, which, by the Bohm criterion, is certainly greater than unity, resulting in equation (13):

$$\kappa\nu^3 - \nu^2 + 1 = 0 \tag{D.1}$$

that has $\kappa = \gamma\left(2\mu\eta_C\right)^{-\frac{1}{2}}\left(1+\frac{1}{2\eta_C}\right) > 0$, but typically less than unity by at least four orders of magnitude. This finding is very important in choosing the solution of equation (D.1) what matters in issue in question, and can be easily confirmed by considering the usual orders of magnitude of the coefficient off emission of secondary electrons $\left(10^{-2} - 10^{-1}\right)$, of the ratio between the mass of a gas discharge ion and electron and the dimensionless potential $\left(10^{1} - 10^{2}\right)$. As explained in this work, a value consistent of the ratio should be slightly larger than unity, and are expected, therefore, that the lower is, the more approaches one. Thus, mathematically, it should be adopted as the value of the zero of the function that satisfies the condition:

$$\lim_{\kappa \to 0^+} \nu = 1 \tag{D.2}$$

The *software Mathematica 5.0* can be confirmed that only one of the three solutions (D.1) satisfies (D.2) and other satisfies and a third, $\lim_{\kappa \to 0^+} \nu = \infty$, not being thus physically consistent with the proposed theoretical model. The solution which satisfies (D.2) has an extensive expression and a function of κ :

$$\nu = \frac{1}{6\kappa}\left[2 - \left(1 - i\sqrt{3}\right)\left(\frac{2}{2 - 27\kappa^2 + 3\kappa\sqrt{81\kappa^2 - 12}}\right)^{\frac{1}{3}} - \left(1 + i\sqrt{3}\right)\left(\frac{2}{2 - 27\kappa^2 + 3\kappa\sqrt{81\kappa^2 - 12}}\right)^{-\frac{1}{3}}\right] \tag{D.3}$$

and is a positive real number since it has $0 < \kappa < \frac{2}{3\sqrt{3}} \approx 0,3849$, which is certainly satisfied in the model. From the equation (D.3), the expression of as a function of the parameters γ, μ, and can also be obtain, by simply replacing the expression of in D.3.

References

[1] N. Hershkowitz, *Physics of Plasmas,* 12, 55502-1 (2005).

[2] R. S. Pessoa, L. L. Tezani, H. S. Maciel, G. Petraconi, M. Massi, *Plasma Sources Science and Technology,* 19 025013 (7pp) (2010).

[3] R. S. Pessoa, B. N. Sismanoglu, J. Amorim, H. S. Maciel, G. Petraconi, Gas Discharges: Fundamentals and Applications vol 1 ed J. Amorim (Kerala, India: Transworld Research Network) pp 175–90 (2007).

[4] L. Bárdős, *Surface and Coatings Technology,* 86–87, 648–56 (1996).

[5] Günterschulze, *Z. Tech. Phys.,* I1, 49 (1930).

[6] P.F. Little, A. von Engel, *Proc. Roy. Sot.,* 224, 209 (1954).

[7] S. Hashiguchi, *IEEE Trans. on Plasma Sci.,* 19(2), 297 (1991).

[8] Z. Donkó, *Phys. Rev., E* 57(6), 7126 (1998).

[9] P. Gill, C.E. Webb, *J. Phys. D: Applied Phys.,* 10, 299 (1977).

[10] V.S. Borodin, Y.M. Kagan, *Sov. Phys. Tech. Phys.,* 11, 131 (1966).

[11] J. P. Boeuf, L. C. Pitchford, *IEEE Trans. on Plasma Sci.,* 19(2), 286 (1991).

[12] V.I. Kolobov, L.D. Tsendin, *Phys. Rev. A,* 46, 7837 (1992).

[13] R.R. Arslanbekov, A.A. Kudryavtsev, *Physical Rev. E,* 58(5), 6539 (1998).

[14] B. Chapman, Glow discharge processes: Sputtering and plasma etching. New York: J. Wiley and Sons, (1980).

[15] Z. Donkó, *Physical Review E,* 64, 26401-1 (2001).

[16] Bogaerts, R. Gijberls, *Plasma Sources Science and Technology,* 11 27–36 (2002).

[17] M. A. Lieberman, A. J. Lichtenberg, Principles of Plasma Discharges and Materials Processing, 2th edition, New York: Wiley (2005).

[18] D. Wang, D. Liu, J. Liu, *J. of Appl. Phys.,* 88 1276 (2000).

[19] F. Chen, Introduction to Plasma Physics and Controlled Fusion, New York: Plenum Press (1984).

[20] R.N. Franklin, Plasma phenomena in gas discharges. Oxford: Clarendon Press (1976).

[21] R.S. Pessoa, Estudos de plasmas fluorados aplicados em corrosão de silício usando simulação de modelo global e diagnósticos experimentais, PhD Thesis, Instituto Tecnológico de Aeronáutica, Brazil (2009).

[22] A.V. Phelps, Z.L. Petrovic, *Plasma Sources Science and Technology,* 8, R21–R44 (1999).

[23] Grill, Cold Plasma in Materials Fabrication: From Fundamentals to Applications, New York: Institute of Electrical and Electronics Engineers Press (1994).

In: Argon: Production, Characteristics and Applications
Editors: B.N. Sismanoglu, H.S. Maciel et al.
ISBN: 978-1-62618-204-2

Chapter 11

SAMPLE PREPARATION USING ARGON ION BEAM MILLING

C. Dahlström[*]
Mid Sweden University, Sweden

ABSTRACT

Surface and cross section analysis can provide important information on material properties. It has, however, been difficult to characterise the internal structure at a micro scale using cross sections prepared with conventional methods, such as e.g. mechanical grinding and polishing, because these procedures can damage the surface and introduce artefacts. A few years ago, a new precision argon ion beam cross section polisher was shown to be a suitable sample preparation technique for microstructure characterisation. It produces high quality cross sections free from artefacts and distortions. This technique can be used for cross section preparation of a wide variety of materials including composites, metals, ceramics and polymers. Argon ion beam milling can also be used in combination with focused ion beam (FIB) as a final step to remove FIB-damaged layers before nano- and microstructure characterisation in high resolution transmission electron microscopy (HRTEM).

1. INTRODUCTION

Material analysis of surface and internal structures has been performed for many years, both for research and engineering purposes. The material structures may affect e.g. absorbency, gloss, optical properties, fluid transport and mechanical properties. Micro- and nanostructure characterisation has been a hot topic the last decades and the growing interest to study smaller structures puts higher demands on both the sample preparation and the resolution of e.g. the electron microscope.

[*] E-mail address: christina.dahlstrom@miun.se.

It has been difficult to characterise the internal structure at a micro scale using cross sections prepared with conventional methods, e.g. resin embedding with subsequent mechanical grinding and polishing [1-4]. Composite material, such as e.g. a coated paper, consists of material with different hardness values, where the softer components are ground faster than the harder components. A drawback with the embedding and mechanical grinding/polishing procedures is that it might damage the surface and introduce artefacts; such as e.g. pigment particle removal during grinding or polishing, resulting in poor image quality. In soft material such as the embedding resin, particles of hard abrasive can be captured in the resin during grinding. Fine details in the structure might be smeared out and some varnish might get dissolved in the resin. Microtoming is another method for cross section preparation and it is a mechanical sectioning [5, 6]. It is commonly used for preparation of soft materials, e.g. biological specimens, which also requires resin embedding.

The focused ion beam (FIB) technique [7, 8] did overcome some of these sample preparation problems and one advantage with this method is that it is site-specific. However, this method is often used in-situ in the electron microscope and is limited to the production of very small sections, e.g. production of ultrathin sections for transmission electron microscopy (TEM).

A few years ago, a new precision broad argon ion beam milling technique was presented to the market and it was suggested as a potential alternative to the FIB technique [9]. Although no measurement data were presented from its images, Robinson *et al.* showed very impressive pictures of coating cross sections that could be used for quantitative analyses [10]. The broad argon ion beam milling technique produces high quality sections similar to FIB, but larger cross sections are obtained, around 2 mm in length. It is often desirable to quantify the microstructure over a larger area and to assess many images for images analysis. For example, 30 images (both backscattered electron images, BEI, and secondary electron images, SEI) of the coating layer were obtained from one single cross section and analyzed with image analysis to obtain pore and binder distribution [11, 12]. The argon ion beam milling in combination with field emission scanning electron microscopy (FESEM) was shown to be suitable techniques for microstructure characterisation. Argon ion beam milling has also been used in combination with FIB and TEM, where the argon ion beam removes damaged layers caused by the FIB [13, 14].

2. Cross Section Polisher via Argon Ion Beam Milling

A cross section polisher typically consists of a specimen chamber with a vacuum system, an optical microscope and a stationary ion beam. The optical microscope is used for sample positioning and this is a critical step in the sample preparation because the ion milling is usually set to etch around 75 μm from the outer edge of the sample, and a sample that is positioned slightly offset will not get milled. Before milling, a masking plate is placed on top of the sample to shield part of the sample from the ion beam. The other part of the incident ion beam gradually removes the sample protruding from the mask and this result in a flat cross section along the edge of the applied mask, see Figure 1. The argon ion beam is uniform in energy and direction, and atoms are sputtered from the sample surface.

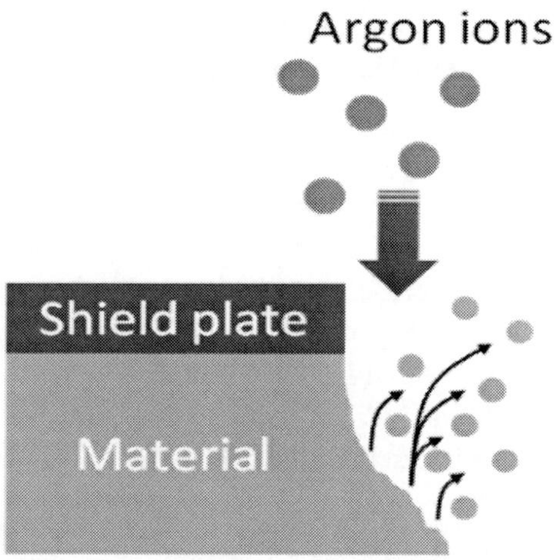

Figure 1. Schematic image of the milling procedure.

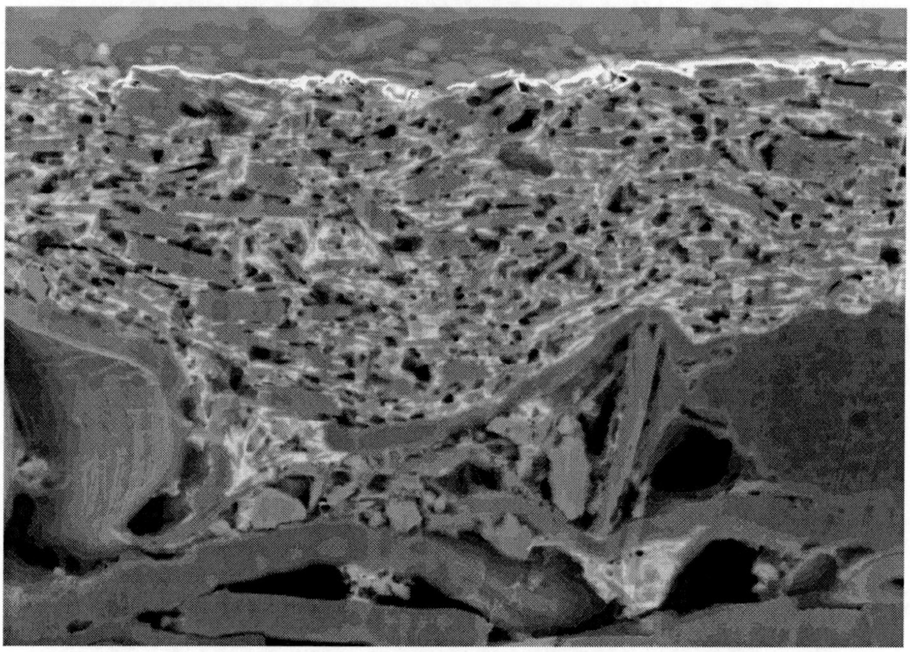

Figure 2. Backscattered electron image (BEI) of a coating layer cross section. The clay particles are platy with different shades of grey, the binders are bright and the pores are dark.

The gas flow is adjusted to optimise the ion beam current, and the ion accelerating voltage can be selected from 0 to 6 kV depending on the sample properties, such as the hardness value. The argon ion beam milling is using a low accelerating voltage and is gentle to the sample compared to the gallium ion beam milling used in FIB. The specimen stage can be rocked during milling to obtain uniform etching and to avoid striations. Special rotation holder has been developed to prepare cross sections of powder or specimens with circular cross sections [15]. The obtained cross sections are around 1.5 - 2.0 mm in length. One advantage is that there is no need for resin embedding.

3. Microstructure Characterisation

TEM and scanning electron microscopy (SEM) has been widely used for morphology studies to assess material structures, e.g. coating layer properties and multilayer of paint [4, 16-19]. Further, FESEM and high resolution TEM (HRTEM) provides even higher resolution than conventional SEM and TEM, and it is possible to detect micro- and nanostructures, such as e.g. pores, crystal structures and crystal lattice fringes [9, 12, 13, 20].

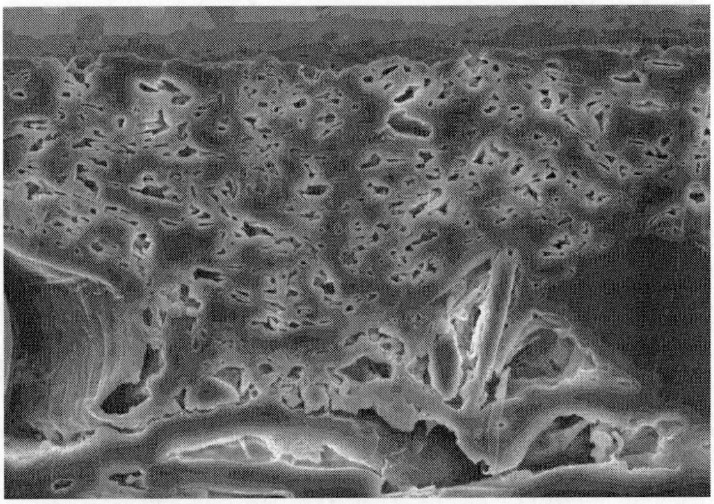

Figure 3. Secondary electron image (SEI) of a coating layer cross section where the pore structure is clearly visualised.

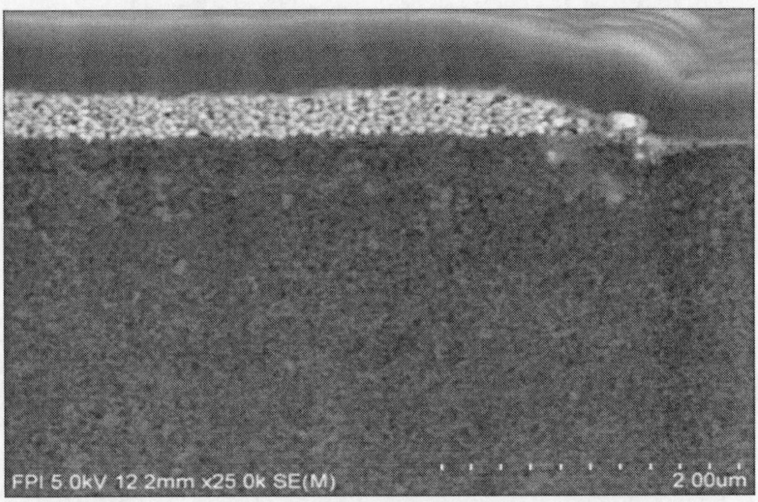

Figure 4. (Continued).

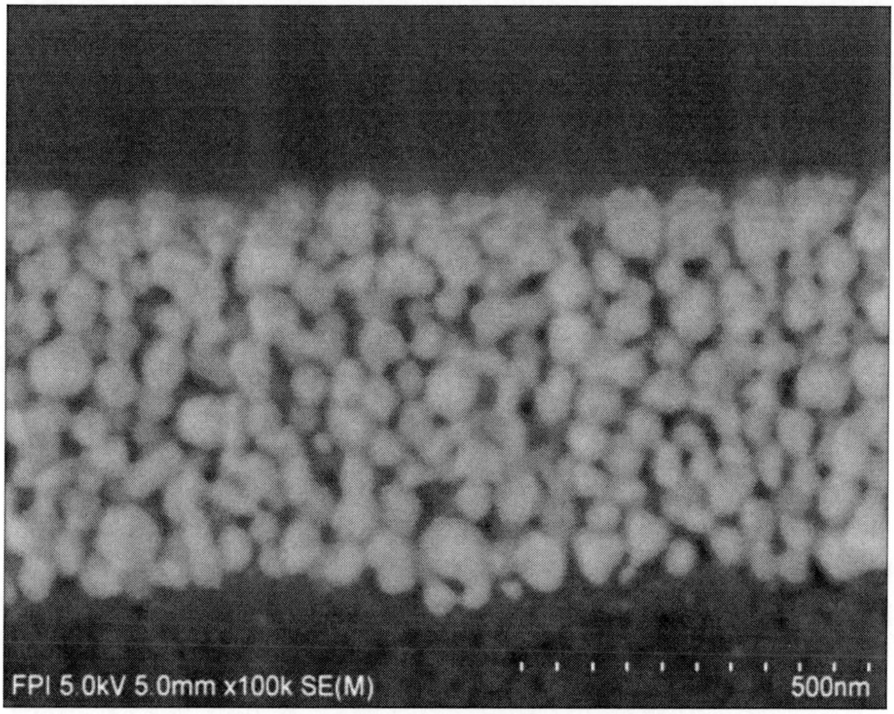

Figure 4. SEI of the electronic print on top of a photo paper (top) and a magnified view (bottom). (Images taken by Dr. Allem and Dr. Le Berre at FPInnovations, Montreal).

Figure 5. (Continued).

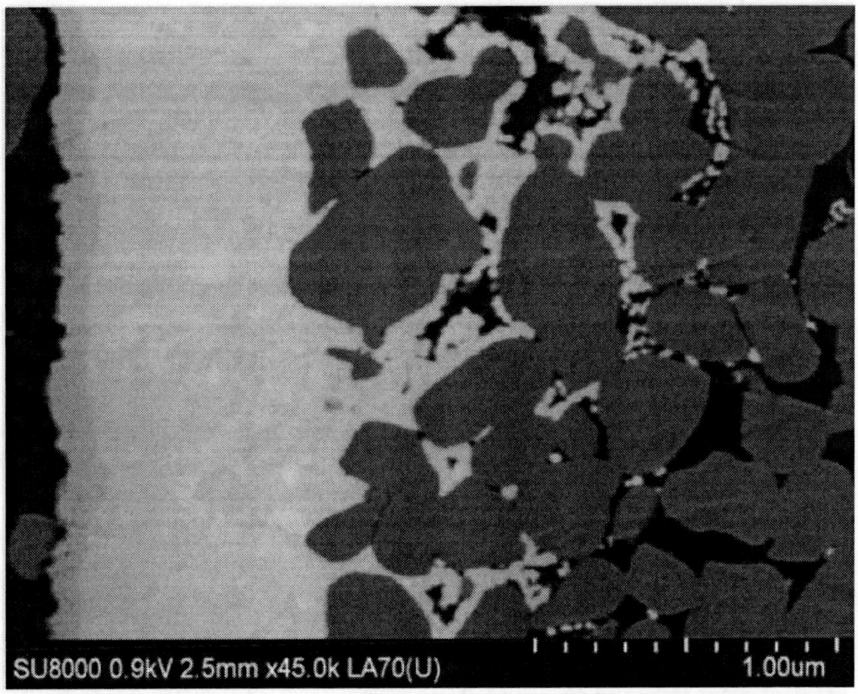

Figure 5. Multilayer ceramics filter with catalyst (top) and a magnified view (bottom) [22].

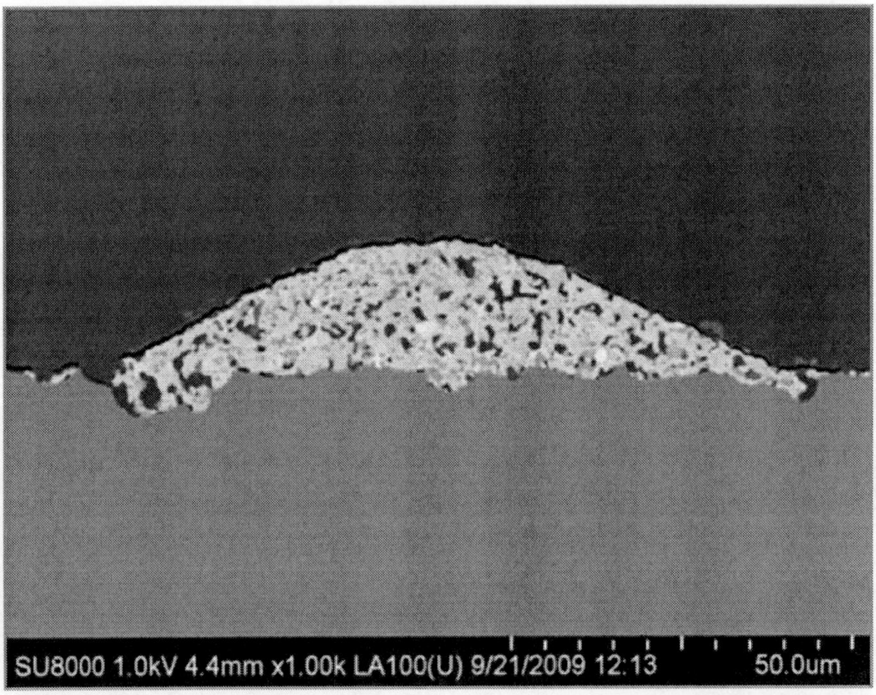

Figure 6. (Continued).

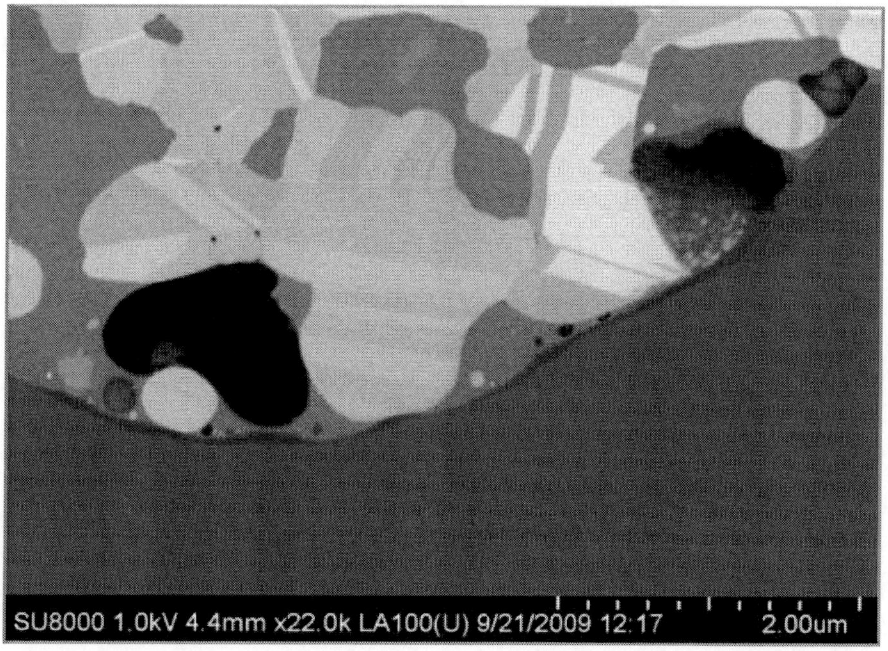

Figure 6. Printed silver electrode on solar cell (top) and a magnified view (bottom) [22].

In paper coatings, the pore structure is of interest since it is important for the transfer and setting of ink in the printing process. Traditionally, the pore measurements are done by image analysis of the BEI, see Figure 2. However, the high quality SEI obtained using the cross section polisher via argon ion beam milling in combination with FESEM made it possible to develop a new method for porosity measurement based on marker-controlled watershed segmentation of the SEI (Figure 3) [12]. Another area of interest in paper coatings is the binder distribution [21], and to detect the binder the paper is stained with OsO_4 [16]. In Figure 2 the binder is seen as the bright material surrounding the particles. The bright layer following the top surface is a thin gold layer that is applied to easier outline the coating layer.

Another interesting research area is printed electronics. These samples are very sensitive due to sintering, despite of that the argon ion beam milling was found to be a suitable method for sample preparation allowing nanostructure characterisation (Figure 4). More examples of materials that were prepared using the argon ion beam milling technique are shown in Figures 5 and 6 [22]. Figure 5 represents a multi-layer ceramics filter with catalyst and Figure 6 shows a printed silver electrode on solar cell.

REFERENCES

[1] Williams G. J. and Drummond J. G., 1994 *TAPPI Papermakers Conference*, 517.
[2] Dickson A. R., *Appita J.,* 53, 292 (2000).
[3] Klein R. and Schulze U., *Int. Papierwirtschaft* 48 (2006).
[4] Allem R., *J. Pulp Pap. Sci.*, 24, 329 (1998).
[5] Peterson A. W. and Williams L. C., *Tappi J,* 75, 122 (1992).
[6] Chinga-Carrasco G., *J. Microsc.*, 234, 211 (2009).

[7] Uchimura H., Ozaki Y. and Kimura M., 1998 *TAPPI International Printing & Graphic Arts Conference*, 121.

[8] Okayasu T., Kusama S., Tozaki E. and Miyagawa T., *2001 International Coating Graphic Arts,* 183.

[9] Erdman N., Campbell R. and Asahina S., *Adv. Mater. Processes*, 164, 33 (2006).

[10] Robinson M. and Rodier J.-M., *2006 TAPPI International Conference on Nanotechnology*

[11] Dahlström C. and Uesaka T., Ind. Eng. Chem. Res., 51, 8246 (2012).

[12] Dahlström C., Allem R. and Uesaka T.*, J. Microsc.*, 241, 179 (2011).

[13] Sasaki H., et al., *J. Electron Microsc.,* 53, 497 (2004).

[14] Huang Z., *J. Microsc*., 215, 219 (2004).

[15] Ogura K., Kamidaira M., Asahina S. and Erdman N., *Microsc. Microanal*., 13, 1518 (2007).

[16] Chinga G. and Helle T., *Nord. Pulp Pap. Res. J.,* 17, 307 (2002).

[17] Dahlström C. and Uesaka T., *Ind. Eng. Chem. Res.,* 48, 10472 (2009).

[18] Climpson N. A. and Taylor J. H., *Tappi J.,* 59, 89 (1976).

[19] Boon J. J. andAsahina S., *Microsc. Microanal*., 12, 1322 (2006).

[20] Pöhler T., Juvonen K. and Sneck A., *2006 TAPPI Advanced Coating Fundamentals Symposium,* 89.

[21] Dahlström C., *Quantitative Microscopy of Coating Uniformity*, Mid Sweden University, Sundsvall, 2012.

[22] Hitachi High-Technologies Europe Gmbh, http://www.hht-eu.com.

In: Argon: Production, Characteristics and Applications
Editors: B.N. Sismanoglu, H.S. Maciel et al.
ISBN: 978-1-62618-204-2

Chapter 12

ARGON PLASMA COAGULATION FOR GASTRIC ANTRAL VASCULAR ECTASIA ASSOCIATED WITH PORTAL HYPERTENSION

T. Sato*

Department of Gastroenterology, Sapporo Kosei General Hospital, Chuo-ku, Sapporo, Japan

ABSTRACT

Gastric antral vascular ectasia (GAVE) is a rare cause of chronic gastrointestinal bleeding. The majority of GAVE patients have underlying cirrhosis and portal hypertension, and GAVE is related to severe liver damage. Argon plasma coagulation (APC) is a new modality of non-contact electrocoagulation. APC is a safe and results in good temporary control of GAVE-associated hemorrhage. However, a high recurrence rate of GAVE has also been reported at medium and long-term follow-up times after APC treatment. Despite the high recurrence rate mentioned above, APC is a first choice for treatment of watermelon type GAVE. The histologic changes of GAVE exist in the mucosal and submucosal region of the stomach, therefore, endoscopic band lagation (EBL) may be more effective for GAVE because of its ability to obliterate submucosal vascular plexus. The best management to treat the diffuse and severe GAVE may be EBL plus APC combination therapy.

1. INTRODUCTION

Gastric antral vascular ectasia (GAVE) is a rare cause of chronic gastrointestinal bleeding, often presenting as iron- deficiency anemia, melena, or rarely, hematemesis [1-4]. The diagnosis of GAVE is achieved on the basis of endoscopic and histological appearance. The majority of GAVE patients have underlying cirrhosis and portal hypertension. The optimal treatment of GAVE is not known. Argon plasma coagulation (APC) is a relatively

* E-mail address: taka.sato@ja-hokkaidoukouseiren.or.jp.

new non-contact electrocoagulation technique. Grund et al. have reported on the first experiences with APC as an option in the treatment of gastrointestinal hemorrhage [5]. In this chapter, we review the clinical features of GAVE and the efficacy of APC in treating GAVE.

2. Etiology of GAVE

The etiology of GAVE remains unknown, and it is thought to encompass a variety of conditions including autoimmune disease, liver cirrhosis, portal hypertension, chronic renal failure, cardiovascular disease, achlorhydria, severe hypochlorhydria, hypothyroidism, and diabetes mellitus [6-13]. In Gostout's article, atrophic gastritis occurred in 19 of 19 (100%) with biopsies, and hypergastrinemia was observed in 25 of 33 (76%) tested [6]. Gastrin may play a role in the pathogenesis of GAVE. However, it may not be fundamental, and also, Ikeda et al. have reported that the development of GAVE associated with cirrhosis was concerned with the time courses [14, 15]. The majority of GAVE patients have underlying cirrhosis and portal hypertension. On the other hand, Spahr et al. have reported that GAVE is not directly related to portal hypertension [16]. Several reports explain the elevated antral motility for GAVE [17, 18]. Whether GAVE requires therapy seems to be related to the severity of liver dysfunction, the majority of GAVE patients had previously received endoscopic injection sclerotherapy for esophageal varices. However, portal hypertensive gastropathy was recognized in a few cases [19]. We suggest that GAVE is related not only to portal hypertension but also to hypergastrinemia, and elevated antral motility.

3. Endoscopic and Histological Findings of GAVE

The endoscopic findings demonstrated that GAVE was limited almost exclusively to the antrum, and red spots were either aggregated in linear stripes (watermelon stomach) (Figure 1-a) or diffuse, punctate and vascular telangiectasia in the entire antral area (diffusely spread) (Figure 1-b), in which the dilatation of mucosal capillaries results in gastrointestinal bleeding.

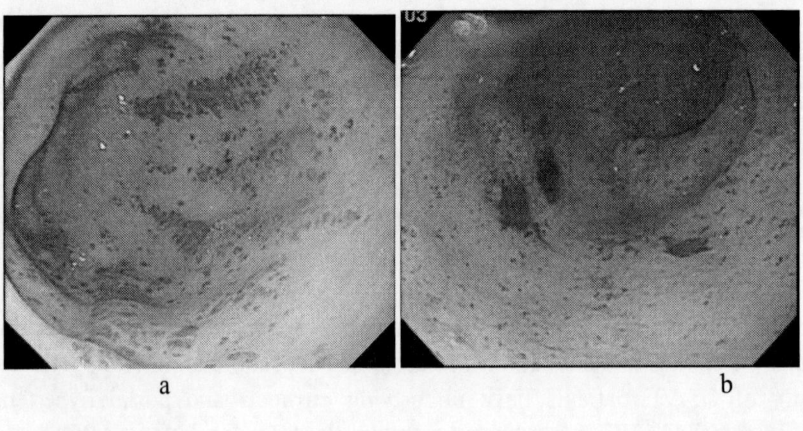

a b

Figure 1. (a) Watermelon stomach type gastric antral vascular ectasia: red spots were either aggregated in linear stripes. (b) Diffusely spread type gastric antral vascular ectasia: diffuse, punctate and vascular telangiectasia in the entire antral area.

The typical histopathological appearance of GAVE consists of hypertrophy of the antral mucosa, dilatation of mucosal capillaries with focal thrombosis, and fibromuscular hyperplasia of the lamina propria [1, 6, 13, 20, 21]. The diagnosis of GAVE is achieved on the basis of endoscopic and histological appearance. The histologic changes of GAVE exist in the mucosal and submucosal region of the stomach.

4. Treatments of GAVE

The optimal treatment of GAVE is not known. Before 1990, surgical anterectomy was the treatment of choice [1]. However, surgery carries a significant risk of mortality in cirrhotic patients. Recently, endoscopic techniques have become the first treatment of choice for GAVE and several endoscopic treatments have been used successfully to treat GAVE patients, including laser [22], electrocoagulation [23, 24], heat probe [25], and APC [26-30]. In the second session of treatment, pharmacological therapy with corticosteroids [1, 31] or others [32, 33] is an alternative. Pharmacologic treatment may be used as adjunct therapy in patients resistant to endoscopic treatments.

5. Argon Plasma Coagulation for Gastric Antral Vascular Ectasia

APC has been performed successfully to treat GAVE, and results in good temporary control of GAVE-associated hemorrhage. However, it is not very effective for medium and long-term treatment, and is associated with a high recurrence rate of GAVE in the medium and long-term term [29, 34, 35].

In our ward, standard APC equipment was used, consisting of a high-frequency electrosurgical generator, (ICC 350, Tuebingen, ERBE Germany), an automatically regulated argon source (APC 300) and a flexible APC probe. The APC probe was a 2.3 mm Teflon-coated catheter with a heat-resistant ceramic tip, which could be passed through the working channel of an endoscope. The 60 W electrical power and argon gas flow was 2 L/min. In this case, APC was applied to the lesions of active oozing bleeding from GAVE (Figure 2).

APC was performed in 22 diffusely spread type GAVE patients, all with iron-deficiency anemia, 14 of whom also had a history of tarry stools. After APC, endoscopy revealed ulcers and shrinkage of the GAVE in all 22 patients, and no complications were reported. During follow-up of the 22 APC patients (mean 16.6 months), endoscopies revealed the recurrence of GAVE in 15 patients requiring further treatment by APC (recurrence rate: 68.2%). Then, we experienced on a case of multiple large polyps with GAVE recurrence after APC. APC was applied to the lesion beginning at the pylorus and proceeding proximally (Figure 3-a). The fibergastroscopic examination after APC revealed multiple hyperplastic polyps with GAVE recurrence (Figure 3-b). While the exact pathogenetic mechanism of polyp formation is unclear, it seems likely that the polyps developed in the gastric antrum after APC treatment, concurrent with healing of acute gastric inflammations and/or ulcers. Previous reports also describe development of hyperplastic polyps as a complication associated with APC treatment for GAVE [36, 37].

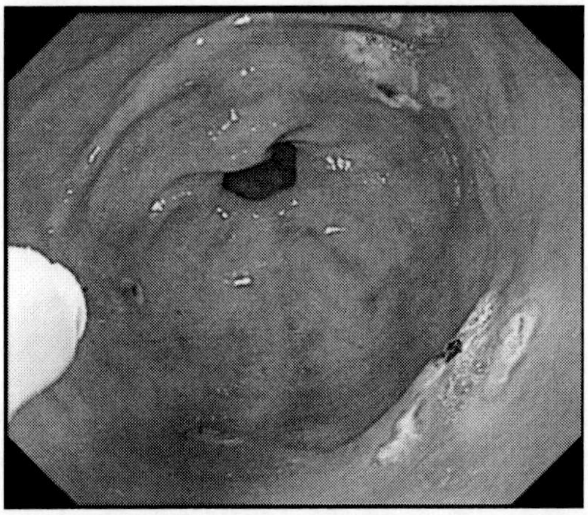

Figure 2. Argon plasma coagulation was applied to the lesion of active oozing bleeding from gastric antral vascular ectasia.

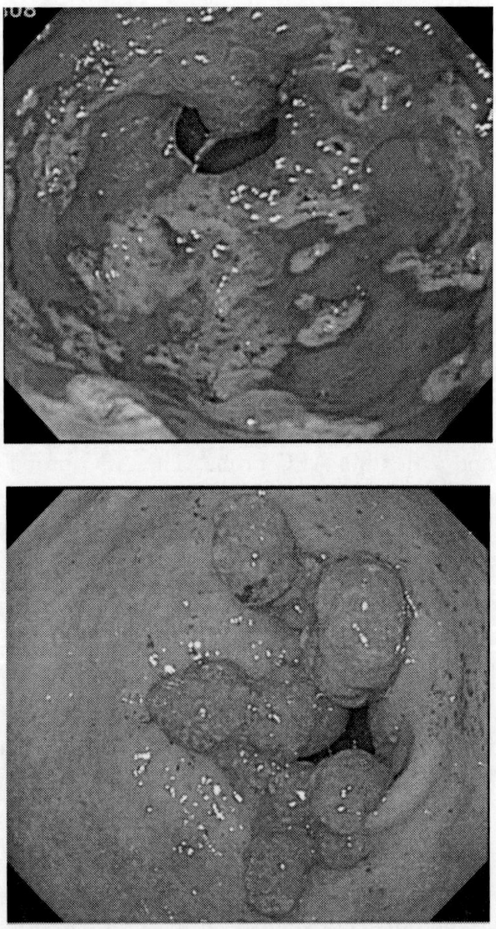

a)

b)

Figure 3. (a) After argon plasma coagulation for gastric antral vascular ectasia, endoscopy revealed ulcers. (b) Multiple large polyps with gastric antral vascular ectasia recurrence after argon plasma coagulation.

Despite the high recurrence rate mentioned above, APC is a first choice for treatment of watermelon type GAVE. The histologic changes of GAVE exist in the mucosal and submucosal region of the stomach, therefore, endoscopic band lagation (EBL) may be more effective for GAVE because of its ability to obliterate submucosal vascular plexus [19, 38-40]. The best management to treat the diffuse and severe GAVE may be EBL plus APC combination therapy.

CONCLUSION

In conclusions, the results suggest that GAVE is related to severe liver damage in cirrhotic patients and portal hypertension. APC is a safe and results in good temporary control of GAVE-associated hemorrhage.

REFERENCES

[1] M. Jabbari, R. Cherry, J. O. Lough, D. S. Daly, D. G. Kinnear and C. A. Goresky, *Gastroenterology* 87, 1165 (1984).

[2] R. H. R. Park and R. I. Russell, *Br. J. Surg.* 78, 395 (1991).

[3] J. E. Gretz and S. R. Achem, *Am. J. Gastroenterol.* 94, 2909 (1999).

[4] Szold, L. B. Katz and B. S. Lewis, *Am. J. Surg.* 163, 90 (1992).

[5] K. E. Grund, D. Storek and G. Farin, *Endosc. Surg.* 2, 42 (1994).

[6] C. J. Gostout, T. R. Viggiano, D. A. Ahlquist, K. K. Wang, M. V. Larson and R. Balm, *J. Clin. Gastroenterol.* 15, 256 (1992).

[7] F. I. Lee, F. Costello, N. Flanagan and K. S. Vasudev, *Gastrointest. Endosc.* 30, 87 (1984).

[8] T. Arendt, M. Barten, V. Lakner and R. Arendt, *Endoscopy* 19, 218 (1987).

[9] D. Kamberoglou, M. Kakkak and J. R. Bennett, *Gut.* 33, 964 (1992).

[10] G. R. Zuckerman, G. L. Cornette, R. E. Clouse and H. R. Harter, *Ann. Intern. Med.* 102, 588 (1985).

[11] P. Dave, J. Romeu, A. Antonelli and A. R. Eiser, *Arch. Intern. Med.* 114, 1781 (1984).

[12] J. T. Cunningham, *Gastroenterol.* 81, 1131 (1981).

[13] G. Borsch, *Am. J. Gastroenterol.* 82, 1333 (1987).

[14] M. Ikeda, H. Ishida, E. Nakamura, Y. Shirai, T. Michida, N. Hayashi, et al, *Endoscopy* 28, 390 (1996).

[15] M. Ikeda, N. Hayashi, E. Imamura, A. Kaneko, T. Michida, K. Yamamoto, et al, *J. Gastroenterol.* 32, 587 (1997).

[16] L. Spahr, J. P. Villeneuve, M. P. Dufresne, D. Tasse, B. Bui, B. Willems, et al, *Gut.* 44, 739 (1999).

[17] J. R. Lowes and J. Rode, *Gastroenterol.* 97, 207 (1989).

[18] J. Charneau, R. Petit, P. Cales; A. Dauver and J. Boyer, *Gut.* 37, 488 (1995).

[19] T. Sato, K. Yamazaki and J. Akaike, *Dig. Endosc* 2012 (in Press).

[20] J. H. III Gilliam, K. R. Geisinger; W. C. Wu, N. Weidner and J. E. Richter, *Digest. Dis. Sci.* 34, 885 (1989).

[21] D. C. Wolf and W. S. Brooks, *Gastrointest. Endosc.* 32, 139 (1986).

[22] M. J. Bourke, R. L. Hope, P. Boyd, P. E. Gillespie, M. Ward, A. E. Cowen, et al. *J. Gastroenterol. Hepatol.* 11, 832 (1996).

[23] M. Yamada, M. Ichikawa, O. Takahara, T. Tsuchida, A. Matsubara, M. Yamada, *Dig. Endosc.* 11, 241 (1999).

[24] K. F. Bimmoeller and R. M. Karon, *Gastrointest. Endosc.* 36, 399 (1990).

[25] J. L. Petrini and J. H. Johnson, *Gastrointest. Endosc.* 35, 324 (1989).

[26] Probst, R. Scheubel and M. Wienbeck, *Z. Gastroenterol.* 39, 447 (2001).

[27] P. J. Wahab, C. J. Mulder, G. den Hartog and J. E. Thies, *Endoscopy* 29, 176 (1997).

[28] J. M. Canard and B. Vedrenne, *Endoscopy* 33, 353 (2001).

[29] Yusoff, F. Brennan, D. Ormonde and B. Laurence, *Endoscopy* 34, 407 (2002).

[30] S. Sou, T. Nagahama, H. Nomura, F. Matsubara, H. Tanaka, T. Matsui, et al, *Dig. Endosc.* 16, 166 (2004).

[31] S. F. Moss, P. Ghosh, D. M. Thomas, J. E. Jackson and J. Calam, *Gut.* 33, 715 (1992).

[32] P. Disdier, N. Schleinitz, M. Perreard, D. Monges, L. Swiader, A. Gerolami, et al, *Am. J. Gastroenterol.* 90, 1009 (1995).

[33] E. Cabral, J. M. Pontes, M. Toste, E. Camacho, M. C. Leitao, D. Freitas, et al, *Am. J. Gastroenterol.* 86, 927 (1991).

[34] S. Roman, J. C. Saurin, J. Dumortier, A. Perreira, G. Bernard and T. Ponchon, *Endoscopy* 35, 1024 (2003).

[35] S. Nakamura, A. Mitsunaga, H. Konishi, I. Oi, K. Shiratori and S. Suzuki, *Dig. Endosc.* 18, 128 (2006).

[36] Fuccio, R. M. Zagari, M. Serrani, L. H. Eusebi, D. Grilli, V. Cennamo, et al, *Digestion* 79, 143 (2009).

[37] J. S. Baudet, D. Diaz-Bethencourt, M. Soler, M. Vela, S. Morales and J. Aviles, *Med. Clin. (Barc)* 133, 217 (2009).

[38] S. K. Sinha, H. P. Udawat, S. Varma, A. Lal, S. S. Rana and D. K. Bhasin, *Gastrointest. Endosc.* 64, 1028 (2006).

[39] R. Kumar, S. Mohindra and H. S. Pruthi, *Endoscopy* 39 (Suppl 1) E56 (2007).

[40] C. D. Wells, M. E. Harrison, S. R. Gurudu, M. D. Crowell, T. Byrne, G. DePetris, et al, *Gastrointest Endosc* 68, 231 (2008).

INDEX

#

A

B

C

D

E

F

G

H

I

J

K

L

M

N

O

P

Q

R

S

T

U

V

W

X

Y

Z